Springer Proceedings in Mathematics & Statistics

Volume 81

For further volumes:
http://www.springer.com/series/10533

Springer Proceedings in Mathematics & Statistics

This book series features volumes composed of select contributions from workshops and conferences in all areas of current research in mathematics and statistics, including OR and optimization. In addition to an overall evaluation of the interest, scientific quality, and timeliness of each proposal at the hands of the publisher, individual contributions are all refereed to the high quality standards of leading journals in the field. Thus, this series provides the research community with well-edited, authoritative reports on developments in the most exciting areas of mathematical and statistical research today.

Helen MacGillivray • Michael A. Martin
Brian Phillips
Editors

Topics from Australian Conferences on Teaching Statistics

OZCOTS 2008-2012

 Springer

Editors
Helen MacGillivray
Science and Engineering Faculty
Queensland University of Technology
Brisbane, Australia

Michael A. Martin
Research School of Finance
Actuarial Studies and Applied Statistics
Australian National University
Canberra, Australia

Brian Phillips
Faculty of Life and Social Sciences
Swinburne University of Technology
Victoria, Australia

ISSN 2194-1009   ISSN 2194-1017 (electronic)
ISBN 978-1-4939-0602-4   ISBN 978-1-4939-0603-1 (eBook)
DOI 10.1007/978-1-4939-0603-1
Springer New York Heidelberg Dordrecht London

Library of Congress Control Number: 2014943925

Printed on acid-free paper

Springer is part of Springer Science+Business Media (www.springer.com)

Introduction

Statistics and statistical thinking have become increasingly important in a society that relies more than ever on information and demands for evidence. Hence the importance of developing statistical thinking across all levels of education has grown, and continues to grow, in a century which will place even greater pressure on society for statistical thinking throughout industry, government and education. However, although handling and interpreting information, data and uncertainty pervade everyday living, disciplines, workplaces and research, they are remarkably challenging skills to learn and teach, and developing the appropriate mindset is difficult. Many reasons have been—and will be—given for this, but it is essentially due to the nature of statistics as the science of variation and uncertainty, together with its pervasiveness and its importance across all disciplines and all evidence-based endeavours. From statistical literacy for citizenship to addressing highly complex and challenging problems, the statistical sciences are servants, collaborators and leaders in information and technological societies.

The roles of servant, collaborator and leader are not distinct, but constantly interact in a continuum which must be acknowledged and embedded in statistical education efforts and systems. Statistical education is education of future consumers, users and clients of statistics, as well as that of future statisticians and statistical collaborators. Thus the entire spectrum of statistical education across all educational levels and disciplines is of vital significance to all statisticians and the statistical profession as well as all involved in statistical education. The supply of statisticians is just one aspect of this spectrum, and the education of future consumers, users, producers, developers and researchers of statistics is both challenging and important for a modern information society and hence to the statistical profession.

Statistics is a living discipline, ever developing to meet new challenges and needs, interacting constantly with users and collaborators. Statistical education has also been, and must continue to be, similarly active, always reflecting, accommodating and anticipating the diversity of students' and clients' current and future needs. Progress in statistics education comes from a combination of deep understanding of statistics and statistical thinking with what is learnt from observing and analysing students and their work. This volume helps to pay tribute to all those who are dedicated to bringing the power of statistics to life in teaching, but embedded in students' and clients' perceptions and prior learning so that they can experience statistical thinking for themselves, and hence continue to learn and develop. This task requires constant interaction with students, colleagues and practitioners, with the statistics teacher always seeking to see inside the minds of students and statistical users to understand how to make the connections and build confidence. It is less demanding to teach statistics in a teacher-centred or surface way, either of which may provide immediate gratification. But such approaches can fail to develop the understanding and confidence needed for ongoing learning and for the data and contexts of real and complex problems.

No matter what aspect of statistics or educational level we are discussing, progress in statistical education is continuous, complex and multi-dimensional, with 'reforms' happening at the frontlines in classes, with students across disciplines and levels, in workplaces, careers and professions. Guidelines, bullet points and slogans reflect developments and help to raise wider awareness of progress underway, but can also risk trivialisation and substitution of new dogma for old.

It is a characteristic of statistics and of its education that developments often start at similar times around the world, and that key steps forward in statistics education reflect the problem-solving processes, thinking and advocacy of statistical practitioners. However, despite ongoing endorsement by statistical practitioners and educators over many years, such approaches are perennially challenging to maintain and embed. Advancing and maintaining improvements in statistics education need collaboration, cooperation and dedication across the community of statisticians and statistical educators, and the Australian Conferences on Teaching Statistics (OZCOTS) have contributed to providing a forum and bringing this community together. Australians and New Zealanders have been part of the international statistical education vanguard for many years, and, since 2008, there has been good international representation at OZCOTS, especially by New Zealanders.

Some Background to OZCOTS: ISI, IASE and ICOTS

The International Association for Statistical Education (IASE) is one of the Associations of the International Statistical Institute (ISI) which was founded in 1885, and whose overall mission is to promote the understanding, development and good practice of statistics worldwide. The ISI Education Committee was set up in 1948, as part of a series of constitutional changes aimed, among other things, at increasing the ISI's mandate to undertake educational activities in statistics and to collaborate for this purpose with UNESCO and other UN agencies. Although statistical education had been a concern of the ISI since its inception in 1885, it was the creation of the Education Committee which marked the beginning of a systematic education programme.

The first international roundtable on statistics education was held in 1968, and the first International Conference on Teaching Statistics (ICOTS) was held in 1982. After the third ICOTS in 1990, the IASE was formed to take over and spread the work of the ISI Education Committee. The seven Associations of the ISI cover a wide spectrum of statistical sciences and applications, facilitating international communication among groups of individuals with common interests, and having their own governance structure, conferences, journals and other activities. The ISI and its family of Associations work closely together to serve the needs of the international statistical community.

Apart from ICOTS, which is held every 4 years, the IASE organises satellite conferences to the biennial ISI World Statistics Congresses (WSC), statistical education strands within the WSC, and international roundtables every 4 years. For its conferences, IASE offers an optional double-blinded refereeing process and publishes proceedings, now online. Papers accepted as meeting refereed status are designated as such in the proceedings.

OZCOTS: Australian Conference on Teaching Statistics

The first OZCOTS was inspired by the fifth ICOTS which had been held in Singapore earlier that year. OZCOTS 1998 was organised by Brian Phillips with papers by Australian speakers from ICOTS5. Its success in bringing together Australians involved in teaching statistics resulted in Brian and his Melbourne colleagues organising annual OZCOTS gatherings from 1999 to 2002. These were quite informal, having no formal proceedings.

In 2006, Helen MacGillivray was awarded one of the first Australian Learning and Teaching Council's National Senior Teaching Fellowships, with her fellowship programme to run throughout 2008. The Australian Learning and Teaching Council (ALTC) (formerly the Carrick Institute of Learning and Teaching in Higher Education), was established in 2005 as an initiative of the Australian Government Department of Education, Science and Training. Its mission was to promote and advance learning and teaching in Australian higher education. In 2011, the ALTC was replaced by the Australian Government's Office for Learning and Teaching which promotes and supports change in higher education institutions for the enhancement of learning and teaching. The Senior Fellowship Scheme supports leading educators to undertake strategic, high profile fellowship activities in areas that support the ALTC/OLT mission. Senior Fellows undertake a full time programme of highly strategic fellowship activities over 1 year. Helen was one of the three inaugural Senior Fellows. Such recognition of the importance of teaching statistics in higher education is of significant value to the statistical and statistical education communities.

As part of her fellowship programme, Helen revived OZCOTS with Brian's help, and ran it as a two-day satellite to the 2008 Australian Statistical Conference (ASC), with a one-day overlap open to all ASC delegates who could also choose to register for the second day. The OZCOTS 2008 invited speakers were all funded as part of Helen's fellowship. OZCOTS 2008 was modelled on the successful IASE satellite conferences to ISI Conferences, with papers in proceedings and an optional refereeing process offered to authors. The success of OZCOTS 2008 lead to further equally successful OZCOTS meetings in Perth, 2010, and in Adelaide, 2012. OZCOTS 2010 and then OZCOTS 2012 were again run as two-day satellites to the 2010 and 2012 ASCs, with a one-day overlap open to all ASC delegates who could also choose to register for the second day, and with a joint keynote speaker on statistics education on the overlap day.

The OZCOTS 2008 programme consisted of six keynote papers, eighteen contributed papers and two forums. The six keynote speakers at OZCOTS were members of the Senior Fellowship's international collaborative team, and their participation in OZCOTS 2008 was fully funded by the Fellowship. OZCOTS was also supported by the ASC and the Victorian Branch of the Statistical Society of Australia Inc (SSAI). The arrangement of the first day of OZCOTS being available to ASC delegates at no extra charge was highly successful with standing room only available on the first day with many ASC delegates attending all or part of the OZCOTS programme. A number of ASC delegates who had not pre-registered for the second day of OZCOTS were so interested by the first day that they stayed on for the second.

The OZCOTS 2008 Conference Committee consisted of Helen MacGillivray (joint chair, joint editor), Brian Phillips (joint chair, local arrangements), Michael A. Martin (joint editor), Kay Lipson and Peter Howley, with entertainment provided by local balladeer extraordinary, Peter Martin.

OZCOTS 2010, held in Perth, had the theme of *Building Capacity in Statistics Education.* The programme consisted of two keynote speakers, twenty-two contributed papers and three forums. The OZCOTS 2010 Conference Committee consisted of Helen MacGillivray (joint chair, joint editor), Brian Phillips (joint chair, joint editor) and Alexandra Bremner (local arrangements). OZCOTS 2010 acknowledged the support of the ASC 2010, SAS and the NSW and Victorian Branches of SSAI. OZCOTS 2008 and OZCOTS 2010 Proceedings are available on the IASE website under Publications > Regional Publications.

For many years, and internationally, statistical consultants and statistical educators have been advocating a holistic approach in statistics education that reflects the thinking and problem-solving of the *practice* of statistics. This was reflected in the theme of OZCOTS 2012, held in Adelaide, namely *Statistics Education for 'Greater' Statistics*, with the 'greater' statistics of the theme reflecting the whole process of the practicum of statistics. The programme consisted of two keynote speakers, eighteen contributed papers and three forums. The OZCOTS 2012 Conference Committee consisted of Helen MacGillivray (joint chair, joint editor), Brian Phillips (joint chair, joint editor), Olena Kravchuk (local arrangements), with ASC 2012 providing significant support and close collaboration.

This Book

Authors of papers from OZCOTS 2008, 2010 or 2012 were invited to develop chapters for refereeing and inclusion in this volume. The chapters in Parts A to E were all developed from papers presented at one of these OZCOTS, and have been grouped under the headings of keynote topics, undergraduate curriculum, undergraduate learning, professional development and postgraduate learning. Each chapter includes identification of the OZCOTS at which the authors presented a paper on the topic. The papers in Part F were also accepted for, and presented at, OZCOTS 2012 and have not appeared elsewhere; they appear here in their OZCOTS 2012 form.

Part A: Keynote Topics

The five chapters in this section are indicative of the diversity of demands on, and the richness of, statistical education. The core enabling power of technology in statistics makes it equally important in statistics education, as well as providing educational platforms for experiential learning. Adrian Bowman's chapter emphasises these in the bringing together of data, concepts and models, which is essential in developing statistical thinking. This is also at the heart of Michael A. Martin chapter, which also demonstrates the extent to which successful teaching strategies in statistics depend on the harmony of deep statistical understanding with far-reaching reflection and analysis, and 'seeing' through student eyes. Larry Weldon's 2008 keynote chapter, reproduced (with permission), is a demonstration of how to bring together experiential learning, the full data investigation process as advocated by statistical practitioners and the underpinning models and structures that make statistics the powerful interdisciplinary problem-solver. Delia North's chapter with co-author Temesgen Zewotir is a tribute to what the combination of dedication to students and statistical understanding can achieve, an indication of the challenges of building statistical capacity in developing countries and a sobering reminder of the societal consequences of prejudice and neglect. The value of national and international interactions among and between statisticians, statistical educators and user disciplines is emphasised in Kaye Basford's overview and personal reflections from a statistician long immersed in an applied discipline, a past president of the International Biometrics Society and a colleague of John Tukey.

Other keynote presentations at OZCOTS 2008-2012 were given by Rob Gould on technological literacy and the nature of modern data (2008), Peter Petocz on student conceptions of statistics and statisticians (2008), Chris Wild on cooperation and competition (2008) and visualisation and conceptualisation of statistical inference (2010) and Jessica Utts on beliefs, choices of analyses and educating psychologists (2012).

Part B: Undergraduate Curriculum

The four chapters in this section are all focussed on that critical and vital aspect of statistics education that is the teaching of statistical thinking and foundational skills and understanding across disciplines at the introductory tertiary level. However, not only is this of central importance to statisticians, the whole statistical community and indeed all of society but the messages in these chapters are also of relevance and assistance for all of statistics education. In a sense, this aspect of statistics education is a vital cusp, informing curricula and pedagogies below and above this level, and linking to the education of future statisticians, statistical collaborators, producers, users and consumers. University systems in many countries, including Australia and New Zealand, are not generally geared to allow or support university-wide statistical literacy courses, and Sue Finch and Ian Gordon's account of taking the opportunities and challenges in designing, implementing and maintaining such a course is both a testament to the authors' efforts and achievements, and an excellent example of linking current statistical education principles with the world and work of professional statistical consulting. Simulation is an essential tool in modern statistical analyses and it also provides a range of educational tools in teaching statistics. Graeme Barr and Leanne Scott's work is as much about helping students to understand distribution as it is about simulating data and demonstrating sampling variability. But Chap. 7 also emphasises the need for readily available software and hands-on experience in accessible everyday contexts, particularly in developing countries.

Student diversity and use of technology are also highlighted by Martin Gellerstedt, Lars Svennson and Christian Östlund who emphasise teamwork and analysis of student learning styles in tackling the challenges of online and mixed online/on-campus courses, demonstrating also how developments often happen simultaneously around the world and how guidelines are built on reflective analysis of frontline experiences. Communication with students, seeing through student eyes and meeting the diversity of student backgrounds and learning styles also feature in the strategies of Małgorzata Korolkiewicz and Belinda Chiera to improve the attitudes of staff and students in large service courses to Business and Life Science students.

Part C: Undergraduate Learning

Student learning is at the heart of all chapters, but the chapters in this section are particularly focussed on similarities and contrasts in student learning and effects of backgrounds, disciplines and conditions due to the practicalities of systems and environments. As in Part B, the focus is again on the extensive and varying challenges of large introductory tertiary classes across disciplines. In Chap. 10, the qualitative component of carefully designed in situ research by James Baglin and Cliff Da Costa investigates how psychology students acquire statistical technological literacy, including whether it is possible to separate this from learning statistical concepts and skills.

An ambitious study by Ayse Bilgin and a team of international collaborators in three very different countries focuses on student learning approaches and how environmental and student characteristics might encourage students to utilise certain approaches, especially students who do not major in statistics. Such a study must therefore carefully identify and allow for the practicalities of the different systems and environments, and Chap. 11 includes validation in a very large Australian cohort of the Approaches and Study Skills Inventory for Students (ASSIST). In Chap. 12, Joanne Fuller also reports on research into the backgrounds of students in a very large cohort, concentrating on the quantitative skills Business students bring to an introductory course and strategies to tackle the associated effects on attitudes and confidence, providing further insight into issues raised in Chap. 9.

Part D: Workplace Learning and Professional Development

As the demand and need for statistical problem-solving, analyses, modelling and know-how continue their rapid growth across disciplines and workplaces, so too do the challenges of helping users and consumers of statistics. The diverse workplace and professional settings of the chapters in this section all demonstrate a reality long known by professional statisticians and statistics educators—that statistical consulting, teaching and the communication of statistics are inherently strongly intertwined and that this interlocking underpins the practice of statistics.

Peter Martin's extensive experience in assisting industry in quality control or process improvement is apparent in Chap. 13 on training programmes in industry settings. In addressing the specific challenges of building client relationships, customer responsiveness and relating to real workers and organisations in order to meet immediate needs and promote ongoing learning, there are also messages on the benefits of such work for all statistics teaching.

National Statistical Offices also have significant interest and involvement in statistical education of the general community, of government users and producers and of future statisticians. The work of Sharleen Forbes et al. demonstrates the extent to which Statistics New Zealand and the Australian Bureau of Statistics have developed advanced partnerships with universities, crafted to progress capabilities in one or more of these diverse but equally important groups.

An extensive review by Kristen Gibbons and Helen MacGillivray of the literature demonstrates the advocacy from statistical consulting on experiential learning of the whole statistical investigation process for future statisticians, and the parallels between student-centred teaching of statistical thinking and consulting, collaboration and communication with statistical users and clients. These links are further exhibited by the benefits for future statistical consultants of participating in an undergraduate mentored developmental programme in learning to tutor statistics, which blends training and apprenticeship stages, including student-centred tutoring in statistical investigations and constructivist teaching approaches.

Chapters 16 and 17 both focus on work-specific statistical training for people who are not statistics graduates and have operational statistical needs. Helen Chick, Robyn Pierce and Roger Wander report on improving teachers' understanding of official data on student performance. Ian Westbrooke and Maheswaran Rohan discuss workplace courses on design of investigations and statistical analysis for scientists in New Zealand's Department of Conservation. Although the workplaces are very different, both groups of users need to understand data handling and relate data reports to their work. The training of both groups is also via workshops or short courses, and is evaluated by how much it helps participants in their work. Both groups also need to deal with the products of statistical technology and concepts of models, even if there are great contrasts in the level of sophistication in both software and models.

Part E: Postgraduate Learning

An area of remarkable growth and increasingly unmet demand is statistical assistance in postgraduate study across many disciplines. Rapidly increasing information technology facilities for gathering and analysing data of all types add to pressures on postgraduates and to the chronic challenges in universities of how to efficiently and effectively meet pleas for statistical assistance for them. Postgraduates are both students and new researchers, and their needs are to understand the statistics they require in their study/research and to develop statistical capabilities for their future careers across diverse workplaces and/or research areas.

Statisticians involved in helping postgraduate students in other disciplines are keenly aware that their needs usually include improving or extending their foundational understanding and skills in statistics, but from a more mature point of view than they may have experienced in an undergraduate introductory course, even if the pedagogy of this was modern and centred on experiential learning of statistical investigations. Glenys Bishop reports on meeting the challenge of providing such foundations in a fully flexible online environment which also offers different levels of attainment, with discussion on indicators of success in online environments. Olena Kravchuk and David Rutley discuss a programme for growing the statistics capabilities of advanced undergraduate agricultural science students in preparation for postgraduate research or workplace decision-making. The programme builds on earlier undergraduate statistical learning and is an integrated package of strategies embedding collaboration with statisticians and student-centred learning in context.

Sue Gordon, Anna Reid and Peter Petocz describe innovative qualitative research into how academics and postgraduate students use quantitative approaches to carry out research in creative and qualitative disciplines, such as music, design and art.

Although the diversity of experiences and attitudes seen across all disciplines is perhaps accentuated by perceptions of less quantitative underpinnings, customs and pathways of research in such disciplines, the research demonstrates the extent of statistical needs and further investigation of these needs and how to support them.

A quote from a statistician interviewed in the research of Chap. 20, provides a comment relevant to the growing challenge of statistical provision for postgraduate students across disciplines:

Statistics is a massive subject with so many pitfalls people commonly fall into—such as confounding. Really if you are doing quantitative research, you need the input of a statistician. ... I am always happy to offer advice to research students, regardless of their subject; however, I am one person and there are not many other statisticians about either. I am sure that if their supervisor doesn't know a 'proper' statistician, and also if they don't understand the subject of statistics sufficiently, the student won't even know they have to look for a statistician, let alone where to find one.

Part F: Papers From OZCOTS 2012

These four short chapters are papers which were also accepted for, and presented at, OZCOTS 2012 and have not appeared elsewhere; they appear here in their OZCOTS 2012 form. Although short, they cover a range of topics which reinforce and link with a number of the preceding chapters. Imma Guarnieri and Denny Meyer consider evaluation of learning in an online environment for postgraduates. Chapters 22 and 23 discuss resources to support holistic experiential learning of the statistical investigation process embedded in data and context, with Robert Brooks et al exploiting sporting interests and Howard Edwards, Sarah Edwards and Gang Xie having a discipline-specific focus. In Chap. 24, Emma Mawby and Richard Penny also touch on online learning challenges and links with Chap. 14 in considering just some of the broad spectrum of statistics educational roles of national statistics offices.

Refereeing Process

Chapters and papers referred to in the proceedings as refereed were reviewed by at least two referees selected from a panel of international peers approved by the editors, with full referee reports and editorial comments provided to authors. The review process was 'double blind', with identification of both authors and referees removed from all documentation during the reviewing process. The refereeing process also provided a mechanism for peer review and critique and so has contributed to the overall quality of statistics education research and teaching. Revisions were reviewed with respect to referee reports and editorial guidance, and chapters marked as refereed have met criteria consistent with the accepted norms for reporting of research. Criteria included that the work would represent a significant contribution to knowledge about statistics education or the research processes in statistical education. Chapters developed from papers in OZCOTS 2008 and 2010 proceedings represent significant new work or progress or analysis.

Brisbane, QLD, Australia   Helen MacGillivray
Canberra, ACT, Australia   Michael A. Martin
Melbourne, VIC, Australia   Brian Phillips

Acknowledgements

The editors are grateful for the assistance of the following referees, whose knowledge and insight were invaluable in preparing the contributions to this book: Stephanie Budgett, Murray Cameron, Matthew Carlton, Len Cook, Tony Croft, Peter Dunn, Michael Forster, Iddo Gal, Gary Glonek, Sue Gordon, Rob Gould, John Harraway, Peter Howley, Helen Johnson, Sandra Johnson, Irena Krizman, Gillian Lancaster, Katie Makar, Maria Gabriella Ottaviani, Maxine Pfannkuch, Jackie Reid, Jim Ridgway, Eric Sowey, Helen Thompson, Neville Weber, Bronwen Whiting, Chris Wild, Emlyn Williams, Therese Wilson, Richard Wilson, Ian Wood.

Thanks are also given for support for OZCOTS 2008-2012, in particular

- The Australian Learning and Teaching Council (now the Australian Government's Office for Learning and Teaching) for the award of Professor Helen MacGillivray's Senior Teaching Fellowship, whose programme included support for OZCOTS 2008
- The Statistical Society of Australia Inc (SSAI) and the Australian Statistical Conferences of 2008, 2010 and 2012, for support of, and collaboration with, OZCOTS 2008, OZCOTS 2010 and OZCOTS 2012
- The Victorian Branch of the SSAI for support of OZCOTS 2008 and 2010
- The NSW Branch of SSAI for support of OZCOTS 2010
- SAS for support of OZCOTS 2010
- All speakers, authors and participants of OZCOTS 2008-2012

Contents

Part F Papers from OZCOTS 2012

Contributors

James Baglin School of Mathematical and Geospatial Sciences, RMIT University, Bundoora, VIC, Australia

Graham Barr Department of Statistical Sciences, University of Cape Town, Cape Town, South Africa

Kaye E. Basford The University of Queensland, Queensland, Australia

Ayse Aysin Bombaci Bilgin Department of Statistics, Faculty of Science, Macquarie University, North Ryde, NSW, Australia

Glenys Bishop Statistical Consulting Unit, Australian National University, Canberra, ACT, Australia

Ross Booth Department of Economics, Monash University, Clayton, VIC, Australia

Adrian W. Bowman School of Mathematics and Statistics, The University of Glasgow, Glasgow, UK

Robert Brooks Department of Econometrics and Business Statistics, Faculty of Business and Economics, Monash University, VIC, Australia

Helen Chick Faculty of Education, University of Tasmania, Hobart, TAS, Australia

Belinda Ann Chiera School of Information Technology and Mathematical Sciences, University of South Australia, Adelaide, SA, Australia

Francesca Chiesi Department of NEUROFARBA—Section of Psychology, University of Florence, Florence, Italy

James O. Chipperfield Australian Bureau of Statistics, Belconnen, ACT, Australia

Cliff Da Costa School of Mathematical and Geospatial Sciences, RMIT University, Bundoora, VIC, Australia

Howard Edwards Massey University at Albany, Auckland, New Zealand

Sarah Edwards Massey University, Albany, New Zealand

Maria del Carmen Fabrizio Department of Quantitative Methods and Information Systems, Faculty of Agriculture, University of Buenos Aires, Buenos Aires, Argentina

Sue Finch Statistical Consulting Centre, University of Melbourne, Parkville, VIC, Australia

S.D. Forbes Statistics New Zealand and Victoria University of Wellington, Featherston, New Zealand

Joanne Elizabeth Fuller School of Economics and Finance, Queensland University of Technology, Brisbane, QLD, Australia

Tamas Gantner Department of Statistics, Faculty of Science, Macquarie University, NSW, Australia

Martin Gellerstedt School of Business, Economics and IT, University West, Trollhättan, Sweden

K.S. Gibbons Mater Research, South Brisbane, QLD, Australia

Ian Gordon Statistical Consulting Centre, University of Melbourne, Parkville, VIC, Australia

Sue Gordon Mathematics Learning Centre, The University of Sydney, NSW, Australia

Petra L. Graham Department of Statistics, Faculty of Science, Macquarie University, North Ryde, NSW, Australia

Imma Guarnieri Faculty of Health Arts and Design, Swinburne University of Technology, Hawthorn, VIC, Australia

J.A. Harraway University of Otago, Dunedin, New Zealand

Małgorzata Wiktoria Korolkiewicz School of Information Technology and Mathematical Sciences, University of South Australia, Adelaide, SA, Australia

O. Kravchuk Biometry Hub, School of Agriculture, Food and Wine, University of Adelaide, Adelaide, SA, Australia

Maria Virginia Lopez Department of Quantitative Methods and Information Systems, Faculty of Agriculture, University of Buenos Aires, Buenos Aires, Argentina

Helen MacGillivray School of Mathematical Sciences, Queensland University of Technology, Brisbane, Australia

Michael A. Martin Research School of Finance, Actuarial Studies and Applied Statistics, Australian National University, Canberra, Australia

Peter Martin University of Ballarat, Ballarat, VIC, Australia

Emma Mawby Statistical Methods, Statistics New Zealand, Wellington, New Zealand

Denny Meyer Faculty of Health Arts and Design, Swinburne University of Technology, Hawthorn, VIC, Australia

Delia North University of KwaZulu-Natal, Durban, South Africa

Christian Östlund School of Business, Economics and IT, University West, Trollhättan, Sweden

Richard Penny Statistical Methods, Statistics New Zealand, Christchurch, New Zealand

Peter Petocz Department of Statistics, Macquarie University, Sydney, Australia

Robyn Pierce Melbourne Graduate School of Education, University of Melbourne, Parkville, VIC, Australia

Caterina Primi Department of NEUROFARBA—Section of Psychology, University of Florence, Florence, Italy

Veronica Frances Quinn University of Sydney, Sydney, NSW, Australia

Macquarie University, North Ryde, NSW, Australia

Anna Reid Sydney Conservatorium of Music, The University of Sydney, Sydney, Australia

Maheswaran Rohan Department of Conservation, Hamilton, New Zealand

D.L. Rutley School of Animal and Veterinary Science, University of Adelaide, Adelaide, SA, Australia

Leanne Scott Department of Statistical Sciences, University of Cape Town, Cape Town, South Africa

Nishta Suntah Department of Econometrics and Business Statistics, Monash University, Berwick, Australia

Lars Svensson School of Business, Economics and IT, University West, Trollhättan, Sweden

Siu-Ming Tam Australian Bureau of Statistics, Belconnen, ACT, Australia

Roger Wander Melbourne Graduate School of Education, University of Melbourne, Parkville, VIC, Australia

K.L. Weldon Simon Fraser University, Burnaby, BC, Canada

Ian Westbrooke Department of Conservation, Christchurch, New Zealand

Jill Wright Department of Econometrics and Business Statistics, Monash University, Berwick, Australia

Gang Xie CRC for Infrastructure and Engineering Asset Management (CIEAM), Queensland University of Technology, Brisbane, Australia

Temesgen Zewotir University of KwaZulu-Natal, Durban, South Africa

Part A
Keynotes

Interacting with Data, Concepts and Models: Illustrations from the `rpanel` Package for R

Adrian W. Bowman

Abstract The rich statistical computing environments which are available for data analysis also provide very flexible tools for practical work in the teaching and learning of statistics. R is an outstanding example of this, with an enormous variety of data exploration and modelling tools which are easily accessible. However, there is also a need to support students in the understanding and intuition of key statistical concepts and models. The tools available within R to create interactive graphical controls can be used for this to very good effect. This is explored through the `rpanel` package, which aims to make the creation of this type of control as straightforward as possible. Examples of interacting with data (in windowing time series), concepts (random variation and correlation) and models (analysis of covariance and logistic regression) are described. The mechanism for coding this type of interactive display is also outlined.

Keywords Animation • Graphics • Interaction • R

1 Introduction

Computing technology has had a major influence on the teaching and learning of virtually all subjects within higher education and not least in Statistics. The ability to explore data graphically and fit a wide variety of statistical models has made real practical applications feasible to tackle in a lecture or laboratory setting and, where appropriate, has allowed the focus to be placed on issues of modelling and interpretation rather than on more technical aspects. In particular, the arrival of the open-source

A.W. Bowman (✉)
School of Mathematics and Statistics, The University of Glasgow,
Glasgow G12 8QQ, UK
e-mail: adrian.bowman@glasgow.ac.uk

H. MacGillivray et al. (eds.), *Topics from Australian Conferences on Teaching Statistics:*
OZCOTS 2008-2012, Springer Proceedings in Mathematics & Statistics 81,
DOI 10.1007/978-1-4939-0603-1_1, © Springer Science+Business Media New York 2014

system R (R Development Core Team 2006) has provided data and modelling tools, from the elementary to the state of the art, which are easily accessible. In designing courses, the availability of suitable statistical computing tools is no longer a barrier.

However, while the process of statistical modelling is very well supported computationally, this is much less true of the process of understanding the underlying concepts and methods. There have been many projects which have produced illustrative graphical software directed at the teaching and learning of concepts, but these have often been stand-alone tools written in languages which allow flexible, and often interactive, graphical tools to be created. The *STEPS* (Bowman and McColl 1999), *ActivStats* (Velleman 2004) and *CAST* (Stirling 2012) projects are all examples of this type of material, including multimedia resources, which all remain available.

Many users of R remain unaware that systems for providing GUI (graphical user interface) tools have been available within R for some time. These do not provide the full range of facilities provided by multimedia authoring systems but they do provide a very useful set of tools for adding a degree of interactivity to R operations. While R has, for very good reasons, been constructed with the philosophy of command-driven control, there are some specific operations where GUI control is very helpful and the teaching and learning context provides a large number of examples. Systems which can provide this include iPlots (Urbanek and Theus 2003) based on Java and RGtk2 (Lang and Lawrence 2006) based on the GTK tools. Verzani (2007) provides the gWidgets interface which provides access to several different GUI systems in R.

This paper focusses on the rpanel package (Bowman et al. 2006, 2007) for R which has two aims. The first is to provide access to GUI tools in as simple and direct a manner as possible. The rtcltk package (Dalgaard 2001) is used because of its native presence in R for many years. rpanel carries out the management of communications behind the single function calls required to add individual controls. The second aim is to provide higher level functions which use these lower level tools to create useful interactive operations, with a particular emphasis on teaching and learning. An illustration of this is given in Sect. 2, focussing on a simple example involving data exploration and plotting. Sections 3 and 4 extend this into interaction with concepts and models. Some final discussion is provided in Sect. 5.

2   Interacting with Data

Time series data offer a simple, easily understood structure which has many applications and creates interest in the setting of teaching by raising many interesting questions. In the context of climate change, the Central England Temperature data provide a remarkably long documentation of monthly temperature (°C), from 1659 to the present day. The data are available from the Hadley Centre (www.metoffice. gov.uk/hadobs/hadcet/) but are also available in the multitaper package (Rahim and Burr 2012) in R.

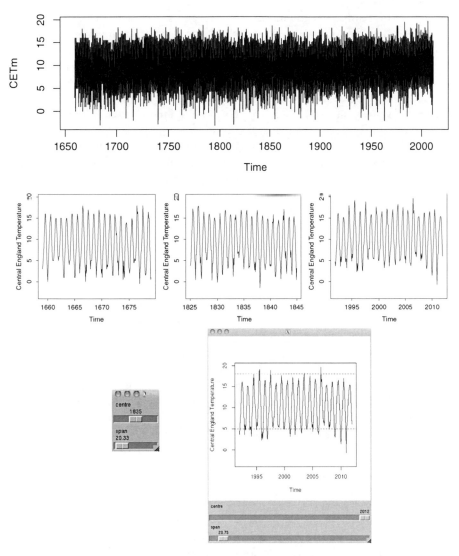

Fig. 1 The *top plot* shows the Central England Temperature from 1659 to 2011. The three plots in the *second row* show the effects of setting the centre of this window to 1659, 1835 and 2011, respectively, with the span set to 20 years in all three cases. The *bottom row* shows the two slider controls in a separate panel and then integrated with the plot into a single display

The top panel of Fig. 1 plots the entire time series but, with such a substantial amount of data, it is very difficult to assess anything other than the broadest features. It is a simple matter to plot a subset of the data which corresponds to any particular time window. However, repetition of this is rather cumbersome. A very attractive alternative is to use two sliders, one for the centre of the time window and the other for its span, as shown in the bottom left panel of Fig. 1. The three graphs in the middle row show spans of 20 years of data centred on different locations. It is hard to communicate the effect of this in static, printed plots but the advantage of animation and direct interaction is considerable, allowing easy inspection across the entire range of the time series, with simple graphical identification of the seasonal effect and cases of unusually high or low summer and winter temperatures. The physical process of using the slider substantially enhances the psychological effect of flexible interaction with the data. The bottom right panel shows the further convenient step of integrating the sliders and the plot into a single window, using the tkrplot package (Tierney 2005).

As R is in such widespread use, it is worthwhile indicating the mechanism by which this type of plot can be constructed, to encourage those with some knowledge of R coding that the addition of interactive controls is a very straightforward step. The starting point is code to plot the data y (a time series object) in a time window defined by its centre and span.

```
w   <- centre + c(-0.5, 0.5) * span
w   <- w + max(0, tsp(y)[1] - w[1])
        - max(0, w[2] - tsp(y)[2])
yw <- window(y, w[1], w[2])
plot(yw, ylim = range(y))
```

The second line of this code simply adjusts the window near the ends of the range of the time series to ensure that the window is always of length span. The code segment as a whole can easily be made into an *action* function, as the code below shows. The list object panel is the mechanism used by rpanel for communication and each action function should return the panel object. The lines of code at the end of the section below simply create a control panel window and add the two sliders with nominated start and end values. As each slider is moved, the action function is called with the new setting and the repeated redrawing which this invokes creates the animation.

```
subset.draw <- function(panel) {
  with(panel, {
    w   <- centre + c(-0.5, 0.5) * span
    w   <- w + max(0, tsp(y)[1] - w[1])
            - max(0, w[2] - tsp(y)[2])
    yw <- window(y, w[1], w[2])
    plot(yw, ylim = range(y))
    abline(h = mark, col = "red", lty = 2)
  })
    panel
}
panel <- rp.control(y = y)
rp.slider(panel, centre, tsp(y)[1], tsp(y)[2],
subset.draw)
rp.slider(panel, span, 2 / tsp(y)[3],
diff(tsp(y)[1:2]),
            subset.draw, initval = 20)
```

This simple mechanism allows a wide variety of controls to be added easily to code which may have been created for a limitless variety of graphical and other tasks. A further small step is to encapsulate this kind of code in new functions which offer further flexibility, such as specifying axis labels or adding horizontal lines for reference values. Bowman et al. (2006, 2007) describe the `rpanel` tools in detail and give a wide variety of examples, while Bowman et al. (2010) discuss spatial examples in particular.

3   Interacting with Concepts

The concept of random variation is fundamental to an understanding of probability and statistics, but this can be a difficult concept to grasp when it is met for the first time. A classic mistake is to over-interpret the detailed shape of histograms, scatterplots and other data displays, attributing meaningful structure to features which are simply manifestations of random variation. A simple device to counteract this is to use repeated simulations of data. A simple example is to simulate several groups of data from the same population and observe the apparent differences which can arise when the data are plotted. The upper panels of Fig. 2 use an `rpanel` control to do this, with radiobuttons to select the sample size, a button to create a new

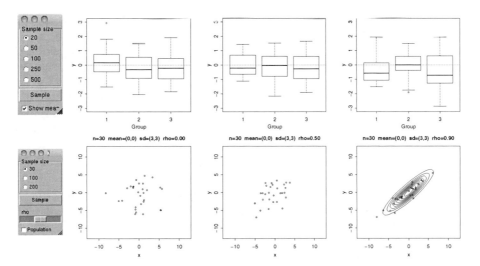

Fig. 2 In the *top row*, the *left hand panel* shows radiobutton controls for sample size, a button to simulate new data and a checkbox to control whether the common mean of the groups is displayed. The following three panels show the results of different simulations. In the *bottom row*, the *left hand panel* shows radiobutton controls for sample size, a button to simulate new data, a slider to control the value of the correlation coefficient and a checkbox to control whether the contours of the true bivariate distribution are displayed. The following three panels show the results of different simulations, with the correlation coefficient set to 0, 0.5 and 0.9, respectively

simulation and a checkbox to control whether or not the underlying true mean is superimposed. The resulting boxplots can sometimes show marked differences, especially for small sample sizes, despite the fact that the underlying means of the groups are identical. Of course, plots like this can be created by repeated execution of a small segment of code, but the GUI controls are extremely convenient and allow rapid and easy investigation of the effects of sample size. This is particularly convenient in a lecture or classroom setting but it also has advantages for student use as it focusses attention on the concept of interest and avoids the distraction of repeated direct execution of code.

The concept of correlation is another case where intuition and experience are very valuable in interpreting the strength of association in observed data. The lower panels of Fig. 2 show a control panel which allows easy repetition of sampling with specified values of sample size and true underlying correlation coefficient. The contours of the true bivariate normal distribution from which the data are samples can also be superimposed. These are elementary operations in R and, as indicated in Sect. 2, the addition of GUI controls is very straightforward. The advantage of this relatively small amount of additional effort is that an effective and easily used display tool can be created.

4 Interacting with Models

The idea of a model is central to statistical analysis and it is very helpful to be able to plot and compare models and their suitability for observed data. Again, there are potential advantages in encapsulating this in tools which use GUI controls, for example allowing interactive specification of the particular terms involved in the model. Analysis of covariance provides a good example of this. It is straightforward to write code to display the fitted models graphically on a scatterplot of the data. The addition of an interactive control to specify these terms enhances the meaning of each model by giving immediate graphical feedback on the associated changes.

Figure 3 illustrates this on data, available in the `rpanel` package, from a study of the weight changes in herring gulls throughout the year. Some birds were caught in June (coded as month 1) and others in December (month 2). Since weight is dependent on the size of the bird, this information is recorded in the form of the head and bill length, `hab` (in mm), the distance from the back of the head to the tip of the bill. The first graph displays the weight data, plotted against `hab` and colour coded by month. The following two graphs show two fitted models of particular interest, one corresponding to additive effects of weight and month and therefore producing parallel regression lines, while in the other the interaction model relaxes this constraint. The GUI control panel allows terms to be specified simply by checking the appropriate boxes, with immediate graphical feedback in terms of the fitted model. This helpfully reinforces the meaning of the models. There is also an opportunity to give feedback on inappropriate models, such as the presence of an interaction term without both main effects. In a small way, the software then plays the role of a tutor, giving appropriate prompts which encourage suitable modes of thinking in the

Fig. 3 The *left hand panel* shows checkbox controls to specify the terms to be fitted in an analysis of covariance model. The following three panels show the data, a model with *parallel lines* and a model with *different lines*, respectively

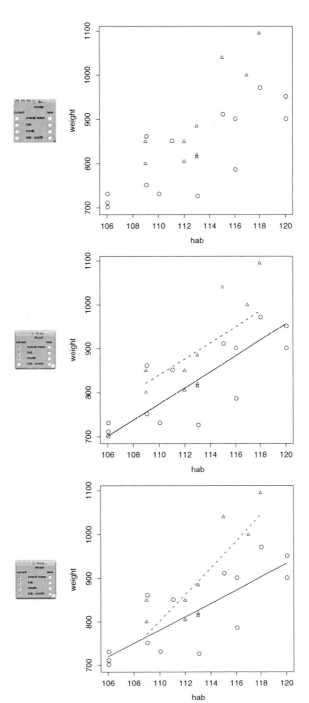

student. These facilities are available in the rp.ancova function in the rpanel package, where a comparison model can also be specified and an F-test used to assess its suitability.

A second illustration is of logistic regression, another model which is a very standard tool in statistical analysis and yet which, in the experience of many students, requires some time and effort to understand when it is first met. Figure 4 shows a well-known set of data on budworms, discussed in Collett (1991) and used as an illustration in Venables and Ripley (1994), where R code is also provided. Data on the numbers killed (from groups of 20) which were exposed to different doses of a chemical are plotted here, for males budworms only. The grouped nature of the data helps in motivating the shape of the logistic link function. The ability to superimpose a logistic regression, with the values of the intercept and slope parameters in the linear predictor controlled by "double-buttons", allows the effects of changing these parameters to be investigated. This promotes intuition on the meaning of the parameters and the process of selecting suitable values to describe the observed data leads naturally to a discussion of scientific principles which can be used for model fitting. Where appropriate within the syllabus, likelihood can be introduced, with further opportunities for graphical display and GUI interaction. Bowman (2007) discusses the uses of rpanel in a likelihood setting. The right-hand panel of Fig. 4 shows the fitted model which is produced by maximising the likelihood. These facilities are available in the rp.logistic function in the rpanel package.

5   Discussion

Opportunities to support the understanding of statistical data, concepts and models have been explored and the use of interactive GUI controls has been advocated. This has been discussed in the context of the R statistical computing environment, which is now very widely used and which provides an enormous variety of data and modelling tools.

The ability to add interactive controls in this rich computational environment allows very useful teaching tools to be constructed. These can be used in a lecture or classroom setting by teachers, to illustrate topics in a convenient but dynamic manner. This mode of use allows discussion to move beyond the scope of static diagrams and communicates the fact that analysing data is a dynamic and exploratory process. The use of animation in particular supports more formal presentations of concepts by illustrating the meaning of parameters or models in a more intuitive manner. The liveliness of this form of presentation also often has a beneficial effect on the attention levels of the audience.

This type of material can also be used by students in laboratory or self-study mode. Again, the aim is to promote more intuitive and conceptual understanding but now with the additional use of reinforcement and interaction. We all know from our own experience of the use of software that users are most comfortable when they have a sense of active control of activity and speed, rather than being placed in the passive role of an observer. Learning is promoted when some degree of self-direction is present, as the learner takes on a degree of responsibility for the process.

Fig. 4 The *top panel* shows checkbox and button controls for fitting in a logistic regression model. The following two panels show a model for specified parameters and a model fitted by likelihood

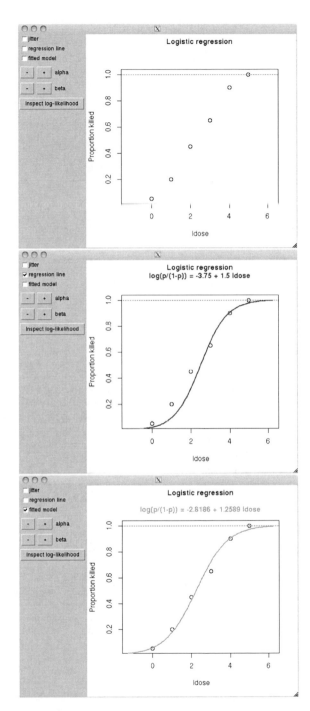

The particular illustrations discussed in the paper have been at the elementary end of the statistical syllabus, but the tools and techniques described can be applied to more sophisticated data, concepts and models with similar ease and to similar good effect.

6 Software

The rpanel package for R is available at cran.r-project.org/web/packages/rpanel. Further information on rpanel is available at www.stats.gla.ac.uk/~adrian/rpanel.

7 Note

Developed from a keynote presentation at the Sixth Australian Conference on Teaching Statistics, July 2008, Melbourne, Australia.

This chapter is refereed.

References

Bowman, A. W. (2007). Statistical cartoons in R. *MSOR Connections*, 7(4), 3–7.

Bowman, A., & McColl, J. (1999). *Statistics and problem solving*. London: Arnold.

Bowman, A., Crawford, E., & Bowman, R. (2006). rpanel: making graphs move with tcltk. *R News*, 6(4), 12–17.

Bowman, A., Crawford, E., Alexander, G., & Bowman, R. W. (2007). rpanel: Simple interactive controls for r functions using the tcltk package. *Journal of Statistical Software*, 17(9), 1–18.

Bowman, A., Gibson, I., Scott, E. M., & Crawford, E. (2010). Interactive teaching tools for spatial sampling. *Journal of Statistical Software*, 36(13), 1–17.

Collett, D. (1991). *Modelling binary data*. London: Chapman and Hall.

Dalgaard, P. (2001). The R-Tcl/Tk interface. In K. Hornik & F. Leisch (Eds.), *Proceedings of the 2nd International Workshop on Distributed Statistical Computing*, 15–17 March, 2001. Technische Universität Wien, Vienna, Austria.

Lang, D., & Lawrence, M. (2006). Rgtk2: R bindings for gtk 2.0. http://cran.r-project.org.

Rahim, K., & Burr, W. (2012). multitaper: Multitaper spectral analysis. R package version 1.0–2. http://cran.r-project.org/web/packages/multitaper.

R Development Core Team (2006). R: *A language and environment for statistical computing*. Vienna: R Foundation for Statistical Computing.

Stirling, D. (2012). Cast: Computer-assisted statistics textbooks. http://cast.massey.ac.nz.

Tierney, L. (2005). tkrplot: Simple mechanism for placing r graphics in a tk widget. http://cran.r-project.org.

Urbanek, S., & Theus, M. (2003). iPlots: High interaction graphics for R. In K. Hornik, F. Leisch, & A. Zeileis (Eds.), *Proceedings of 3rd International Workshop on Distributed Statistical Computing (DSC 2003)*, Vienna.

Velleman, P. (2004). *ActivStats*. Glenview: Addison Wesley Longman (Pearson).

Venables, W. N., & Ripley, B. D. (1994). *Modern applied statistics with S-Plus*. New York: Springer.

Verzani, J. (2007). An introduction to gWidgets. *R News*, 7, 26–33.

An Elephant Never Forgets: Effective Analogies for Teaching Statistical Modeling

Michael A. Martin

Abstract Analogies are useful and potent tools for introducing new topics in statistics to students. Martin (J Stat Educ 11(2); Proceedings of the 6th Australian Conference on Teaching Statistics) considered the case for teaching with analogies in introductory statistics courses, and also gave many examples of particular analogies that had been successfully used to make difficult statistical concepts more accessible to students. In this chapter, we explore more deeply analogies for statistical concepts from more advanced topics such as regression modeling and high dimensional data.

Keywords Analysis of variance decomposition • Influence • Leverage • Model selection • Multicollinearity • Regression • Sequential sums of squares • Testing multiple hypotheses

1 Introduction

Many students approach their statistics classes with trepidation, perhaps because many of the concepts they encounter seem so foreign. Yet, despite a lexicon steeped in jargon and technical expressions, much statistical thinking has its basis in ideas with which most students will be already familiar—the trick for statistics educators, it seems, is to bridge that gap between existing, familiar ideas and new, forbidding ones. Analogy is an effective tool for bridging this gap, with some particularly evocative uses including the alignment between statistical hypothesis testing and the process of a criminal trial (Feinberg 1971; Bangdiwala 1989; Brewer 1989; among

M.A. Martin (✉)
Research School of Finance, Actuarial Studies and Applied Statistics,
Australian National University, Canberra, Australia
e-mail: michael.martin@anu.edu.au

H. MacGillivray et al. (eds.), *Topics from Australian Conferences on Teaching Statistics:*   13
OZCOTS 2008-2012, Springer Proceedings in Mathematics & Statistics 81,
DOI 10.1007/978-1-4939-0603-1_2, © Springer Science+Business Media New York 2014

many others), and the idea of a sample mean as a "balance point" for the data (Moore 1997, p. 262 as well as on the cover of the text). These famous examples leverage the key features of analogical thinking:

- *Access*—the relevant source idea must be retrieved from memory.
- *Mapping*—alignment between elements (both objects and relationships) in the source and the target must be identified.
- *Evaluation and adaptation*—the appropriateness of the mappings needs to be assessed and adapted where necessary to account for critical features of the target.
- *Learning*—the target is understood, and new knowledge and relevant items and relationships are added to memory. The "transfer" from old to new domains is completed and the new situation can be accessed without reference to the source domain.

These elements are described in detail in the monograph by Holyoak and Thagard (1995), in which is presented a comprehensive, modern overview of analogical thinking. Martin (2003) explored the use of analogies in teaching statistics and offered many examples of analogies that had been effectively used in his statistics classes, including the legal analogy for hypothesis testing and the balance point analogy for the average. Martin later presented this work at the OZCOTS 2008 conference (Martin 2008). In the original 2003 paper and in the OZCOTS presentation, Martin focused on analogies and examples useful in a first course in statistics—the critical time when students first encounter our "mysterious" discipline. In this chapter, we consider examples and analogies specific to statistical concepts from more advanced topics from regression modeling and high-dimensional data analysis. We explore in more detail the mappings—for both items and relationships—that exist between the source and target ideas, critique the strengths and weaknesses of the analogies, and offer some new ideas that have been found useful in describing these more advanced topics.

In describing and critiquing the examples below, we utilize the "teaching-with-analogies" framework developed by Glynn (1991) (see also Glynn and Law 1993; Glynn 1994, 1995, 1996; Glynn et al. (1995), for further discussion and refinements). This framework identified six steps: introduce target concept; access source analog; identify relevant features of both source and target; map elements and relationships between source and target; assess breakdowns of the analogy; adaption and conclusion. These six steps essentially give form to the four features (access, mapping, evaluation, learning) listed above, and allow the construction of powerful analogs for thinking and learning.

This chapter is designed to be read in combination with the earlier article by Martin (2003), in which a formal argument is made supporting the use of analogies in teaching statistics, so the focus of this chapter is principally descriptive. The focus of that paper was largely on analogies for teaching a first course in statistics, while this chapter gives more consideration to and provides more detail for topics covered in a later course on statistical modeling.

2 Analogies for Describing Regression Modeling

Martin (2003) introduced several analogies useful in the context of describing regression models. We explore some of these examples in greater detail here, including a couple of analogies not included in the 2003 article or the OZCOTS Martin (2008) presentation.

2.1 Analogy 1: Signal-to-Noise and F-Ratios

Most students become familiar with hypothesis testing by considering tests for means and proportions, and so come to associate testing with the location-scale structure of Z and t tests. Similarly, in regression contexts, tests for β coefficients also work in this familiar way. So, when F tests are introduced, the immediate reaction is that this test is somehow different, as it is now based on a ratio rather than a scaled difference. Worse still, that ratio is "tampered with" through degrees of freedom adjustments! To motivate the use of a ratio-based test statistic, the analogous concept of a "signal-to-noise" ratio is a useful one. Almost without exception these days, students use Wi-Fi technology every day, so the idea that signals emanate from some central server and then are degraded by noise as the wireless device moves further from the source is a very familiar one. Most devices measure "signal strength" using bars—a rudimentary graphical display. The idea of a signal-to-noise ratio is thus a natural one, and the further idea that as the signal-to-noise ratio drops, the ability of the receiver to satisfactorily recover the true signal drops with it. In this analogy, the correspondence between *objects* is strong (signal/model; noise/error), and a key relationship (the use of a multiplicative measure of distance) also holds. As a result, the analogy has strong appeal and good memorability. On the other hand, there are some unmapped elements: in the Wi-Fi example, the notion of distance from the server is not represented in the target domain, and the role played by degrees of freedom in the F test has no direct map back to the source domain. As a result, the map is incomplete, but good enough to serve to motivate further discussion.

The way that degrees of freedom impact the definition of the F test statistic is often a difficult one for students to understand. To elucidate this idea, one approach that has been successful is the notion of "a fair fight". In comparing the signal with the noise, we wish to make this comparison as "fair" as possible, but the numerator in the F statistic is based on only *one* piece of information (the location of the line), while the denominator is based, essentially, on $n-2$ pieces of information (this having been established when degrees of freedom were discussed), and so in order to make the comparison "fair" we must scale each of the numerator and denominator by the number of pieces of information on which each is based. This argument seems to resonate with some students, though the idea attempts to use knowledge about degrees of freedom that may be too "new" for students to readily access initially. This problem leads to inevitable questions: why is the line based on only one piece of information? Why $n-2$? Why isn't the regular ratio (unadjusted by degrees of freedom) good enough? These are tough questions—and the analogy is not strong enough to provide accessible answers. Of course, the questions have reasonable answers, but the answers lie outside the map implied by the analogy. The double-edged sword of analogies remains that while they can produce in students that "eureka" moment, when the map is incomplete they can instead produce frustration.

2.2   Analogy 2: The Undiscovered Island and Partitioning Variability with Sequential Sums of Squares

Martin (2003) introduced the analogy of the "undiscovered island" to explain how the order in which variables are fit in a model changes their sequential sums of squares. Here, the analogy is explored more deeply, with a view to more clearly incorporating the notion of multicollinearity and its effect on the sequential breakdown of explained variation in the analysis of variance. The analogy describes an uninhabited, unexplored island in the days of the great exploration of the oceans by colonial powers. The source idea is that the exploration and the claiming of territory depended critically on which explorer arrived first. So, as explorers arrive at the island one after another, they are only able to explore and claim territory that has not already been claimed. Further, some parts of the island are impenetrable jungles, so some territories cannot be explored (remembering that the colonial powers did not have access to Google Earth!). Mapped objects (source/target) exist in both domains (explorers/covariates; explored territory/explained variation; impenetrable jungle/ unexplained variation; sequence of arrival of explorers/sequence of fit in model). The map is fairly strong, and the story sufficiently engaging that students can readily transfer the idea from the source domain to the target domain. Further, other notions such as multicollinearity and marginal explanatory power can also be integrated into the analogy with strongly mapped elements. Figures 1 and 2 show how two great explorers coming from the same direction can each look "marginal" if they happen to arrive after the other one. This situation is an analog of two "good" variables that are roughly collinear—i.e., they are both carrying much the same information—so, the order in which they are fit determines which of them seems most important in terms of explaining variation in the response. Figure 3 shows the situation when variables (explorers) are roughly independent (coming from completely different directions)—in this case, the order of fit (arrival) doesn't matter as the way the variation is explained (island is partitioned) does not change. In either case, the total amount of explored island (variation explained) is the same irrespective of the order of arrival (order of fit), so the fitted model itself does not change, only the way that the territory has been divided up (explained variation has been partitioned).

Fig. 1 X_1 and X_2 come from roughly the same direction (collinear), X_1 arrives first (gains largest sequential sum of squares), leaving little for X_2

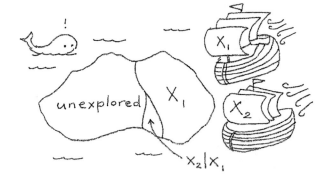

Fig. 2 X_1 and X_2 come from roughly the same direction (collinear), X_2 arrives first (gains largest sequential sum of squares), leaving little for X_1

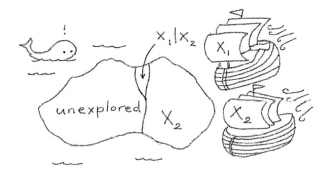

Fig. 3 X_1 and X_2 come from opposite sides (roughly independent), order of arrival does not materially affect territory claimed (sums of squares)

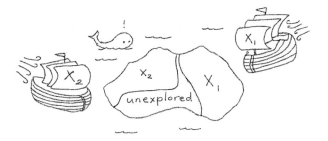

2.3 Analogy 3: Symptoms Versus Disease and Leverage Versus Influence

Perhaps the most frustrating experience when teaching regression is that of having students who confuse high leverage with influence. Over many years, this one concept seems to have been the hardest of all to reliably communicate in my regression courses. Why this should be is hard to pinpoint, as the distinctions could hardly have been made clearer, with many facts and examples used as evidence of the difference between the two concepts. For instance, leverage is a function only of the covariates, not the response, so it simply *cannot* be the same as influence which must, necessarily, involve *some* consideration of the response variable. Yet in almost every assessment item, students routinely (and cheerfully) declared points with high leverage to be influential. For many years, despite my intense efforts to clarify this issue, the confusion between these two concepts continued—until I began using a simple but powerful analogy that has radically addressed this confusion. In the analogy, high influence is aligned with a disease, while high leverage is aligned with a symptom of that disease. The analogy is strong because along with the strong map between objects in the analogy, there is also a strong map of structural relationships in the source domain to the target domain. In general, in human health, symptoms are fairly readily detected, just as leverage is easily calculated. Disease, on the other hand, may be hard to *directly* detect, and so it is often the case with high influence. Diseases are often signaled by symptoms, just as high influence is often signaled by high leverage, but as everybody knows, a sneeze *may* be a sign of a cold, but to actually detect the virus causing a cold would require a visit to a laboratory, and, in any event, not every sneeze is associated with a cold. In this way, the tendency I noticed for students to declare a point as influential in many ways resembled the tendency for people to declare that they had a cold when, in fact, they were simply sneezing! Even more complex concepts such as masking are well accommodated within the analogy—the presence of a disease may well be masked by the presence of additional symptoms beyond those classically associated with the disease. Since I began using this analogy, the tendency for points to be routinely assessed as having high influence simply because they have high leverage has dropped markedly. Just as the realization that symptoms and diseases are associated but not synonymous is part of what people commonly understand, now this realization has been transplanted to the context of leverage and influence. Even more powerfully, the context of both the source domain and the target domain is diagnostic, the former medical and the latter statistical, making this a very appealing analogy.

2.4   Analogy 4: Competition Between Sporting Teams and Combining p-Values

A very common sight within journals in just about any field is a large table with columns labeled "variable", "estimate", "SE", "*t*-value", and "*p*-value" under which sits row after row of figures, typically festooned down the right-hand edge with an array of daggers and stars representing significance at 10 %, 5 %, and 1 % for each of the listed variables. This table is typically the result of a model fitting exercise, the ultimate intention being to make a judgment—and choice—of which explanatory variables are important in describing the response variable, and in many instances, this choice is made by simply retaining those variables "starred" in the table and removing the others as extraneous. What this exercise does not explicitly take into account, of course, is that the multiple tests on which this aggregate judgment is made *cannot* be simply combined to produce this outcome because each individual test is marginal—that is, each test is based on an assumption that is incompatible with every other test. At first blush, most students find this situation utterly confusing: if the first line in the table suggests that β_1 is plausibly zero, and the second line in the table suggests that β_2 is plausibly zero, why can't I just take both of the corresponding variables out of the model? The answer lies, of course, in considering the actual models being compared in each of the tests being conducted (this discussion leaving aside for the moment the vexed problem of multiple testing). For simplicity, suppose there are only two covariates, X_1 and X_2. The test summarized by the first line (corresponding to X_1 and β_1) in the table reflects a comparison between the two models

$$E(Y_i) = \beta_0 + \beta_1 X_{1i} + \beta_2 X_{2i}$$
$$\text{and}$$
$$E(Y_i) = \beta_0 + \beta_2 X_{2i},$$

while the test summarized by the second line in the table (corresponding to X_2 and β_2) reflects a *completely different comparison*, between

$$E(Y_i) = \beta_0 + \beta_1 X_{1i} + \beta_2 X_{2i}$$
$$\text{and}$$
$$E(Y_i) - \beta_0 + \beta_1 X_{1i}.$$

Meanwhile, the proposed action suggested by combining the two tests is that of removing both variables from the model, which amounts to making a comparison between yet another pair of models, between

$$E(Y_i) = \beta_0 + \beta_1 X_{1i} + \beta_2 X_{2i}$$
$$\text{and}$$
$$E(Y_i) = \beta_0;$$

that is, between the full model and a model featuring only an intercept. You cannot combine the tests summarized by the first two lines, the separate tests for β_1 and β_2 — as the underlying comparisons are inherently incompatible. Further, you certainly cannot infer the result of the final comparison you wish to make by considering the first two tests. By now, most students' heads are spinning. Compare what model with what model? How can you say β_1 is plausibly zero and β_2 is plausibly zero, but they are not both plausibly zero? Huh?? One of the difficulties rests with the way in which null hypotheses are expressed, typically only explicitly referencing one parameter under an implicit assumption that all other parameters are present. But students often interpret these statements as *absolute* statements and ignore the implicit assumptions and underlying models, leading to the misunderstanding related above.

Here, a simple analogy can help. Think of each of the competing models in the above description as sporting teams engaged in a round-robin contest. Call the teams Sydney $(E(Y_i)=\beta_0+\beta_1X_{1i}+\beta_2X_{2i})$, Melbourne $(E(Y_i)=\beta_0+\beta_2X_{2i})$, Brisbane $(E(Y_i)=\beta_0+\beta_1X_{1i})$, and Canberra $(E(Y_i)=\beta_0)$. Then run the student through the following sequence of games for Sydney:

Game 1: Melbourne plays Sydney, and Melbourne wins
Game 2: Brisbane plays Sydney, and Brisbane wins
Game 3: Sydney plays Canberra, and Sydney wins

Then ask, is there anything about this set of results that is inherently contradictory? The answer is invariably no, sets of results like this are commonplace in sports. Even if stronger teams always beat weaker teams, this set of results is completely unsurprising, indeed expected, if Melbourne and Brisbane are strong teams, Sydney is a medium-ranking team, and Canberra is weak. But this system of games is analogous to the sequence of tests described above, a series of hypothesis tests for which students typically assume that the results of the first two tests imply the result for the third. The strong identification between teams and models is a useful mapping, and the key to understanding in the target domain rests in the realization that the set of results in the source domain is also unsurprising because the three contests (tests) are not as related as the null hypothesis statements make them seem.

2.5   Analogy 5: Choosing the "Right" Meal from the Menu and Model Selection

Model selection is a process many students find difficult to understand, particularly when there is a large number of covariates. Having been warned of selecting combinations of variables based on large tables of marginal p-values, they know they cannot proceed that way, and in the presence of many covariates, the sheer number of available models is formidable. Automatic procedures such as stepwise procedures are a seductive alternative, but remembering the algorithm has proven difficult for many students (particularly when the process is completely automatic—"black box"—in software). To motivate the algorithm, the following analogy has proven useful. Imagine a restaurant with a large, diverse menu. Obviously you want the "optimal" meal. So, begin by selecting from the menu the food you most like. Having eaten that morsel, you gaze again at the menu, at the next step choosing the food you like next best provided, of course, it goes well with what you have already eaten. The process continues until either there is nothing on the menu that complements what you have already eaten or you are full. This process is like the forward selection method of model selection—each step is conditional on the previous step, and the process cannot step backwards (since you eat the courses as you progress).

In the model selection process, the same sequence is followed, with variables chosen at each step depending on their contribution to the model given what has already been added. The meal (model) is built one item at a time until the contributions from additional menu items (variables) have diminished below some acceptable threshold. Refinements such as moving to a forward stepwise procedure that incorporates successive add–delete variable phases can be accommodated by the analogy by simply removing the requirement that courses are eaten as they arrive at the table—instead, the order is built sequentially with menu items added—and potentially deleted—as their suitability is assessed in the context of what else has already been ordered at the preceding step. The analogy is very simple but in my experience students find it very motivating. The experience of ordering food and thinking of pleasant combinations of food is both a common experience and, generally, a positive and pleasant one. These factors, plus the strength of the maps between objects and relationships between the source and target, create a positive environment for understanding the new algorithm, and my experience has been that this analogy is a particularly effective way to describe stepwise regression procedures.

2.6 Analogy 6: The Blind Men and the Elephant and Understanding High-Dimensional Data

Visualization is an incredibly useful tool in statistical modeling. Every student of statistical modeling has to have seen Anscombe's quartet (Anscombe 1973), the collection of four datasets that all yield identical numerical regression output but which could scarcely be more different when plotted. This powerful example immediately convinces all students of the wisdom of visualizing data, but visualization is a seriously difficult problem when data is high dimensional. Explaining *why* visualization in high dimensions is so problematic can be difficult—many graphical displays, for instance scatterplot matrices and trellis displays, offer a glimpse at high-dimensional data, but the truth behind the data can remain well hidden.

One approach to demonstrating this truism is to carefully construct a multivariate dataset that effectively defeats all attempts to discover its real structure by looking in the obvious directions. This approach can work well, but it has a considerable downside—it casts the teacher as an illusionist, a trickster, even a huckster.

Yet here an analogy—the brilliant fable of the blind men and the elephant—illustrates the situation wonderfully. The history of this story is long, and it has been used to teach a wide range of lessons, from the need for effective communication to the idea of tolerance for those who have different perspectives. Perhaps the best-known rendering of the tale is the poem by John Godfrey Saxe (1816–1887), a work now in the public domain as Saxe has been deceased for over a century:

> It was six men of Indostan,
> To learning much inclined,
> Who went to see the Elephant
> (Though all of them were blind),
> That each by observation
> Might satisfy his mind.

The First approach'd the Elephant,
And happening to fall
Against his broad and sturdy side,
At once began to bawl:
"God bless me! but the Elephant
Is very like a wall!"

The Second, feeling of the tusk,
Cried, "Ho! what have we here?
So very round and smooth and sharp?
To me 'tis mighty clear,
This wonder of an Elephant
Is very like a spear!"

The Third approach'd the animal,
And happening to take
The squirming trunk within his hands,
Thus boldly up and spake:
"I see," quoth he "the Elephant
Is very like a snake!"

The Fourth reached out an eager hand,
And felt about the knee:
"What most this wondrous beast is like
Is mighty plain," quoth he,
'Tis clear enough the Elephant
Is very like a tree!"

The Fifth, who chanced to touch the ear,
Said "E'en the blindest man
Can tell what this resembles most;
Deny the fact who can,
This marvel of an Elephant
Is very like a fan!"

The Sixth no sooner had begun
About the beast to grope,
Then, seizing on the swinging tail
That fell within his scope,
"I see," quoth he, "the Elephant
Is very like a rope!"

And so these men of Indostan
Disputed loud and long,
Each in his own opinion
Exceeding stiff and strong,
Though each was partly in the right,
And all were in the wrong!

MORAL
So, oft in theologic wars
The disputants, I ween,
Rail on in utter ignorance
Of what each other mean;
And prate about an Elephant
Not one of them has seen!

Apart from the teacher having the delightful experience of reciting a poem in a statistics class (so you already have everyone's attention), the reward is that the final line of the poem states *exactly* the critical problem with high-dimensional data—it simply *cannot* be seen, at least not in the low-dimensional space in which humans live. The time to consider the scatterplot matrix for *that* trick dataset is right *after* the poem has been read. Even despite the enormous advantages conferred by the use of small multiples allowing so many directions in the data to be assessed at once, the students realize very quickly that they are no better off than the committee of blind men standing before the elephant. It is then that, as a class, the journey to understand high-dimensional data begins, acknowledging that we all begin with the same basic problem—we are all essentially blind.

It is interesting to note also that visualization of data and relationships within data—a basic tool for statisticians—is itself a classic example of analogical thinking, one that is so embedded that it is now a completely automatic process. Statistical graphics all embed a very simple metaphor—the size of a visual element (e.g., length, area, angle) must be proportional to the number it represents. As long as this metaphor is satisfied—and, remarkably, this rule is broken very frequently—then the simple analogy allows our visual comparisons of size to transfer seamlessly and quickly to an understanding of the difference between the underlying numbers. The metaphor is extraordinarily powerful, and the effects when the metaphor fails can be catastrophic. Edward Tufte even has a name for the effect when the metaphor breaks—he calls it the "lie factor" (Tufte 2001, p.57). The effect on decision-making when graphics misrepresent the numbers they are supposed to communicate further demonstrates the power of analogical thinking—when the relationship map behind the analogy fails, the whole house of cards can come tumbling down.

Analogies are a potent bridge between what is familiar and comfortable and what is new, uncharted territory. Analogical structure—mappings from the old to the new, along with the preservation of critical relationship maps—can be used to acquire new knowledge, and thus explore new vistas. Once the new knowledge is transferred from the source to the target domain, it becomes itself accessible. Analogies are also evocative, so their use promotes students remembering concepts far better than rote memorizing of formulas ever could—as the folklore says of elephants, they never forget.[1]

As a postscript, it is also prudent to remind students of that other lesson from the fable of the blind men and the elephants: the value of considering differing perspectives. In that vein, I close with the following tale …

Six blind elephants gathered together and the discussion turned to what humans were like. After a gentle discussion (elephants dislike heated argument), it was decided that they should each feel a human and then they could meet again to discuss their findings. After a careful examination of a human, the first blind elephant returned to the group. One by one the elephants went and made their own assessments, and when the group assembled again, the first blind elephant announced that she had determined what humans were like. A brief discussion ensued, with each elephant describing its findings. The verdict was unanimous. Humans are flat.

---

[1] http://www.scientificamerican.com/article.cfm?id=elephants-never-forget

3 Note

Developed from a keynote presentation at the Sixth Australian Conference on Teaching Statistics, July 2008, Melbourne, Australia.

This chapter is refereed.

References

Anscombe, F. J. (1973). Graphs in statistical analysis. *American Statistician, 27*(1), 17–21.

Bangdiwala, S. I. (1989). The teaching of the concepts of statistical tests of hypotheses to non-statisticians. *Journal of Applied Statistics, 16*, 355–361.

Brewer, J. K. (1989). Analogies and parables in the teaching of statistics. *Teaching Statistics, 11*, 21–23.

Feinberg, W. E. (1971). Teaching the type I and II errors: The judicial process. *The American Statistician, 25*, 30–32.

Glynn, S. M. (1991). Explaining science concepts: A teaching-with-analogies model. In S. M. Glynn, R. H. Yeany, & B. K. Britton (Eds.), *The psychology of learning science* (pp. 219–240). Hillsdale, NJ: Lawrence Erlbaum.

Glynn, S. M. (1994). *Teaching science with analogies: A strategy for teachers and textbook authors* (Research Report No. 15). Athens, GA: University of Georgia.

Glynn, S. M. (1995). Conceptual bridges: Using analogies to explain scientific concepts. *The Science Teacher, 1995*, 24–27.

Glynn, S. M. (1996). Teaching with analogies: Building on the science textbook. *The Reading Teacher, 49*(6), 490–492.

Glynn, S. M., Duit, R., & Thiele, R. (1995). Teaching science with analogies: A strategy for transferring knowledge. In S. M. Glynn & R. Duit (Eds.), *Learning science in the schools: Research reforming practice* (pp. 247–273). Mahwah, NJ: Lawrence Erlbaum.

Glynn, S. M., & Law, M. (1993). *Teaching science with analogies: Building on the book [Video]*. Athens, GA: University of Georgia.

Holyoak, K. J., & Thagard, P. (1995). *Mental leaps: Analogy in creative thought*. Cambridge, MA: MIT Press.

Martin, M. A. (2003). It's like, you know—The use of analogies and heuristics in teaching introductory statistics. *Journal of Statistics Education, 11*(2) online. http://www.amstat.org/publications/jse/v11n2/martin.html. See also Letter to the Editor: http://www.amstat.org/publications/jse/v11n3/lesser_letter.html and response: http://www.amstat.org/publications/jse/v11n3/martin_letter_response.html.

Martin, M. A. (2008) What lies beneath: inventing new wheels from old. In: MacGillivray, H. L., & Martin, M. A. (Eds) *OZCOTS 2008, Proceedings of the 6th Australian Conference on Teaching Statistics* (pp. 35–52). Statistical Society of Australia Incorporated (SSAI) http://iase-web.org/documents/anzcots/OZCOTS_2008_Proceedings.pdf.

Moore, D. S. (1997). *Statistics: Concepts and controversies* (4th ed.). New York: W. H. Freeman.

Saxe, J. G. (1816–1887) *Six men of Indostan*. Poem is in the public domain and the work is no longer in copyright.

Tufte, E. R. (2001). *The visual display of quantitative information*. Cheshire, CT: Graphics Press.

Experience Early, Logic Later

K.L. Weldon

Abstract The motivational value for students of problem-based immersion in the process of data collection, data analysis, and interpretation is accepted by many. However, the culture of instruction through technique-based courses is still used at the tertiary level in many universities. The coverage of topics seems to trump guidance through the process of data analysis. In this chapter, I suggest how to complement a problem-based experiential presentation of statistical methods with a presentation of the abstract structures necessary for future applications. A series of problem-based courses might fail to highlight the general and transferable concepts and principles that help to bring coherence to the toolbox of statistical techniques. To overcome this shortcoming one can present the logical structure—that is definitions, strategies, theoretical frameworks, and justifications— to unify the collection of problem-specific methods, but only after extensive immersion in practical problems. Once students have experienced the effectiveness of the practical statistical approach, they may be better prepared to absorb the abstract generalizations.

Keywords Experiential learning • Statistics education • Mathematical statistics • Graphical methods in statistics • Data analysis • Reform of statistics education

1 Introduction

Statistics educators have been trying to improve undergraduate statistics instruction for decades. Some progress has been made but the forces of the status quo are formidable. One of the most frustrating constraints relates to the economics of textbook publication: few publishers will accept a script that is much different from the current market.

K.L. Weldon (✉)
Simon Fraser University, Burnaby, BC, Canada
e-mail: weldon@sfu.ca

H. MacGillivray et al. (eds.), *Topics from Australian Conferences on Teaching Statistics:*
OZCOTS 2008-2012, Springer Proceedings in Mathematics & Statistics 81,
DOI 10.1007/978-1-4939-0603-1_3, © Springer Science+Business Media New York 2014

Another constraint is the human effort required by both instructors and students to blaze a new path. Moreover, the disincentives to teaching effort, especially at influential universities, are well known. In spite of these impediments to reform, a small group of reformers is motivated to keep trying.

If students as a group had a keen interest in statistics, both teaching and learning would be more successful. In this chapter, I want to encourage course designers and instructors to focus on the student motivation for the subject, even at the expense of short-changing the student with the usual list of inferential tools. I will argue that guided immersion in real data-based problems, in contexts of interest to students, is a more effective way to produce useful learning of statistics basics than to present a logical sequence of techniques, even if the techniques are illustrated with applications as they are introduced.

The organization of the chapter is as follows: The first section considers an overview of the progress of statistics education over the last quarter century. Next, the style and content of textbooks is used as a proxy for the teaching style and course content of many undergraduate courses in statistics—that the advances in textbooks have not solved the pedagogical problems. The important role of context-based motivation, "experience-based instruction," is then discussed. Next, some suggestions are presented concerning the year-levels at which context-based instruction is appropriate, and the related issue of class size is considered. Three examples of context-based teaching of statistics theory are then outlined. The final sections of the chapter discuss the implications of context-based instruction for both undergraduate and graduate statistics courses.

2 Reform in Statistics Education

The ICOTS conferences that began in 1982 initiated a continuing international focus on the issues of teaching statistics. OZCOTS, USCOTS, ICME, and the ISI/ IASE Satellite Conferences have also been a part of this activity. An unofficial theme of all the early conferences seems to be that instruction in the subject had not adapted appropriately to the expansion of statistics audiences from math majors to all majors. An additional theme of the more recent conferences seems to be that the changes associated with statistical software availability have not been adequately absorbed into undergraduate curriculum and pedagogy. In fact, an overarching theme is the lack of adaptation to changes in statistics instruction to reflect the changing practice in the discipline. As a participant in ICOTS 2, I joined the rising voices asking for change, and there were many good ideas being proposed in 1986. Consider the following quotes from ICOTS 2 in Victoria, BC, 1986:

> The development of statistical skills needs what is no longer feasible, and that is a great deal of one-to-one student-faculty interaction … (Zidek 1986)
>
> The interplay between questions, answers and statistics seems to me to be something which should interest teachers of statistics, for if students have a good appreciation of this

interplay, they will have learned some statistical thinking, not just some statistical methods. (Speed 1986)

Using the practical model [of teaching statistics] means aiming to teach statistics by addressing such problems in contexts in which they arise. At present this model is not widely used. (Taffe 1986)

To take advantage of these developments, one must recognize that, while most statistics professors like statistics for its own sake, most students become interested in statistics mainly if the subject promises to do useful things for them. I believe that even the seemingly limited goal of developing "intelligent consumers of statistics" is best attained if students try to produce statistics on a modest scale. Only then do most students seem to become sufficiently intrigued with statistics to want to learn about statistical theory. (Roberts 1986)

These ideas span different parts of the problem: the need for interaction of students with experts in statistics, the need for students to learn the whole process of statistics from verbal questions to verbal answers, the need to incorporate context into students' experiences in statistical analysis, and the need to excite students about applications before presenting the theory. As an index of the extent to which these suggestions have been adopted, consider their impact on current textbooks. I suggest that the impact has been very slight, partly because the suggestions all relate to the process of teaching rather than the techniques to be learned. In fact, it is hard to imagine how a textbook would be written that would help the instructor with the above recommendations. One might conclude that a good start to implementing the recommendations is to contemplate abandoning the dependence on a textbook, at least for the sequencing of course topics. For undergraduate courses, the current textbooks could be used as a reference resource for students, rather than as a course outline. The text assigned to the course could remain as in a traditional course, but the instructor could change completely the role of the text.

I cannot summarize all the recommendations of the series of ICOTS, ISI, and ICME conferences any better than to quote from Brian Phillips report of David Moore's Invited address to the ICME 1996 conference (Phillips(1996)):

In discussing what helps students learn, [David Moore] listed the following:

Hands-on activities	Explaining reasoning
Working in small groups	Computer simulations
Frequent and rapid feedback	Open questions in real settings
Communicating results	Learning to work co-operatively

Even though many of these ideas were discussed in earlier statistics education conferences, they were still "new" in 1996 as they are today. The question I wish to ask readers to consider is how well modern textbooks incorporate these strategies, and also whether it is possible for a textbook to provide for all these strategies. I suggest experiential learning of the kind proposed in this chapter is one way to incorporate all these strategies, and that textbooks should pursue an important but limited role as reference agents for students, and not as lecture guides for instructors and students.

3   More than Textbook Reform

Lovett and Greenhouse (2000) review and update the psychological research on course design in statistics to make recommendations for curriculum reform. The reforms they highlight are listed in their paper as "Collaborative Learning, Active Learning, Target Misconceptions, and Use of Technology." "Collaborative learning" refers to learning in teams and peer discussion. "Active learning" includes exploratory investigation and data collection, the "target misconceptions" item is described more fully as "confront students with their misconceptions", and "use of technology" means allowing students to use statistical software for both calculations and exploration. Few instructors would find these suggestions startling. However, it should be noted that textbooks do not help much with any of these reforms. What do textbooks say about teamwork, learning based on exploration, or confronting students with their misunderstandings? More and more textbooks do encourage "use of technology" although even in this reform area, exploratory use of software is less often proposed than are demonstrations or prescribed calculations with software. The implementation of the recommended reforms requires much more than reforming the textbook.

Moving away from the textbook, or a sequenced curriculum based on a list of techniques, raises many questions for the course designer. Some of the things we want students to experience are unobservable, and the final examination that tests the outcome may not be, by itself, the perfect instrument for guiding the learning. One suggestion to improve the situation has been proposed by Wessa (2008): providing an archive that captures students' calculations as well as their calculation outcomes. In fact, the facility provided by Wessa facilitates communication between instructor and student, and among students as well, about the actual calculations under discussion. This removes the concern about arithmetic, and replaces it with a focus on method and interpretation. A pedagogical benefit is that the instructor does not have to force the student into one particular mode of calculation, and this recognizes that there are often a variety of ways to extract information from data. Wessa's facility is based on R but the user does not need to know R to use it. There is no charge to use the facility. To fully understand the potential of the Wessa facility, it is necessary to explore the website www.wessa.net. However, Wessa (2008, Personal Communication) has reported informally some quantitative evidence that involving students in discussion of quantitative strategies does actually improve their score on objective examinations, in which the examinations aim to assess conceptual understanding.

Another way to assist the instructor in moving away from the textbook as a lecture outline is to provide a reference-friendly electronic textbook for student access. Students depend increasingly on clickable sources. One excellent example of this is the freely downloadable text called CAST (Sterling 2002). With searchable key words and a detailed table of contents and index, this provides easy access to text material. Because it is electronic, optional links for further information are available. Another benefit of the electronic source is the multitude of java-based animations and parameter sliders. It is just more fun to use than a paper text!

Chance and Data Analysis

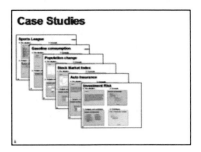

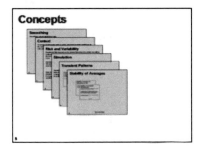
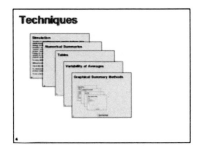

Fig. 1 Display from Roadmap Tools version of STAT 100. Each icon provides links to a more detailed display, and cross links are facilitated

An interesting development making an experience-based course feasible is the Excel-based Roadmap Tools (Carr 2008). This software allows the instructor to prepare notes outlining the "experiences" and at the same time capturing the techniques and concepts illustrated as they are met in the explorations of the experiences. Figure 1 gives a hint of the display: embedded slides are brought forward by clicking, and communication with the instructor or other links are also enabled through this medium. In reviewing a technique, it is possible to return with a click to the case study in which the technique was introduced, or more than one case study if appropriate. Similarly for the concepts arising in the case studies. Or, if the student wants to know what techniques and concepts were supposed to have been learned from a case study, the linkage is there to provide the information. Students can use the display initially to access the case studies, and subsequently to ensure that they have mastered the intended techniques and concepts. This is one way technology can help students stay organized within a case-study or experiential approach.

An example of an electronic textbook designed for reference rather than as a lecture guide is the handbook provided jointly by NIST and SEMATECH at http://www.itl.nist.gov/div898/handbook/

With such a helpful electronic resource, students should be less dependent on a traditional textbook.

The benefits of experiential learning have been recommended for many years, notably as a common theme of ICOTS 4 (1994) sessions in Marrakesh. Don Bentley's "Hands-On and Project-Based Teaching" (Bentley 1994) was so popular that it had to be broken into three sessions. However, the adaptation of this strategy into the current curriculum context seems to be problematic, since it is still not a common strategy at the tertiary level. An example of the creative atmosphere surrounding these sessions is the following quote from an abstract of Allan Rossman (1994): "In this presentation, I describe a project which takes this approach to the extreme of abandoning lectures completely." His abstract of the paper titled "Learning statistics through self-discovery" concludes with "The goals of these activities are to create a more enjoyable and productive learning environment as well as to deepen student's understanding of fundamental statistical ideas." However, university traditions seem to require lecture schedules and student–faculty interaction, and so a practical problem is how to incorporate the widely-recommended project device into a traditional lecture course. The proposals of this chapter suggest a way of incorporating experiential learning into the undergraduate statistics curriculum.

4   Experience-Based Instruction

To a mathophile, logic is beautiful, but most practitioners of statistics are not mathophiles. We need to keep that in mind when we are directing our pedagogical efforts toward students of statistics. We want to attract future practitioners of statistics to our statistics courses. What device can we use to show the charm of statistics without losing the underlying logical structure? I will argue that the logic of statistics can be instilled subversively by seducing students through immersion in the process of context-specific, data-based "discovery", and only later providing the logical framework that is more generally applicable.

Of course, this approach is not new. Not only the ICOTS 4 concentration on the idea in 1994, but an intriguing compilation of student-conceived projects is recorded in the locally published volume in 1997 by MacGillivray and Hayes "Practical Development of Statistical Understanding." The report records the results of student-selected projects which satisfied the criteria for a problem-based statistics course. Although each student would have primary responsibility for only one project, and not a sequence of projects, the resource does suggest the richness of student-selected problems for motivating learning of the entire process of statistical analysis.

Even earlier, Tukey (1977) emphasized the importance of involving students in data analysis unencumbered by assumptions of parametric models. His emphasis on visualization and an exploratory approach was revolutionary at the time. He felt that students needed experience with data more than knowledge of formal methods of parametric inference. But the project-based approach that would provide this experience tends not to be used—most undergraduate courses still follow closely textbooks organized by parametric technique. When the project-based approach is used in a text-dependent course, it is thought to be an add-on rather than the main driver of exposure to statistics. The suggestion here, as it was in Rossman(1994), is that with adaptation, it can be the main driver, and that there are compelling reasons to consider doing it this way.

The mathematical culture of statistics instruction is pervasive. In this culture we think it is obvious that, to teach statistics, we need to start with basic definitions, follow up with basic tools, and build on these basics to construct the commonly used strategies of statistics practice. However, if we apply this seemingly obvious approach to other disciplines, it does not seem so obvious. For example, to teach conversational Spanish, we would start with vocabulary, grammar, and pronunciation, and after a long period of becoming familiar with these skills, encourage students to converse. But immersion programs show that the formal phase of instruction works best after a lengthy immersion in motivated oral practice. Likewise, English grammar is best taught to children after they can speak English! Or, as another example, consider the math approach to teaching social geography: start with definitions of urban, climate, transportation corridor, enumeration area, etc. and get to the human impacts much later. To engage students' interest, it might work better to talk about human impacts first, and get to the formal definitions later. This same approach might not work for mathematics instruction, but it may well work for statistics instruction.

Whether a statistics course is designed as a service course or a mainstream course, the content tends to be technique-based rather than problem-based. Textbooks encourage this approach, and both students and instructors find textbooks a useful guide. Within the style of technique-based courses, many strategies have been devised to increase interest in the presentation of the techniques: data-collection projects, personal-data comparisons, in-class presentations, computer-based games, computer-quizzes with feedback, and simulation by applets or statistical software programs. These strategies certainly improve the likelihood that students will learn the techniques, and in some cases will increase interest in the techniques. But absent from these many strategies is the thrill of discovery: unexpected findings or anomalies that may have a bearing on the information gained from the data. What is often missing from traditional courses is the opportunity to use general intelligence, in combination with techniques learned from past experiences, to uncover information from data. How many students of statistics are aware of the fact that most applications do not use the "standard" techniques without adaptation?

An alternative to the technique-based course is the project-based course. The obvious argument against a project-based course is that the students will find the collection of techniques associated to be a jumble of unrelated tricks, rather than a logical sequence of strategies. But just as with geography or language, the formalities are a lot more useful to students after the students have an in-depth exposure to some examples. If students have been motivated to wonder what it takes to decide if a differential group effect is consistently reproducible, or transient, then they will be interested in the concept of a hypothesis test to tidy up the confusion. But the understanding of the information dilemma really needs to be internalized before this tidying is appreciated. Most instructors would agree that both theory and application need to be covered in a statistics course for the course to be useful to students.

However, the mathematical approach of theory-first, application-after may not work as well as application-first, theory-after. The reason is that for the vast majority of students, it is easier to arouse interest in an application than in a theoretical concept.

Some instructors routinely use application examples to introduce statistical techniques, which is a reasonable way to motivate theory. But if the guiding theme of such a course is a series of application-technique pairs, this has not really deviated from the technique-oriented course. For application immersion, one needs an application that engages students for several days or weeks, with data to analyze, external background information to track down, feedback on analytical techniques suggested by students, opportunities to verbalize what the question is and what the findings are, and time to read up on similar studies that have been done. Classes are an opportunity for the instructor to suggest multiple approaches, invite criticisms, provide feedback from past suggestions of students, and ask questions for students to work on outside class. Only after two or three such projects are completed (maybe over 15 h of instruction) would the instructor have enough technique material to outline the logical links among the techniques: types and numbers of variables, formal and informal analyses, estimation vs. testing, prediction vs. modeling, and experiments vs. experiences. The theory would be presented as a *simplifying* device rather than as an perplexing abstraction.

Students in "mainstream" courses of statistics (courses that would not be called service courses) generally have some tolerance for a mathematical approach to statistics. In particular, students of engineering, natural science, business, or psychology are required to have some math basics in their programs. But do these students actually like the mathematical, logical approach to their subjects? Is the chemistry student more interested in the periodic table or the relationship of sugar and alcohol? Is the business major more interested in the definitions of credits and debits or the earnings growth rate of Google? Mathematically-oriented students might actually be more interested in the periodic table and definitions of debits and credits, than the more applied topics, but such students are a small minority of the ones we want to introduce to statistics. As Cleveland (1993) says:

> A very limited view of statistics is that it is practiced by statisticians. … The wide view has far greater promise of a widespread influence of the intellectual content of the field of data science.

If we accept that the target audience, even in mainstream courses, is users of statistics, rather than the statisticians that are a subset of this group, then we need to think about how to make statistics courses attractive to future users. While some advantages can be had from making the course process a pleasant social experience, with team assignments, and convenient venues, an important draw is to have very interesting content. To say that the logical, sequential approach to the introduction of statistical techniques is the only viable one is to deny the importance of motivation in learning. It is important to make the introduction of statistical ideas fall out of stimulating investigations into real data-based questions. The logical structure of statistical strategies can come later.

Improved learning as a result of heightened motivation is just one reason to use in-depth exposure to applications. Another reason is that students need to learn the *process* of statistical investigation, and not only the strategies and tools. This process involves many attitudes that need to be learned: the presumption that more subject-matter knowledge may help; the realization that modeling and analysis is to some extent a trial-and-error process, the appreciation of the dangers of overfitting or overanalysis, the importance of graphical methods for identifying anomalies and summarizing results, the availability of resampling methods when standard methods fail, and the appreciation of power in summarizing findings. These things are hard to reduce to single lessons—demonstration and repetition are required. Guiding students through immersion in real data analysis exercises, right from the problem formulation stage to the ultimate report of findings, is one way to "indoctrinate" students in this process.

The efficacy of experiential learning has support from the psychological literature. The "constructivist" approach to learning involves social collaboration and communication and individual responsibility and experimentation. Although the concept of constructivism is quite old, course designers are still working toward courses which incorporate all the aspects identified by recent researchers (Moreno et al. (2007)). Another recent example of support for this approach comes from Konold (2007). He argues for "bottom-up" instructional design rather than "top-down." The idea is that for effective teaching, we need to start with the student's current context, and use this as a starting point for introducing new ideas. Experiential learning, including exploratory data analysis, student-choice projects, and verbal reports, does seem suited to bottom-up instruction.

A recent and extensive bibliography of the contribution of educational psychology to pedagogy in statistics is given by Garfield and Ben-Zvi (2007). In reviewing this bibliography, one is struck by the long list of difficulties experienced by students in learning from traditional technique-based courses. It does seem that experience-based pedagogy deserves a greater emphasis than has been usually practiced so far.

An important contribution to the role of experiential learning in the computer age is explored extensively in the series of articles recently published in the International Statistical Review. A lead-in to this series is provided by Seneta and Wild (2007). Although the emphasis is on computer-based learning environments, the important role of experiential learning is highlighted.

5   Integrating Experience-Based Courses into Undergraduate Education

Where does such an "experience" course fit in the undergraduate curriculum? I think the approach can be used at all levels. In recent years I have initiated a few new courses into the offerings at my university. Although they were proposed for various levels of student, they all have the feature that they are a series of examples rather than a logical sequence of techniques. I'll give a brief overview of these courses, since their style is close to the style I am recommending.

STAT 100: "Chance and Data Analysis." At the first year level, data relating to accidents by young drivers, sports leagues, blue whales, and the stock market, and others introduced ideas of causality, time series smoothing, simulation, sampling surveys, and survival analysis. Of course, the basic definitions of means and standard deviations, frequency distributions, sampling variation, and scatter diagrams are introduced and repeated many times, in the context of the examples discussed. By the end of the course, the student has had all the techniques normally included in a first course although perhaps less drill than usual.

STAT 300: "Statistics Communication." After 2 years of basics, students can begin to comment verbally on what they have learned. More specifically, they should be learning how to explain why certain techniques are appropriate in a particular data analysis, and what the analysis really shows. This course asks students to critique or defend criteria like "unbiasedness" or "minimum variance," to comment on the use of hypothesis tests when the sampling frame is uncertain, to discuss how to report anomalous data, and how to present orally or in writing what the real findings are in an instance of data analysis.

STAT 400: Data Analysis. The approach in this course is to ask students to suppose that the information in the data is more important than the techniques normally used to analyze the data. This helps the student understand that statistical practice is problem-based, and students are expected to use all their knowledge and intelligence to get at the information in the data. Of course, it helps if the content of the examples is of intrinsic interest to students. I have used a badly designed internationally funded agricultural study to draw attention to the value of good design as well as to provide an opportunity for students to try to rescue some information from the study in spite of its bad design. Another example uses a profit maximization strategy for a wholesale distribution network, in a situation where demand data is censored by inventory. Other examples use the data sets from Cleveland's *Visualizing Data* Book for which multivariate graphing provides a visual approach to information retrieval. Creativity in analysis is encouraged. These experiences lead to the use of resampling techniques, simulation to assess trial-and-error solutions, graphical smoothing in one or more dimensions, context-guided strategies to avoid the pitfalls of stepwise regression, and more.

Students find these courses both challenging and rewarding, judging from the feedback provided as part of the department's routine evaluation procedures. They are challenging because they require the student to move away from mere textbook knowledge, and they are rewarding because they confirm that a student can integrate their intelligence with the techniques they have learned to produce useful information from data.

Of course, the big question concerning this type of course is: When do the students find out about the logical structure of the discipline? There are different approaches for students with different needs. For students who take only one course, it may be futile to try to convey the logical structure. Perhaps for this group it is better to convey an appreciation for the utility of statistical strategies, rather than the basic tools and concepts themselves. For students who take more than one course, the subsequent courses can supply the logical structure—if the students appreciate the utility of the subject from the problem-based course, they will be both motivated and receptive to the more formal approach. However, an alternative is to include this step within a problem-based course. For example, each module of say, 10 contact hours, can be followed by a "what tools have we learned" session with the logical structure emphasized. As mentioned earlier, this phase can be described as a simplification of the apparently chaotic collection of tools introduced for a particular problem. The website for STAT 100 at www.stat.sfu.ca/~weldon includes some examples of this approach. STAT 300 and STAT 400 are embedded in course sequences that include more formal courses, and so the logical structure is, to some extent, left to these other courses.

6   The Class Size Constraint

The idea of 1–1 instruction is clearly impractical at the undergraduate level. But small classes that allow discussion can sometimes be afforded. Over the last few decades, with the ubiquitous spread of data-based research into most disciplines, undergraduate class sizes have grown to 100 and more, making discussion during a class meeting a rare event. How can students be exposed to the whole process of data analysis in a setting of large-class lectures? The efficiency of large classes may be an illusion in the case of statistics.

Various strategies have been used to try to solve the lack of student–faculty interaction that occurs with large classes. Small group tutorials are one common approach. Tutor-assisted group projects such as the ones documented by MacGillivray and Hayes (1997) are another. Group assignments in which students help each other and thus reduce the need for faculty help is a third approach. But the guidance in the complete process of data analysis is most effective if an instructor experienced in both tools and applications has frequent interaction with students. Instructor-to-student lectures, as are common with large classes, do not provide this interaction. Strategies involving group work outside of lectures (as just mentioned) do provide some benefit. But the small class idea would be best at allowing the instructor to balance the motivation of student exploration with the provision of guidance in the most effective tools and strategies.

One recent report (Carnell 2008) which attempted to gauge the impact on learning of a single project in an introductory stats course, found that this project addition did not make an appreciable difference in learning outcomes. Perhaps several projects are necessary. If a project is seen as an extra, it may not be treated the same way as if projects are the main drivers of the course.

In a world of large classes in statistics, how does one move in the direction of experience immersion as a teaching device? One way is to have whole courses that are taught to smaller classes at the advanced stage (like STAT 400). Another is to try to create the experience in the large class by abandoning the "technique-coverage" approach and instead describe for students the process of development from idea to report. In such an approach, the assumption is that enough tools and strategies will be covered incidentally to the case-studies described to satisfy the programs that require one course in statistics (like STAT 100). The justification for this shotgun approach is that it is better to have an understanding of a few common tools, than little understanding of a complete toolkit. However, it may be necessary to convince university administrations that the best statistics education requires small classes. For this to happen, the public view of the discipline of statistics may have to change from a necessary evil to that of a creative and vital subject! If students become excited about the small-class statistics course they are taking, then perhaps the message will get through to administrators that the subject is worth the higher price. So this is another reason to focus on drawing students into the subject matter with material that seems immediately interesting and useful.

7   Exploratory Contexts

These days the internet provides a wealth of good examples for teaching material. In fact, there are some conventionally published sources as well: the text mentioned previously, MacGillivray and Hayes (1997), details the teaching experiences in 19 different application scenarios: from fishing to motorcycle accidents to Murphy's Law. The availability of ideas for projects is useful, but the pedagogic effectiveness of an example really depends on how it is presented. Consequently, the instructor's role is still key to the learner's outcomes.

The suggestion in this chapter is that exploratory data analysis, suitably guided, will lead a student to understand basic statistical strategies, and the student will learn the basic statistical strategies more thoroughly in a given timeframe than if the same strategies are presented in the more conventional way. The reason is that the student will be motivated by the obvious relevance of the strategies since they will be introduced as the data exploration requires them. But will the student be able to apply the strategies to new contexts? This is where the instructor's role is crucial. After several data exploration examples have been worked through, the instructor needs to make sure that the student has the big picture. This is where the logical relationships of the techniques need to be presented.

For students to be able to use their learning of statistical tools and concepts in new contexts, they do need the logical structure clearly in mind. For example, they need to be aware of the different scales of measurement, of the different ways comparisons can be made, and of the difference between parameter estimation and testing parameter credibility. But, as we have argued, to try to teach these in a sequence of techniques has not worked well—better to have them as a framework for techniques motivated by data-based projects, when there is a readiness for aggregating the pieces learned.

To further illustrate the potential of teaching techniques through experience immersion, I will briefly describe three examples. The first one should appeal to students since they get to choose an activity from their own lives. The second one has the advantage of relating to local conditions. The third one relates to a personal characteristic that students hold subconsciously and would often be of interest to a student in comparison with others. All of the examples could be used in a first course, or in an advanced course. Of course, statistics courses with a subject area focus (e.g. life sciences, business, engineering, or psychology) would likely include examples more closely related to the subject area, and would build on a more specialized student background. But these examples will suffice to illustrate the point of conveying useful statistical theory through comprehensible applied projects with real-world contexts.

7.1   Example 1: Sports Leagues

Students often have at least one sport they are interested in, either as a participant or a spectator. Team sports have the feature that game results are accumulated throughout the season and teams are repeatedly ranked using some points system. Suppose students are given the task of finding the accumulation table of a currently operating league, and commenting on the relative quality of the teams suggested by the table. Students should be advised to choose a sport that interests them, if possible. Questions for discussion might be:

1. If team A has more points than team B, does team A have a better than even chance of winning the next contest with team B?
2. Is there any evidence of a home team advantage?
3. If all the teams have the same chance to win each game, what would the league ranking look like?

Note that many students will have an opinion regardless of their statistics knowledge so far, so a discussion should be easy to stimulate. Where would the discussion lead? Here are just a few of the possibilities:

A better understanding of "better than and even chance"
A realization that current rankings are, at least in part, subject to "luck" or "random variation"
An opportunity to test a hypothesis by observing data
Consideration of conditional probability
An appreciation that randomness can deceive and often does
An opportunity for answering a question via simulation (by coin or computer)
A need to define a measure of variability (in point status)

The point of this example is that a context of interest to students can be the platform for introducing many important statistical tools and strategies, and because the answers to the questions are of interest to the students, the tools and strategies that help to get at the answers will also be of interest to the students. The motivation for learning statistics is based on a genuine interest and not only on the need for a good mark in the course. The learning will include the entire process of data-based study and not only an artificially simplified context. Moreover, the freshness of the discussion should make the process stimulating for both student and instructor.

7.2   Example 2: Auto Fuel Consumption

An example that I like to use is based on some personal data I accumulated on gas consumption for my car during a 5-year period. The graph shows the data:

Students are asked if they find any interesting or useful information in this graph, and are asked to analyze the data to see if there are any "trends" or anomalies. The initial response is usually negative. It turns out that fitting ordinary polynomial regression reveals nothing much, just as a visual scan would suggest. But any kind of nonparametric smoothing, even a moving average, shows a nice sinusoidal pattern in sync with the annual seasons. This pattern raises the question of the likely cause, and the many potential explanations include usage, temperature, precipitation, traffic, and tire pressure. Note also the likelihood of a negative serial correlation as a result of the way gas consumption is measured. What does the student learn from this experience?

Not all regular patterns in time can be discerned by eye
A correlation can have many causes, and often further data collection is suggested
Nonparametric smoothing is a useful exploratory technique
Not all interesting data is a sample from a population (time series)
The measurement context must be examined as part of the analysis

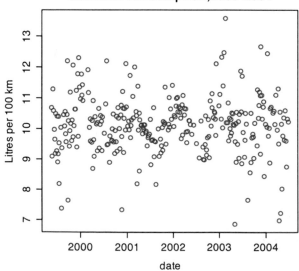

Gasoline consumption,1999-2004

7.3   Example 3: Crossing of Arms and Clasping of Hands

MacGillivray (2007) reports a project in which a large class of students were asked to report their normal way of crossing their arms and clasping their hands. Students are initially surprised that there is a "natural" way to do these things that is personal for each student. There is the question of whether gender, or handedness, explain the differences, and whether arm crossing is related to hand clasping. An advantage of this example over the other two is that, even if one is an unusual student with no interest in sports or driving, one is still likely to practice arm crossing and hand clasping. The referenced paper gives the full detail of this example, but some of the obvious lessons one learns from the experience are:

The procedure of developing an idea into a data-based study is a non-trivial exercise
Careful definitions and protocols need to precede data collection
Two features can be related even when exceptions exist
Categorical data is summarized by frequencies
Descriptive techniques need to be used before inferential techniques
Descriptive techniques can often reveal unexpected findings
"Obvious" relationships are sometimes not confirmed by data
Apparent relationships can be deceiving and require testing for reproducibility

Looking over the 20 "lessons learned" from these three examples, even though they are a partial list, should suggest the richness of the learning experience with respect to statistical practice. Are these lessons as useful to students as the ability to fit a line to x–y data, or test if two groups have the same mean? While students do need to learn about the calculation methods, they need these meta strategies as well to have a useful education in statistics.

These three examples suggest the broad spectrum of statistical tools and strategies that can be conveyed through immersion of students in stimulating applications. One only has to consider how the lessons-learned from the examples would be taught one-at-a-time in a logical sequence to see that the logical approach would lack the charm of the experience immersion. If we want to attract and retain students' interest in statistics, we need to consider charm! And, if we want students to understand the whole process of data analysis, we need to give them experience with the whole process of data analysis.

8   Target Audience for Statistics Strategies

The mathematics model of teaching statistics is a product of the twentieth century, math-based history of statistics. It seems to work for those few students who relish mathematical abstraction, and the current cadré of statistics instructors is mostly drawn from this group. But today modern statistics is practiced by a wide spectrum of engineers, scientists, and social scientists, and these users need more than a superficial knowledge of statistical strategies. STAT 100 does not adequately prepare these users. These users must be able to identify opportunities for data-based studies, plan data collection, explore data, extract valid information from data, and defend what they have done in words. It is this large group of future practitioners of statistics that needs the most attention at the undergraduate level, not the stat major. The stat major can benefit from math courses, even math courses with little statistical overlap. But the future statistics practitioners need to know how to use statistical software to explore data, and how to allow for study design shortcomings in coming to conclusions.

Practitioners educated in applied disciplines will not be able to handle all data-based problems they meet, and will find it necessary to seek the help of expert statisticians. However, to take advantage of this expertise, they must recognize the opportunity. Do the technique-based courses give students the insight needed to know when an expert can help? A student who has met a situation in which the perfect test is unknown, such as would be likely to happen in an exploratory study, may be willing to admit the need for expert help in later practice.

To become expert statisticians, graduate work would usually be required. This is where the full confluence of mathematics and statistics should be explored. Instead of designing a special undergraduate education for stat majors (future statisticians), it might be efficient to give all the statistical practitioners the same undergraduate education in statistics, but require further mathematical statistics in graduate courses. With this approach, undergraduates learn to appreciate statistics as a vital subject relevant to many careers, and are not deceived into thinking of statistics as a specialized form of mathematics; and graduates will not arrive at the serious study of statistics with a naive view that statistics is a bundle of calculation tools, and will realize that the role of mathematics in statistics is to allow adaptation of available methods to suit particular application contexts.

Mathematics is a powerful technology for clarifying complex ideas, and is essential for expertise in many "applied" disciplines. Theory in engineering, management science, environmental science, and many other fields is assisted by mathematics, but these theories are not only mathematics. The same is true of statistics. It is now clear that statistics is a separate discipline from mathematics, even though this was not always recognized in the past. We need to judge statistics expertise by its relevance to the extraction of useful information from data, and not by the mathematical expression of its tools. Our teaching of statistics should reflect this criterion.

9 Summary

The content of undergraduate courses in statistics has not changed very much in spite of the reform movement of statistics educators. Better textbooks, and the provision of computer software, have changed the tasks of the student, but objectives as revealed through tests and examinations have not kept up with the recommended reforms. What seems to be missing is the immersion of students in the entire process of data-based research along with frequent interactivity with the instructor. The motivation provided by interesting projects (suggested by the instructor or the students themselves) is an important factor in shaping the image of the discipline, as well as being a powerful stimulus to learning. It may be necessary to depart from large classes to accomplish a useful exposure to statistics: higher education administrators need to see undergraduate statistics as concepts and strategies rather than procedures and formulas. Statistics courses need to include more experience and creativity and less coverage and dogma.

Of course, balance of the old with new is probably the optimal strategy, but change has been so slow that extreme measures need to be contemplated. The basic recommendation in this chapter is to provide students with exciting experiences in extracting information from data, and after the student is completely amazed by the many surprising and useful strategies of statistics, proceed to provide the formal structure of the techniques used, including the mathematics as required. We need to recognize that the main audience for statistics at the undergraduate level is future practitioners of statistics, and the principal secondary market is for statistics appreciation courses. The training of future statisticians should not be considered an undergraduate mission. Future statisticians need the practical knowledge of the undergraduate education, as well as graduate work in mathematical statistics. Our focus in undergraduate education should be on experiential immersion in data-based discovery. The conveyance of the logical structure of formal statistical inference will to some extent be achieved simultaneously, but in any case should be relegated to the status of a secondary goal.

10   Note

Reprinted with permission from Weldon (2008) in MacGillivray, H. and Martin, M. *Proceedings of Sixth Australian Conference on Teaching Statistics*, available online at http://iase-web.org/documents/anzcots/OZCOTS_2008_Proceedings.pdf.

This chapter is refereed.

References

Bentley, D. (1994). Sessions 4 A,B,C. In *Proceedings of the Fourth International Conference on Teaching Statistics* (Vol. 1: pp. 17–31). The National Organizing Committee of the Fourth International Conference on Teaching Statistics, Rabat, Morocco.

Carnell, L. J. (2008). The effect of a student-designed data collection project on attitudes toward statistics, *Journal of Statistical Education 16*(1). Retrieved from www.amstat.org/publications/jse/

Carr, R. (2008). Roadmap tools for excel. Retrieved from www.deakin.edu.au/~rodneyc/roadmap-tools.htm

Cleveland, W. S. (1993). *Visualizing data.* Summit, NJ: Hobart Press.

Garfield, J., & Ben-Zvi, D. (2007). How students learn statistics revisited. A current review of research on teaching and learning statistics. *International Statistical Review, 75*(3), 372–396.

Konold, C. (2007). Designing a tool for learners. In M. Lovett & P. Shah (Eds.), *Thinking with data* (pp. 267–291). New York: Taylor & Francis.

Lovett, M. C., & Greenhouse, J. B. (2000). Applying Cognitive Theory to Statistics Instruction. *The American Statistician, 54*(3), 196–206.

MacGillivray, H. (2007). Clasping hands and folding arms: A data investigation. *Teaching Statistics, 29*(2), 49–53.

MacGillivray, H., & Hayes, C. (1997). *Practical development of statistical understanding: A project based approach*. Brisbane, QLD, Australia: Queensland University of Technology.

Moreno, L., Gonzalez, C., et al. (2007). Applying a constructivist and collaborative methodological approach in engineering education. *Computers & Education, 49*(2007), 891–915.

Phillips, B. (1996). Plenary lecture given by David S. Moore (USA) New pedagogy and new content: The case of statistics. *Proceedings of ICME 8*, Seville, Spain.

Roberts, H. V. (1986). Data analysis for managers. *ICOTS 2* (pp. 410–414). Victoria, BC, Canada.

Rossman, A. (1994). Learning statistics through self-discovery. Abstract in *Proceedings of the Fourth International Conference on Teaching Statistics* (Vol. 1: p. 30).

Seneta, E., & Wild, C. (2007). Preface. *International Statistical Review, 75*(3), 279–280.

Speed, T. (1986). Questions, answers, and statistics. *ICOTS 2* (pp. 18–28). Victoria, BC, Canada.

Sterling, D. (2002). *Computer assisted statistics teaching (Version 2.1)*. For an up-to-date version go to http://cast.massey.ac.nz/

Taffe, J. (1986). Teaching statistics: Mathematical or practical model. *ICOTS 2* (pp. 332–336). Victoria, BC, Canada.

Tukey, J. W. (1977). *Exploratory data analysis*. New York: Addison-Wesley.

Wessa, P. (2008). Free Statistics Software. Office for Research Development and Education, version 1.1.22-r5. Retrieved from http://www.wessa.net/. See also www.freestatistics.org/index.php?action=10 for more information.

Zidek, J. V. (1986) Statistication: The quest for a curriculum. *ICOTS 2* (pp. 1–17). Victoria, BC, Canada.

Transforming Statistics Education in South Africa

Delia North and Temesgen Zewotir

Abstract Challenges faced by Statistics Education in developing countries are similar, though often of larger magnitude, and of a more critical nature, than in more developed countries. This chapter focuses on the status of Statistics Education at school and tertiary level in South Africa. The authors give a historical overview, followed by a discussion of the current status, mentioning challenges and successes to building statistics capacity in the country, and emphasizing the importance of facing realities. Finally, mention is made of a few key projects, aimed at statistics capacity building, which have the potential to change the face of Statistics Education in a country grappling with legacies of the past, whilst balancing risks and opportunities of the future.

Keywords Statistics education • Developing country • Statistics capacity building projects

1 Introduction

As the world becomes more connected due to the economic, social, and political interdependence, it is vital that all citizens are able to orient themselves in the information and data-driven age, where decision-making is likely to call for skills of data collection, organization, analysis, and interpretation. Accordingly, there is an ever increasing need to disseminate more data, more accurately, in shorter times and in forms desired by users for further analysis (Wallman 1993). The training of statisticians and the raising of levels of statistical literacy is accordingly a challenge faced by countries all over the world. In developing countries however, these processes

D. North (✉) • T. Zewotir
University of KwaZulu-Natal, Durban, South Africa
e-mail: northd@ukzn.ac.za; zewotir@ukzn.ac.za

H. MacGillivray et al. (eds.), *Topics from Australian Conferences on Teaching Statistics:*   43
OZCOTS 2008-2012, Springer Proceedings in Mathematics & Statistics 81,
DOI 10.1007/978-1-4939-0603-1_4, © Springer Science+Business Media New York 2014

have to be carefully planned, in order to make the most use of the limited resources available. In South Africa, there are *further* confounding factors, the legacy of Apartheid, severe shortage of mathematics and statistics educators, serious lack of adequate professional development of teachers, a mixture of cultures and multiple languages being spoken in one class room and many more issues that negatively affect statistics capacity building in the country.

This chapter takes the reader on a tour of statistics education in South Africa, from the historical beginnings to the current scenario, mentioning the challenges and successes at the various levels and then concluding with a brief discussion of two key projects aimed at statistics capacity building in the country.

2 The Legacy of Apartheid

South Africa is the 25th largest country in the world, with around 51 million citizens, of which roughly 79.2 % are Black, 8.7 % are White, 9 % are of Mixed Race, and 3 % are of Other races (mostly Indian). The country has a very diverse population, with the Black African population alone having nine major ethnic groups. Accordingly, South Africa has 11 official languages (Brand South Africa 2012), of which English as mother tongue ranks only fifth according to the 2011 national census. Schooling is conducted in home-language at primary school level, but the language of instruction at university is either English or Afrikaans (the two official languages during the Apartheid era). English is by far the dominant language of instruction at university level, with Afrikaans instruction-based universities generally switching over to English at senior post graduate level, for projects, theses, and seminar presentations.

During the Apartheid era (1948–1994), Education Policies in South Africa were designed to assert white domination and African race inferiority (Badat 2004, formalizing the misguided notions of racial superiority and inferiority (Note that hereafter African shall mean Black/Bantu South African citizens, and does not include Colored or African citizens). The Report of the Inter-Departmental Committee on "Native Education," 1935–1936, is a highly instructive document, summing up the Educational Policy of the then South African Government (Wilcox 2003). In this report we read that "The Education of the White child prepares him for life in a dominant society and the Education of the Black child for a subordinate society." Furthermore, the National Party government, which was in office from 1948 to 1994, passed legislations which stripped Africans of political and economic rights, implementing an Education policy deliberately designed to keep them in subservience. The Bantu Education Act, Act No 47 of 1953, established a Black Education Department in the Department of Native Affairs, which was responsible for compiling a curriculum that systematically ensured that the young African scholars of that era were prevented from obtaining an education in keeping with the advances of the twentieth century. The author of the legislation, Dr Hendrik Verwoerd (then Minister of Native Affairs, later Prime Minister (1958–1966)), stated that its aim was to prevent Africans receiving an education that would lead them to aspire to positions they would not be allowed to hold in society. On 17 September 1953, he introduced the Bantu Education Bill in Parliament and stated that "When I have control over the native education I will reform it so that the Natives will be taught from childhood to realize that equality with Europeans is not for them ... . People who believe in equality are not desirable teachers for natives ... What is the use of teaching the Bantu child mathematics when it cannot use it in practice? The idea is quite absurd" (House of Assembly 1953). As a result, the apartheid ideology consciously destroyed a generation of Black African mathematics students, depriving them from access to mathematically based disciplines such as statistics, and in general of having access to an education in keeping with advances of the twentieth century.

3 Statistics Education at School Level: History and Current Status

When Apartheid was abolished in 1991, education and training in South Africa was restructured to reflect the values and principles of a democratic society, leading to the announcement of a new school curriculum with Outcome Based Education as the fundamental building block. A major difference was that the country now had one school curriculum for all learners, in direct contrast to the racially dividing school curricula that had been in place during Apartheid. This curriculum was further intended to overturn the legacy of Apartheid and catapult South Africa into the twenty-first century (Chisholm et al. 2000). The curriculum, known as Curriculum 2005 (DoE 1997), was planned to be fully implemented by 2005. It was later amended and renamed the National Curriculum Statement (NCS) (DoE 2003), to be fully implemented by 2008, then further amended and re-labeled, the Revised NCS, or commonly referred to as RNCS (Chisholm 2003).

In the new curriculum, emphasis was specifically placed on shifting from the traditional aims-and-objectives approach to outcomes-based education. This paradigm shift was seen as a prerequisite for the achievement of the vision of an "internationally competitive country." Outcomes-based curriculum development starts with the formulation of the purposes of learning which are then used as the criteria for further curriculum development and assessment. One of the specific outcomes identified in this school mathematics curriculum was the "use of data from various contexts to make informed judgments" (Steffens and Fletcher 1999). Recognition of the cross-curricular need for statistics as an anticipated outcome, led to the collection of data (methods such as interviews and sampling), the application of statistical tools and communication and critical evaluation of findings (North and Zewotir 2006) being included in the new school curriculum, under the label "Data Handling." This was in total contrast to what had previously been the case as (1) statistics had generally (see following comment) not been taught at school level in South Africa and (2) mathematical-based learning areas, including Statistics, had previously not

Table 1 Number of students registered for Grade 12 mathematics exam

Year	Type of mathematics	Grade 12 learners
2007	Higher grade (main stream)	41,000
	Standard grade (subsidiary)	306,000
2008	Mathematics (main stream)	287,487
	Mathematical literacy (subsidiary)	249,784

been in the curriculum of Black school children. Statistics topics were part of the Additional Mathematics curriculum, which was only offered as an additional subject (over and above the standard load) to the very high-end achievers in Mathematics, and was only available at selected (White) schools. However this exposure to Statistics was only taken up by less than 0.1 % of the White school going children of that era!

The new education system further required that each learner either did mathematics or mathematical literacy in each school year—this was a major shift from what had been the case prior to the adoption of the new curriculum, as it was previously possible for all students (of any race) to complete schooling without doing any form of mathematics in the last 3 years of the schooling system. In 2008, the first school leavers that had been exposed to the new outcome-based education schooling system were ready to enter institutions of higher learning the following year, changing the landscape of higher education in South Africa totally. The first significant difference was the large increase in the number of school leavers with some form of mathematics under the new schooling system, as can be seen from Table 1 above. A huge "wave" of school leavers with at least some mathematics training emerged from schools in South Africa in 2008 (the first year when NCS was fully implemented), when a total of 537,271 students registered for the final school exam in mathematics/mathematical literacy as compared to a total of 347,000 in 2007 (the last school graduates under the previous system), i.e., an increase of more than 50 % in school leavers with some level of mathematical proficiency, in 1 year!

In 1998, the Human Sciences Research Council in South Africa conducted a study under the auspices of the International Association for the Evaluation of Educational Achievement (Howie 1999). A total of 225 secondary schools were randomly selected from the nine provinces, resulting in more than 8,000 learners, 350 teachers, and 190 principals responding to questionnaires and interviews aimed at providing an indication of the status of mathematics training in the country at that time. The average score of all participants turned out to be 275 points out of a possible 800 points, well below the international average of 487 points. The South African pupils' performance was relatively low in every mathematics topic (from 37 % for algebra to 45 % for data representation, analysis, and probability). It is interesting to note that the average score for data representation, analysis, and probability scored 356 points (out of 800 points), making this the scale with the smallest difference from the international average! The results were thus surprising as Data Handling (Statistics) had not yet been introduced as part of the

Mathematics curriculum at school level at that time. One can thus only conclude that the questions must have been very basic (requiring intuition and logic only) and thus did not reflect modern statistics education.

4 Challenges

The Data Handling component of the NCS aims to ensure that each school leaver is statistically literate. Challenges to achieving this aim are to ensure that (1) the school syllabus has the desired content; (2) teachers have the skill and confidence required to promote basic data handling and interpretation skills in the class room.

In-service teachers have generally had no training in Statistics, as this was previously not part of the school syllabus, but the real barrier to meeting the second challenge is the sheer magnitude of the problem. The legacy of Apartheid is that there are simply not enough mathematics teachers to meet the demand.

Not surprisingly, this shortage is most pronounced in rural areas, resulting in non-specialist training of mathematics in many schools (Mail and Guardian Online 2008). All school children now require a level of mathematics (including Statistics) training, but during Apartheid, over 80 % of the nation did not receive any mathematics training, leading to the severe shortage of mathematics teachers today.

The Study conducted by the HSRC in 1998 interestingly showed that school children in South Africa achieved relatively better scores in statistics-related questions in this study, when it was not part of their school curriculum, than they did in the learning areas of Mathematics that they were actually instructed on! Where this came from is unknown, but it indicates either some natural interest or capacity, or that the questions reflected learning in other disciplines and/or learning from everyday living. It is thus a challenge for teachers to build on this, so that they develop proficiency and interest in Statistics which leaves a lasting legacy when they enter further education and training. The challenge is thus ultimately to have developmental training programs for pre- and in-service teachers in line with this objective.

5 Successes

Before the new school syllabus was introduced, school leavers generally had no formal statistics training at all, so the inclusion of statistics in all grades at school may be regarded as a success.

The Statistics component of the new Mathematics curriculum was initially developed by the Department of Education (DoE), taking a formula driven approach, much like the first few chapters of a typical classic university Statistics text book. This prompted intervention from the South Africa Statistical Association (SASA), the mouthpiece of the majority of professional and practicing statisticians in South

Africa, and resulted in the successful rewriting of the school curriculum by the DoE (North and Ottaviani 2002).

The training of teachers to engage with the new curriculum was to be conducted by the subject advisors of the DoE, but they generally had not received any training in Statistics either! In recognition of this dilemma SASA stepped in and since then has been, and continues to be, very actively involved in the teaching of Statistics in schools via its Education Committee, a subcommittee with the specific brief of furthering statistics education at tertiary and preparatory (school) level. This committee has built up a collection of games, projects, newspaper articles, etc. which are used to encourage mathematics teachers to teach statistics in a more meaningful and stimulating way.

Accordingly, the SASA Education Committee initiated an awareness of the dilemma of statistics education of teachers by giving various talks at local conferences, holding workshops and helping with teacher training. It was however only when Statistics South Africa (Stats SA 2012), the national statistics office, launched the maths4stats campaign in 2006, that the human capacity and finances were available to address the dilemma on a national basis. The maths4stats project addresses the dilemma with a roll-out plan to provide statistics training to in-service mathematics educators, from 28,000 schools. The objective of the maths4stats campaign is to create a specialized body of educators with a passion for mathematics, and to instil love and interest for mathematics and statistics in educators and learners. Details of the project can be found in North and Scheiber (2008). The long-term aim of this project is to strengthen the expertise in Statistics at all levels, so that ordinary people have trust in the information they receive from Stats SA (Lehohla 2002).

The author (North) has headed the statistics outreach programs of Stats SA since inception and has had particular success (North and Scheiber 2008) with training teachers who are not qualified to teach mathematics, but are expected to upgrade their knowledge and teach mathematics due to the lack of suitably qualified teachers in this discipline. These teachers in particular have a very positive reaction to Statistics workshops as they find it more practical and easier to teach than the advanced mathematical concepts that they often have not mastered themselves.

Modules in Educational Statistics, including research-based pedagogical content, are now part and parcel of training of pre-service teachers (Wessels 2008) to promote the statistical literacy, thinking, and reasoning abilities called for in the NCS.

6   Statistics Education at University Level: History and Current Status

Statistics was first introduced at South African universities in the 1930s and had a very theoretical focus, deliberately shying away from applied statistics (De Wet 1998). Statistics historically started at second-year level (in a 3-year program

Table 2 Number of passes in statistics modules in the South Africa (2006)

| B.Sc. program | | | | | | |
First year	Second year	Third year	Honours	M.Sc.	Ph.D.	Service courses
4243	1185	945	190	90	22	16,246

leading to a Bachelor's degree in Statistics) as one needed a high-level of mathematics to follow the calculus-based approach to teaching statistics that was in place at that time. Historically statistics training was thus geared towards furthering the discipline in a theoretical way It must however be noted that a few individuals did manage to excel in the applications of statistics through their own self-interest rather than through skills acquired from their tertiary level statistics training.

In recent times, a more balanced view of theory and applications has become apparent in statistics training at South African Universities. Currently 14 universities in South Africa have statistics departments that typically offer a 3-year program leading to a Bachelor's degree in Statistics, a 1-year honors degree, 1- or 2-year Masters' program and four-year Ph.D. programs. Masters and Ph.D. programs are mainly by research dissertation only.

The current structure of the statistics courses in the various South African universities is very similar across academic institutions. Two courses (one per semester) at first year level, two courses at second year level, and four courses at third year level are generally the required courses for the B.Sc. program (major in Statistics). The two courses (one per semester) in the first year offer an introductory approach to the theory, principles, and applications of statistics. The two courses in the second year mainly deal with distribution theories, estimation procedures, and inference. The third year courses are a mix of methods and applications with a theoretical basis. Relatively intermediate level advanced statistical theories and methods with computer practicals are offered at the honors level. A common feature of honors programs amongst the South African universities is the statistics project, where independent research under guidance of faculty is a key factor in spotting talent for subsequent post graduate studies.

Statistics service courses (non-specialists) generally are very big classes as at least one module in statistics is an essential part of almost all programs at South African universities. Classes can be as large as 500–700, making it very difficult to engage in active learning, which is pivotal to conceptual understanding. The result is that many students find Statistics courses difficult with a resulting poor pass rate in such courses (North and Zewotir 2006b). For instance, the average pass rate for Engineering students in a Statistics service course offered at the University of KwaZulu-Natal for the 1997–2005 period is 73.82 % (Zewotir and North 2007).

In 2006 a national survey was conducted by SASA, to determine graduation numbers at the various levels for statistics modules. Table 2 bears evidence of the heavy load that service courses put on statistics departments as well as the small number of graduates in Statistics at the higher post graduate levels.

7   Challenges

Statistics training at tertiary level in South Africa has challenges embedded in South Africa's Apartheid history. The first challenge is the fact that the majority of incoming students still have poor or insufficient mathematical foundation for the theoretically orientated first year statistics courses (North and Zewotir 2006b). This is the direct result of inferior mathematics teaching at school level, generally the result of not having enough adequately trained mathematics school teachers. Instruction at the lower levels is further characterized by large classes with the majority of students typically not having had previous exposure to computers; instruction is in a language other than their home language and financial constraints are experienced by many, so that the buying of text books is also a challenge. It is thus not uncommon to have high failure rates in statistics modules, particularly at level 1 (De Wet 1998).

The second challenge is a shortage of post graduate students and statistics lecturers. Statistics education at the higher levels is characterized by very small numbers, which is a great concern as B.Sc. and Honors graduates in statistics get lucrative offers from industry, business, and government, luring them away from academia. Accordingly, very few South African statistics graduates opt to pursue postgraduate studies. South African universities on average have at least 15 % of their Statistics posts vacant, whilst 20 % of posts at Statistics South Africa, the national statistics office, are vacant. The International Review Panel Report on the Review of Mathematical Sciences Research at South African Higher Education Institutions (Department of Science and Technology 2008) concluded that "the shortage of academic statisticians is so critical that the field is in danger of disappearing through lack of academic capacity," further noting that "the closure of academic departments is a real possibility." The shortage of Statistics lecturers at universities and increasing numbers of first year students, results in ever increasing teaching loads, which has a negative effect on research output—clearly a concern.

8   Successes

A number of universities have introduced access or foundation programs (extended length programs with "catch-up" modules to assist students who have had inferior schooling), supplementary instruction, mentorship programs, and hot seats (private lessons by post graduate students) in first year Statistics and Mathematics modules to deal with the problems discussed earlier. The use of local textbooks, or more effectively, the writing of course packs to replace foreign textbooks do, to a certain extent, overcome the problem of presenting examples that the students can relate to. Each of the 23 universities has programs in place to deal with re-dress and overcome the problems discussed above. For example, at the University of KwaZulu-Natal, special tutor sessions are planned weekly, where instruction is in English, but explanations are given in the home language (Zulu) of local African students. The leaders of these special groups are local African students, who have registered for post graduate degrees and are thus also role models for the junior students on campus.

9 Special Impact Initiatives

The international panel reviewing the status of mathematical sciences research in South Africa (Department of Science and Technology 2008) noted that it was critically urgent that statistics capacity had to be built at all levels, further noting that the country accordingly needed to produce more post graduate students in statistics, so that they could feed the economy and stock statistics departments. But, is this happening? Sadly not! Statistics departments remain under-stocked, with the cream of the statistics graduates not pursuing higher level studies in statistics, due to the lucrative job opportunities in the business sector (banking in particular).

Statistics Education in South Africa faces many challenges—though there have been some successes, there is no doubt that the country urgently needs to take urgent stock of the status of academic Statistics departments, to increase the number of post graduate students and ultimately to grow statistics as a research discipline in South Africa.

There is such acute awareness of this problem that a number of special competitions are held each year to raise the level of awareness of Statistics amongst the youth of the country.

10 South African Statistical Association: Statistics Competitions

In an attempt to create excitement and grow interest in statistics amongst students at institutions of higher education in the country, the South African Statistical Association (SASA) increased the number of national competitions run by the Education sub-committee, over the last 3 years. Prize monies for the various competitions were secured as 5-year sponsorship deals, by means of memorandums of understanding (MOU's), signed between SASA and Statistics South Africa, and also between SASA and SAS, which in broad terms provides student fees, expenses for winners to attend the annual SASA statistics conference, etc. The various competitions, at the different levels of tertiary education, for students of statistics, run by SASA Education committee are as follows:

- Bursary and scholarships competition for second year students, based on merit and need (special category for previously disadvantaged students), prize money goes towards registration fee for third year of study towards a degree, majoring in statistics.
- Bursary competition for third year students, based on merit and need (special category for previously disadvantaged students), prize money goes towards registration fee for an honors degree in statistics.
- Project competition for honors students, winners receive prize money and sponsored attendance of the annual SASA conference.
- Masters and Ph.D. paper competition, winners receive prize money and sponsored attendance of the annual SASA conference.

• Young Statistician's competition (less than 35 years old), based on research papers entered, winners are fully sponsored to attend the WSC. This competition is held every alternate year, in accordance with the hosting of the WSC.

The last competition, introduced in 2013, is an exciting competition for young (South) African statisticians, as winners were fully sponsored to attend the World Statistics Congress in Hong Kong in August 2013. This was initially intended for South African entrants only, but when additional funding became available, the competition was opened up to include young African Statisticians from beyond the borders of South Africa. The excitement this competition created is evidenced by the fact that a total of 149 entries were received from 19 countries in Africa! Unfortunately only 12 of these entries were from South African students … and only one single entry was from a Black African post graduate student, so much work remains to be done! This competition bears further evidence of a national problem—we not only does the country not have enough post graduate students in statistics, but 9 years post democracy, there is still no evidence that local African students are coming through the system at the rate that we would hope to see!

So the question remains, what more can be done to put a structured and politically supported approach to Statistics capacity building in place, which would ultimately have a realistic chance of overcoming the challenges that are deep-rooted in the history of the country, and the more recent challenges of the flow of statistical graduates and academics to business and industry. If this can be achieved, then there would be a realistic chance of preventing the doom and gloom predicted by the International panel that "the closure of Statistics departments is a real possibility," due to sheer lack of capacity!

The international panel made various recommendations, many of which would entail extra budget for statistics departments from government or universities coffers, which has not happened to date, due to other competing urgencies. Academic statistics departments have however, attempted to address the problem of building statistics capacity in various ways that are particular to the drive of individuals that are based in those departments, with approaches very much depending on the profile of students that are on that particular campus.

11   UKZN: Leading the Way to Grow Statistics Capacity

The University of KwaZulu-Natal (UKZN) was formed in 2004, following the merger of the former UND, UNP, UDW (a previously disadvantaged university) and Edgewood Teacher training college, during the government "merger plan" for institutions of higher education in South Africa (Nkoane 2006). UKZN is one of the "big five" research institutions of higher education in South Africa (Boshoff 2009). This university has transformed the racial profile of students attending the institution totally, with the proportion of first year students who are White, declining from 14 % in 2005 to 5 % in 2009, whilst the corresponding figures for Black African students increased from 46 % of first year students in 2004 to 60 % correspondingly

Fig. 1 School children come to University Open Day and are excited to play games and learn more about probability!

in 2009 (Zewotir et al. 2011), and now stands at over 80 % in 2013. The racial profile of students across UKZN campuses is thus in line with the racial profile of the country, making it the ideal institution to address the problem of capacity building in statistics, and in particular, for putting steps in place to increase the number of local African students progressing to the more senior post graduate levels at the institution.

To date the Statistics staff at UKZN have attempted to address the issue of statistics capacity building at ALL levels, embarking on a number of initiatives some examples of which are discussed below.

A series of eye catching posters were designed, which together with the playing of fun games are sparking school children's interest in statistics when they attend UKZN Open Days (Fig. 1).

This university has been instrumental in defining and setting up (together with StatsSA) series of statistics workshops and after-hours classes for in-service teachers as part of the Stats SA maths4stats program (North and Scheiber 2008). Over the last 3 years, just over 1,300 in-service teachers have completed a 5-week statistics module for teachers, run over weekends on UKZN campus, sponsored by Stats SA and presented by UKZN lecturing staff. Note that these classes give preference to teachers from schools identified by the DoE as being schools where the most need is experienced, thus truly attempting to upgrade basic Statistics training of teachers, as well as bringing about awareness of Statistics to pupils from rural areas (Fig. 2).

Fig. 2 In-service teachers eagerly waiting for their Saturday morning statistics lessons given by UKZN staff, sponsored by Stats SA

Extra lessons are presented for students registered for first year statistics modules as students often experience problems adjusting from school to university (Zewotir et al. 2011). The "Hot Seat" system was conceptualized by UKZN Statistics Department and defines a program whereby post graduate Statistics students give one-on-one lessons to first year statistics students, the UKZN tutoring budget sponsors this system (Fig. 3).

Statistics students who are struggling to adapt to university, have a "small class" experience once a week, by dividing the large first year class into smaller weekly tutorial groups. Preference is further given to Black African post grad students to act as role models, in the capacity of Hot Seat tutor leaders, so that they can converse in the home language of the students who are struggling to adjust to university life in a language that is not their home language.

Personal invitations are given to successful students (by one-on-one meetings as well as invitations in writing), at all levels, to encourage them to register for higher level studies in statistics (Fig. 4).

12 Conclusion

The building of statistics capacity to meet the ever increasing demand from business, industry, government, schools, and universities has been a hot topic at conferences all over the world throughout the past decade.

Fig. 3 Some of the UKZN staff and post grad students (2013) who run tutorial groups and "Hot Seat" for first year statistics students

Fig. 4 Successful post grad statistics students at UKZN graduation!

South Africa has a critical shortage of statistics skills, resulting in an urgent need to produce more statistics graduates. The abolishment of Apartheid and recent introduction of Statistics into the school syllabus in South Africa has given the country the opportunity of exposing all school children to basic statistical principles, in direct contrast to what had been the case historically. The extent to which this exposure will result in school leavers with an increased appreciation for, and interest in Statistics, will only be known in the years to come. There is, however, no doubt that the recent joint efforts between Stats SA and SAS, to work collaboratively with SASA, as well as the initiatives of individual universities, such as UKZN, to promote statistics education at all levels, will give the best possible chance for the introduction of statistics at school level to filter through to tertiary level and ultimately to play a meaningful role in statistics capacity building in the country.

13   Note

Developed from a keynote presentation at Seventh Australian Conference on Teaching Statistics, July 2010, Perth, Australia.

This chapter is refereed.

References

Badat, S. (2004). Transforming South African higher education, 1990–2003: Goals, policy initiatives and critical challenges and issues. In N. Cloete, P. Pillay, S. Badat, & T. Moja (Eds.), *National policy and a regional response in South African higher education.* James Currey (Oxford) and David Philip (Cape Town): Published in association with Partnership for Higher Education in Africa.

Boshoff, N. (2009). Shanghai Academic Ranking of World Universities (ARWU) and the 'big five' South African research universities. *South African Journal of Higher Education, 23*(4), 63–655.

Brand South Africa. (2012). *The languages of South Africa. Media South Africa.* Retrieved from http://www.southafrica.info/about/people/language.htm

Chisholm, L. (2003). *The politics of curriculum review and revision in South Africa.* Retrieved from http://repository.up.ac.za/handle/2263/5044

Chisholm, L., Volmink, J., Ndhlovu, T., Potenza, E., Mahomed, H., Muller, J., et al. (2000, May 31). *A South African curriculum for the twenty-first century, Report of the review committee on curriculum 2005.* Pretoria, South Africa.

De Wet, J. I. (1998). Teaching of statistics to historically disadvantaged students: The South African experience. In *Proceedings of the Fifth International Conference on Teaching Statistics.* Voorburg, Netherlands: International Statistical Institute. Retrieved from http://www.stat.auckland.ac.nz/~iase/publications/2/Topic5e.pdf

Department of Education [DoE]. (1997). *Senior phase policy document.* Pretoria, South Africa: Author.

Department of Education [DoE]. (2003). *NCS mathematics grades 10–12.* Pretoria, South Africa: Author. Retrieved from http//www.education.gov.za/Curriculum/SUBSTATEMENTS/Mathematics.pdf

Department of Science and Technology. (2008). *The International Review Panel Report on the review of mathematical sciences research at South African higher education institutions.* Retrieved 18 December 2008 from http://www.nrf.ac.za/files/file/Report.pdf

House of Assembly. (1953, August–September). Debates Vol. 78, No. 3585.

Howie. S (1999). *Third International Mathematics and Science Study Repeat (TIMSS-R)* Executive Summary. Retrieved from http://www.hsrc.ac.za/Document-540.phtml

Lehohla, P. (2002). Promoting Statistical literacy: A South African perspective. In Phillips, B. (Ed.), *Proceedings of the Sixth International Conference on Teaching Statistics.* Cape Town, South Africa: International Statistical Institute and International Association for statistics Education. Retrieved from www.stat.auckland.ac.nz/~iase/publications

Mail and Gaurdian Online. (2008). *Teachers flunk maths.* Retrieved from http://www.mg.co.za/article/2008-08-03-teachers-flunk-maths

Nkoane, M. M. (2006). Challenges facing South Africa's educational advancement. *International Journal of Educational Advancement, 6*(3), 243–252.

North, D., & Ottaviani, G. (2002), Statistics at foundation school level in South Africa—The way forward. In *Proceedings of the Sixth International Conference on Teaching Statistics.* Voorburg, Netherlands: International Statistical Institute. Retrieved from http://www.stat.auckland.ac.nz/~iase/publications/1/2d2_nort.pdf

North, D., & Scheiber, J. (2008). Introducing statistics at school level in South Africa. The crucial role played by the national statistics office in training in-service teachers. In *Proceedings of ICMI Study and IASE Roundtable* Mexico. Retrieved from http://www.ugr.es/~icmi/iase_study/

North, D & Zewotir T (2006), Introducing statistics at school level in South Africa. In Ben-Zvi, D., & Garfield, J. (Eds.), *Proceedings of the Seventh International Conference on Teaching Statistics.* Cape Town, South Africa: International Statistical Institute and International Association for Statistics Education. Retrieved from www.stat.auckland.ac.nz/~iase/publications

North, D., & Zewotir, T. (2006). Teaching statistics to social science students: Making it valuable. *South African Journal of Higher Education, 20*(4), 503–514.

Stats S. A. (2012). *Highlight of key results* (Report No 03-01-42). Retrieved from http://www.statssa.gov.za/Census2011/Products/Census_2011_Methodology_and_Highlights_of_key_results.pdf

Steffens, F. E., & Fletcher, L. (1999). Statistics as part of the mathematics curriculum in South Africa. *Proceedings of the 1st International Conference of the Mathematics Education into the 21st Century Project* (Vol. 1, pp. 298–305). Retrieved from http://math.unipa.it/~grim/ESteffensFletcher298-305.pdf

Wallman, K. K. (1993). Enhancing statistical literacy: Enriching our society. *Journal of the American Statistical Association, 88*(421), 1–8.

Wessels, H. (2008). Statistics in the South African school content: Assessment and teacher training. In Batanero, C. Burrill, G., Reading, C., & Rossman, A. (Eds.), *Joint IcMI/IASE study: Teaching statistics in school mathematics challenges for teaching and teacher education, Proceeding of the ICMI Study 18 and 2008 IASE Round Table Conference.* Retrieved from http://www.ugr.es/~icmi/iase_study/Files/Topic1/T1P3_Wessels.pdf

Wilcox D. (2003). *On mathematics education in SA and the relevance of popularising mathematics.* Retrieved from http://www.mth.uct.ac.za/~diane/on_math_ed_in_SA.pdf

Zewotir, T., & North, D. (2007). Focus on the statistical education of prospective engineers in South Africa. *Pythagoras, 65,* 18–23.

Zewotir, T., North, D., & Murray, M. (2011). Student success in entry level modules at the University of KwaZulu-Natal. *South African Journal of Higher Education, 25*(6), 193–199.

Beyond the Statistical Fringe

Kaye E. Basford

Abstract Having attended the First International Conference on Teaching Statistics (ICOTS) in Sheffield in 1982 (and a few since then), I found it informative to look back and see what has happened in statistical education during the past three decades. In this chapter, I comment on some consistent themes and my perception of how the focus has changed over time, particularly from the viewpoint of training beyond the introductory course for students in other disciplines. Some comment is also made on particularly influential publications and other relevant meetings overseen by the International Association of Statistical Education (IASE). My vantage point is from involvement in, and then responsibility for, teaching biometry in an agriculture department for two decades, followed by another decade in university and statistical society management where I was not directly involved in the delivery of such coursework training.

Keywords Statistical education perspectives • Resources • Technology • Advanced courses for other disciplines

1 Introduction

When asked to give a keynote address at the Australian Conference on Teaching Statistics (OZCOTS) in July 2012, I decided to review what had been happening in statistical education during the past three decades; since I had attended the First International Conference on Teaching Statistics (ICOTS) in Sheffield in 1982, I concentrated on some consistent themes and my perception of how the focus

K.E. Basford (✉)
The University of Queensland, Queensland, Australia
e-mail: k.e.basford@uq.edu.au

H. MacGillivray et al. (eds.), *Topics from Australian Conferences on Teaching Statistics:* *OZCOTS 2008-2012*, Springer Proceedings in Mathematics & Statistics 81, DOI 10.1007/978-1-4939-0603-1_5, © Springer Science+Business Media New York 2014

changed over time, particularly with regard to training beyond the introductory course for students in other disciplines. It is not an exhaustive review, but rather a personal perspective. Some comments are made on a few other particularly influential publications and their relevance to advanced courses for such students. Finally, some issues with and impacts of advances in technology and potential issues for the future are presented.

To understand my vantage point, it is helpful to know my background. I was trained as a statistician and worked as a consulting biometrician for more than a decade. After a Fulbright Postdoctoral Fellowship in the Department of Plant Breeding and Biometry at Cornell University, I moved to a teaching and research position in the Agriculture Department at the University of Queensland. For the next decade I was responsible for courses and research training in biometry, followed by university and professional society management (for a further decade). During the latter period, I was not directly involved in the delivery of coursework training.

A brief historical perspective on statistical education in the International Statistical Institute (ISI), which was founded in London in 1885, is also useful. As detailed in Vere-Jones (1995), the ISI set up an Education Committee in 1948 "aimed, among other things, at increasing the ISI's mandate to undertake educational activities in statistics and to collaborate for this purpose with UNESCO and other UN agencies." Responsibilities soon separated, with Vere-Jones stating that "the operating function, that is responsibility for developing and running training facilities, would be taken over by the UN Agencies, while the ISI retained responsibility for the broader, more nebulous task of promoting statistical education." Zarkovich's (1976) reappraisal of the ISI statistical education program argued that it was "the only international body with the competency to guide the future growth of statistical education, and strongly urged that it should take a more active role, at all levels" (Vere-Jones 1995). This was enacted upon by Joe Gani (who became Chair of the Education Committee in 1979), when he set up the Task Force on Teaching Statistics at School Level, led initially by Vic Barnett, and the Task Force on International Conferences in Statistical Education, led initially by Lennart Råde (Vere-Jones 1995). The Education Committee established its successor, the International Association for Statistical Education (IASE) in 1991, as a new section of the ISI and the first international body specifically devoted to statistical education.

2 Conferences

The particular series of conferences concerned with statistical education considered here are as follows:

- IASE Round Tables (from 1968)
- International Conferences on Teaching Statistics (ICOTS) (from 1982)
- IASE Scientific and Satellite Meetings (from 1993)
- Australian Conferences on Teaching Statistics (OZCOTS) (from 1998)

Each of these is now considered in more detail.

IASE Round Tables had the following themes, with location and year in parentheses:

- University teaching of statistics in developing countries (The Hague, 1968)
- New technologies in teaching of statistics (Oisterwijjk, 1970)
- Statistics at the school level (Vienna, 1973)
- Teaching of statistics in schools (Warsaw, 1975)
- Teaching of statistics (Calcutta, 1977)
- Teaching of statistics in the computer age (Canberra, 1984)
- Introducing data analysis in schools: who should teach it and how? (Quebec, 1992)
- Research on the role of technology in teaching and learning (Granada, 1996)
- Training researchers in the use of statistics (Tokyo, 2000)
- Curriculum development in statistics education (Lund, 2004)
- Teaching statistics in school mathematics-challenges for teaching and teacher education (Monterrey, 2008)
- Technology in statistics education: virtualities and realities (Cebu City, 2012)

ICOTS had the following themes (where found), with location and year in parentheses:

- ICOTS 1 (Sheffield, 1982)
- ICOTS 2 (Victoria, 1986)
- ICOTS 3 (Dunedin, 1990)
- ICOTS 4 (Marrakech, 1994)
- ICOTS 5 (Singapore, 1998)
- ICOTS 6 – Developing a statistically literate society (Cape Town, 2002)
- ICOTS 7 – Working cooperatively in statistics education (Salvador, 2006)
- ICOTS 8 – Data and context in statistics education: towards an evidence-based society (Ljubljana, 2010)
- ICOTS 9 – Sustainability in statistics education (Flagstaff, 2014)

IASE Scientific and Satellite Meetings had the following themes (where found), with location and year in parentheses:

- First Scientific meeting (Perugia, 1993)
- First Satellite meeting – Statistical literacy (Seoul, 2001)
- Second Satellite meeting – Statistics and the internet (Berlin, 2003)
- Third Satellite meeting – Statistics education and the communication of statistics (Sydney, 2005)
- Fourth Satellite meeting – Assessing student learning in statistics (Guimarães, 2007)
- Fifth Satellite meeting – Next steps in statistics education (Durban, 2009)
- Sixth Satellite meeting – Statistics education and outreach (Dublin, 2011)
- Seventh Satellite meeting – Statistics education for progress (Macao, 2013)

OZCOTS had the following themes (where found), with location and year in parentheses:

- OZCOTS 1 (Melbourne, 1998)
- OZCOTS 2 (Melbourne, 1999)
- OZCOTS 3 (Melbourne, 2000)
- OZCOTS 4 (Melbourne, 2002)
- OZCOTS 5 (Melbourne, 2003)
- OZCOTS 6 – Teaching and assessment of statistical thinking within and across disciplines (Melbourne, 2008)
- OZCOTS 7 – Building capacity in statistical education (Perth, 2010)
- OZCOTS 8 – Statistics education for greater statistics (Adelaide, 2012)

OZCOTS 1–5 were all coordinated by Brian Phillips and held at Swinburne University of Technology. As detailed on the SSAI website for the Australian Statistical Conference in 2012 (for which ICOTS 8 was a satellite) (http://www.sap-mea.asn.au/conventions/asc2012/ozcots_info.html accessed July 2012), the background to the conferences is as follows:

> The first OZCOTS was run in 1998 by Brian Phillips with papers by the Australian speakers from the 5th ICOTS which had been held in Singapore earlier in 1998. Its success in bringing together Australians involved in teaching statistics resulted in Brian and his Melbourne colleagues organising annual OZCOTS gatherings from 1999 to 2002. In 2006, Helen MacGillivray was awarded one of the first Australian Learning and Teaching Council's Senior Fellowships, with her fellowship programme to run throughout 2008. As part of her fellowship programme, Helen revived OZCOTS with Brian's help, and ran it as a two-day satellite to the 2008 Australian Statistical Conference (ASC), with one-day overlap open to all ASC delegates who could also choose to register for the second day. The OZCOTS 2008 invited speakers were all funded as part of Helen's fellowship. OZCOTS 2008 was modeled on the successful IASE's satellite conferences to ISI Conferences, with papers in proceedings and an optional referreeing process offered to authors. The success of OZCOTS 2008 led to an equally successful OZCOTS 2010 and then to OZCOTS 2012.

Brian and Helen coordinated OZCOTS 6–8 which were held in association with the Australasian Statistical Conferences.

3   Musings on Particular Publications

As ICOTS 1 was the first conference on statistical education which I attended, I shall initially comment on this series. Maria Gabriella Ottaviani (ICOTS6 2002) considered how the scientific content of ICOTS conferences evolved from 1982 to 2002 via textual data analysis (a form of correspondence analysis) of the titles of each ICOTS paper presented at each conference. Her interpretation of the emphases over time was as follows:

1982 – Mainly a matter of teachers and teaching
1986 – Students in evidence as well as teaching/learning problems
1990 – Students right in the center

1994 – Focus on teaching of statistics per se
1998 – About statistics students and statistics teachers (of introductory courses)
2002 – Focus on research (in learning and teaching statistics and through research methods and experience in research)

Ottaviani's conclusion was that during this time period, ICOTS moved from dealing with teaching/learning problems to teaching by real data with suitable computer packages. There was a focus on introductory courses for statistics students and those from applied disciplines, with research methods emphasizing basic statistical concepts. There was also a move to educational research in statistics to better disseminate and teach the discipline. She felt that there was a need for statisticians and statistics education researchers to work together and learn from each other.

I also asked Helen MacGillivray (past President of the IASE and current Vice-President of the ISI) what her impressions were from the various ICOTS she attended. Her informal view of the general feelings of delegates was as follows:

1990 – Very confident on what should be done
1998 – Culture of this is harder than we thought
2002 – We are all in this together and it is hard work
2006 – Let's keep looking at where we are at
2010 – We have done a lot, so where do we go now?

My view of the papers presented from the full ICOTS series is as follows:

- There are abundant general resources available on teaching statistics from these conferences (http://www.stat.auckland.ac.nz/~iase/publications.php, accessed July 2012)
- There is outstanding local expertise available (and these people are willing to share their experiences)

This puts teachers of statistics in Australia in a very good position, particularly for introductory courses. Not only can they access many varied resources but they also can get personal assistance from some of the leaders in the field.

4 Advanced Courses for Other Disciplines

Given my background, I was particularly interested in advanced courses for students from other disciplines. While I am not a fan of the term "service teaching," teaching students from other disciplines is often referred to in this way in the literature. Allen et al. (IASE/ISI Satellite 2009) stated that while great progress had been made in the field of statistics education, a remaining challenge was "to maximize the efficiency and effectiveness of service teaching in order to create successful end-users of statistics and to raise the profile of statistics within the wider community." They demonstrated that "the most efficient and effective approach to service teaching involved close collaboration between all departments involved in order to benefit both those teaching and those learning."

My paraphrasing and reordering of their recommendations are as follows:

- Maintain staff continuity, enthusiasm, and ability
- Use real data to generate interest
- Don't stream students based on ability
- Clarify early why a good understanding and knowledge of statistics is important
- Make innovative changes to teaching practice that students find useful
- Advertise resources and support for non-statisticians
- Integrate quantitative data analysis into subject specific modules
- Spread statistics tuition throughout the postgraduate program

Helen MacGillivray (IASE/ISI Satellite 2011) focussed on teaching statistical thinking foundations for postgraduates across disciplines. In the process, she advocated the use of real data sets in contexts that do not require discipline-specific knowledge, but which are sufficiently complex to allow demonstration of choice, use, interpretation, and synthesis of statistical tools and thinking.

Helen's requirements for undergraduate courses were that they:

- Should be more than introductory
- Include realistic investigations/projects
- Include "focus on essential concepts of statistical inference and data modeling (variables, estimation vs prediction, scientific vs statistical)"

Previously in MacGillivray (IASE/ISI Satellite 2009), Helen had stated that statistical courses in general were to:

- "Align objectives and curricula to maximize learning
- Structure curricula for smooth flow, with incremental steps and connectedness within flow".

I had the good fortune to work with John Tukey and I agree with his views on the application of statistics. Tukey (1977) stated that "A major point, on which I cannot yet hope for universal agreement, is that our focus must be on questions, not models … Models can – and will – get us in deep troubles if we expect them to tell us what the unique proper questions are." He stressed to me that he preferred an approximate answer to the right question rather than the exact answer to the wrong question.

Terry Speed (ICOTS 2 1986) was thinking along similar lines when he made the following statements:

- "The value of statistics … helping us to give answers of a special type to more or less well defined questions"
- "Discern the main question of interest associated with any given set of data, expressing this question in the terminology of the subject area – the "scientific question""
- "Fundamental importance of the interplay of questions, answers and statistics".

David Cox and Christl Donnelly (2011) also have similar ideas, as illustrated by their statements that "It is a truism that asking the 'right' question or questions is a crucial step in success in virtually all fields of research work … The formulation of very focussed research questions at the start of a study may indeed be possible and desirable … In other cases however, the research questions of primary concern may emerge only as the study develops."

In the more general situation of working with students and scientists from other disciplines, my view is as follows:

- There is no substitute for a statistician being immersed in the applied discipline – it enables that person to determine and understand the research questions more easily
- It is worthwhile providing undergraduates in applied disciplines with an introductory statistics course plus an appropriate advanced course (or two)
- It is essential to provide statistical guidance (via short courses or workshops) to postgraduates in applied disciplines throughout their program (particularly when undergraduate statistical education has been lacking).

5 Issues with and Impacts of Technology

Ann Hawkins (IASE Round Table 1996) presented and challenged "myth-conceptions" concerning technology in statistics education. She said that "Innovations in this area tend to be accompanied by a number of myths that have crept into our folklore and belief systems. Myths are not necessarily totally incorrect: they often have some valid foundation. However, if allowed to go unchallenged, a myth may influence our strategies in inappropriate ways."

In her paper, Ann took the opportunity "to recognize and examine the myths that govern innovations and implementations of technology in the classroom, and to establish the extent to which our approaches are justified". Her myth-conceptions were as follows:

- "Technology enhances the teaching and learning of statistics
- Computers have changed the way we do statistics
- Computers have changed the way we teach statistics
- Introducing technology into the statistical teaching process is innovative
- Students learn statistics more easily with computers
- Research is guiding our progress
- People intuitively understand statistics and probability concepts
- Technology will solve students' statistics and probability misconceptions".

Ann felt that her "myth-conceptions about the role of technology in teaching and learning statistics will cease to be misconceptions as current trends toward the amassing of more imperial evidence continue, and our understanding of the processes involved increases." Perhaps she was somewhat optimistic! Nevertheless, I think that most of us will agree with her views that "Our research should be aimed at identifying a broad range of ways in which technology can assist the teaching and learning process. Just as there is 'no one right answer' to a statistical investigation nor is there 'one right way' to teach statistics, there may be many 'wrong' ways! A variety of methods and materials will always be a source of strength and benefit to both teachers and students, provided we have insights into how to use the available resources."

Brian Phillips (IASE/ISI Satellite 2003) gave an excellent overview of online teaching and internet resources for statistics education at the time, noting that developments were "mainly occurring in the teaching of introductory statistics, a subject taken by many students at the post secondary level." He commented on the following:

- Web-based learning was already available and such resources provide a form of distance learning
- Online classrooms provide a variety of experiences
- The web is a place to store and disseminate information
- A variety of sites are especially designed by and for statistics educators
- Online assessment is still in its infancy

Thus even a decade ago, entire courses were offered online and there were extensive resources available for "teachers of statistics at all levels to give their students a greater opportunity than ever to learn and understand statistical concepts." Brian's findings support my earlier viewpoint that there are abundant general resources available on teaching statistics (see http://iase-web.org/Conference_Proceedings. php, accessed February 2013), some provided locally. Note that the website http:// iase-web.org is the update to the previously quoted website http://www.stat.auckland.ac.nz/~iase/publications.php.

6   Potential Issues for the Future

There have been many innovators in visualization, with some well-known ones in statistics including the following:

- Adrian Bowman
- Aaron Kobin
- Hans Rosling (videoed at ICOTS8)
- Markus Gesmann

What about learning innovations? I asked Phil Long, Director of the UQ Centre for Educational Innovation and Technology at The University of Queensland, and he felt that the top two would be the following:

- Inverted or flipped classrooms
- MOOCs

By "inverting" the classroom we mean going from "transmission in class" and "assimilation outside of class" to "assimilation in class" and "transmission outside of class" (Brown 2012; Stannard 2012). As shown in Fig. 1 (used with permission from Robert Talbert, Grand Valley State University), Bloom's taxonomy aligns cognitive complexity, learning and access to help, in that assimilating increases cognitive load while transmission increases accessibility of help. In summarizing evidence

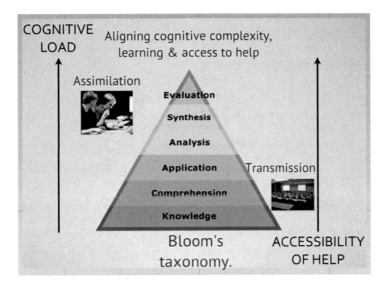

Fig. 1 Bloom's taxonomy (from Talbert 2011)

from randomized trials of interactive learning online at public universities, Bowen et al. (2012) state that "Our results indicate that hybrid-format students took about one-quarter less time to achieve essentially the same learning outcomes as traditional-format students."

Phil also provided me with an embedded 2010 YouTube clip on Massive Open Online Courses (MOOCs) by David Cormier (http://www.youtube.com/watch?v= eW3gMGqcZQc, accessed January 2013) who states that "a MOOC is one way of learning in a networked world." Elements of a collaborative or cMOOC go from text, to blogs, to videos, to Twitter (microblogs), to web pages, and more. They are fundamentally about networking people of common interests around a topic – typically without nearly the formality of assessment and often without certification (though sometimes you can opt in for that and get assessments when doing so). This is sometimes referred to as "students teaching students" (http://epic2020.org/, accessed January 2013). MOOCs have proliferated during the past couple of years and been legitimized in Udacity, edX, and Coursera (to name a few), in that they are supported by consortia of reputable institutions including Princeton University, Stanford University, Penn State University, MIT, and Harvard, to name a few. So far, the courses are free to users, with up to 160,000 enrolling in a single one. They are more course focussed, or xMOOCs. How they will be used commercially for more than the reputational benefit of the provider is yet to be determined.

Interestingly, a free online introductory statistics course is now available through edX (see https://www.edx.org/courses/BerkeleyX/Stat2.1x/2013_Spring/ about, accessed in February 2013).

In the last 50 years, we have gone from mainframes, minicomputers, and personal computers, to desktop Internet and mobile Internet. Everybody expects to be connected to anyone else in the world and to any resources at any convenient time. Units keep getting smaller, more powerful, easier to use, more connected, and cheaper to buy. No wonder education is being delivered, and is expected to be delivered, this way.

As another guide to the future of educational technology, Phil suggested looking at the Horizon Project (http://www.nmc.org/horizon-project, accessed January 2013). The New Media Consortium (NMC) is an international community of experts in educational technology and their Horizon Project is designed to identify and describe emerging technologies likely to have an impact on teaching, learning, and creative inquiry.

Key trends from the Horizon Report (Johnson et al. 2012) are as follows:

- People expect to be able to work, learn, and study whenever and wherever they want
- The technologies we use are increasingly cloud-based, and our notions of IT support are decentralized
- The world of work is increasingly collaborative, driving changes in the way student projects are structured
- The abundance of resources and relationships made easily accessible via the Internet is increasingly challenging us to revisit our roles as educators
- Education paradigms are shifting to include online learning, hybrid learning, and collaborative models
- There is a new emphasis in the classroom on more challenge-based and active learning

The NMC Technology Outlook > Australian Tertiary Education 2012–2017 (found under the publications tab of the above Horizons Project web address, accessed January 2013) suggested that the technologies likely to influence teaching, learning, and the creative arts in the given time frames are as follows:

- Time-to-adoption horizon: 1 year or less

    - Cloud computing
    - Learning analytics
    - Mobile apps
    - Tablet computing

- Time-to-adoption horizon: 2–3 years

    - Digital identity
    - Game-based learning
    - Open content
    - Personal learning environments

- Time-to-adoption horizon: 4–5 years

    - Digital preservation
    - Massively open online courses
    - Natural user interfaces
    - Telepresence

The listing of four technologies per time horizon in the new Technology Outlook format is a departure from past Country Editions where only two items per time horizon were listed.

Perhaps the time-to-adoption horizon for MOOCs in Australia has been overestimated, given that several Australian universities have already joined some of the consortia mentioned above.

7 Conclusion

I found researching the recent history of statistical education over the past three decades and gaining some informed insight into the near future very rewarding. I hope you enjoyed it too.

My final comments are as follows:

- More students are likely to learn fundamentals from fewer outstanding teachers via online courses
- More emphasis will be placed on advanced courses in particular application areas
- There will be continued emphasis on statistical thinking to answer scientific questions
- Personal interaction and campus experience will still add value

It is an exciting time to be a statistician, particularly one involved in statistical education in Australia!

8 Note

Developed from a keynote presentation at Eighth Australian Conference on Teaching Statistics, July 2012, Adelaide, Australia.

References

Allen, R., Abram, B., Folkard, A., & Lancaster, G. (2009). *Next steps in statistics education: Successful service teaching* (9pp.). Retrieved February 2013 from http://iase-web.org/documents/papers/sat2009/2_1.pdf

Bowen, W. G., Chingos, M. M., Lack, K. A., & Nygren, T.I. (2012). *Interactive learning online at public universities: Evidence from randomized trials* (52pp.). Ithaka S+R.

Brown, B. (2012). Flipping the classroom. In *Proceedings of the 43rd ACM Technical Symposium on Computer Science Education* (p. 681). New York: ACM.

Cox, D. R., & Donnelly, C. A. (2011). *Principles of applied statistics*. Cambridge, England: Cambridge University Press. 202pp.

Hawkins, A. (1996). Myth-conceptions (14pp.). Retrieved February 2013 from http://iase-web. org/documents/papers/rt1996/1.Hawkins.pdf

Johnson, L., Adams, S., & Cummins, M. (2012). *The NMC Horizon report: 2012 higher education edition*. Austin, TX: The New Media Consortium. Retrieved February 2013 from http://nmc. org/publications/horizon-report-2012-higher-education-edition.

MacGillivray, H. (2009). *Some principles of flows and steps in designing tertiary statistics curricula for learning* (22pp.). Retrieved February 2013 from http://iase-web.org/documents/ papers/sat2009/1_2.pdf

MacGillivray, H. (2011). *Statistical thinking foundations for postgraduates across disciplines* (21pp.). Retrieved February 2013 from http://iase-web.org/documents/papers/sat2011/IASE20 11Powerpoint5A.3MacGillivray.pdf

Ottaviani, M. G. (2002). *1982-2002: From the past towards the future* (8pp). Retrieved February 2013 from http://iase-web.org/documents/papers/icots6/07_ot.pdf

Phillips, B. (2003). *Overview of online teaching and Internet resources for statistics education* (10pp.) Retrieved February 2013 from http://iase-web.org/documents/papers/sat2003/ Phillips.pdf

Stannard, R. (2012). The flipped classroom or connected classroom? *Modern English Teacher, 21*, 35–37.

Speed, T. (1986). *Questions, answers and statistics* (11pp.). Retrieved February 2013 from http:// iase-web.org/documents/papers/icots2/Speed.pdf

Talbert, R. (2011). Inverting the classroom, improving student learning, SlideShare.com. Retrieved December 2012 from http://www.slideshare.net/rtalbert/inverting-the-classroom-improving-student-learning

Tukey, J. W. (1977). Contribution to the discussion of "A reformulation of linear models" by J.A. Nelder. *Journal of the Royal Statistical Society, Series A, 140*, 72.

Vere-Jones, D. (1995). The coming of age of statistical education. *International Statistical Review, 63*, 3–23.

Zarkovich, S. S. (1976). A reappraisal of the ISI statistical education program. *International Statistical Review, 44*, 289–295.

Part B
Undergraduate Curriculum

The Development of a First Course in Statistical Literacy for Undergraduates

Sue Finch and Ian Gordon

Abstract We describe the development of a statistical literacy subject suitable for undergraduate students from any discipline. Dispositions and fundamental ideas in thinking statistically as a cognizant consumer of media and research reports are identified from the literature. Using this framework, we describe how learning is embedded in a rich context of a diverse range of real-world problems. We illustrate the importance of visual representations in teaching fundamental ideas, including those about variability and statistical modelling, without assuming a strong mathematical background. The development of statistical dispositions is supported by modelling a statistician's enquiring approach to a question informed by data, structured practice at this approach, and checklists and tips to prompt scepticism.

Keywords Statistical literacy • Critical thinking • Statistical graphics

1 On the Need and Opportunity for Statistical Literacy

The great body of physical science, a great deal of the essential fact of financial science, and endless social and political problems are only accessible and only thinkable to those who have had a sound training in mathematical analysis, and the time may not be very remote when it will be understood that for complete initiation as an efficient citizen of one of the new great complex worldwide States that are now developing, it is as necessary to be able to compute, to think in averages and maxima and minima, as it is now to be able to read and write. (H.G. Wells 1903, p 126)

S. Finch (✉) • I. Gordon
Statistical Consulting Centre, The University of Melbourne, VIC 3010, Australia
e-mail: sfinch@unimelb.edu.au; irg@unimelb.edu.au

H. MacGillivray et al. (eds.), *Topics from Australian Conferences on Teaching Statistics: OZCOTS 2008-2012*, Springer Proceedings in Mathematics & Statistics 81, DOI 10.1007/978-1-4939-0603-1_6, © Springer Science+Business Media New York 2014

One hundred and ten years on, the need for increased statistical literacy in our complex world remains. Introductory statistics subjects in the mathematical sciences and allied disciplines have become more applied in the last two decades, with increasing emphasis on design, interpretation of analyses in context, and statistical thinking. Moore and McCabe's (1999) *Introduction to the Practice of Statistics* was one of the first of a new era of introductory texts with an applied emphasis. Outside the traditionally allied disciplines, liberal arts style courses teach the fundamentals of statistical thinking with less emphasis on mathematics. Such subjects are exemplified by the subject 'Lies, damned lies and statistics', at the University of Auckland. Utts (2005) book 'Seeing through statistics' skilfully guides learners and teachers through the first important topics in statistical literacy.

What should H. G. Wells' efficient citizen know? In 2003, Utts suggested that the teaching of introductory statistics subject had not kept pace with increasing everyday exposure to the findings of empirical studies, availability of statistical tools, and expectations that graduates will be statistically literate. She articulated seven important topics for the statistically educated citizen of today. We considered this broad agenda in the development of a statistical literacy subject for university undergraduates.

1.1   The Institutional Context

The traditional structure of undergraduate degrees in universities, including Australia, limits the opportunity to teach statistical literacy to a broad audience. Macquarie University (Australia) offers a general education unit 'Gambling, Sport and Medicine' and Auckland University offers 'Lies, Damned Lies and Statistics'. These introduce important statistical ideas in context and have no prerequisites. Other courses, such as Taplin's (2003) consulting subject at a second-year undergraduate level to students of mathematical statistics, included statistical literacy skills.

In 2008, a restructure of the undergraduate degree programme at The University of Melbourne introduced a requirement for students to take one-quarter of their degree as 'breadth', defined as subjects that are not part of a core degree. This was an unusual step for an Australian university although the structure is seen in degrees offered in other countries.

Students could take existing, traditional subjects as 'breadth' if they were outside their core degree. In addition, the new structure offered students freshly designed subjects in a special category of 'breadth' with a strongly interdisciplinary focus and no prerequisites. This new type of subject is broad in the sense of taking perspectives from many disciplines on a topic, involving teaching from different faculties and teaching students generic skills as well as subject-specific content. 'Critical thinking with data' (CTWD) is in this category of breadth subject; it aims to teach undergraduate students from across the University (predominantly, first-year students) some fundamentals of statistical science.

A requirement of being in this category of 'breadth' is that the statistical literacy subject must be able to be taken with any other subject. Clearly the content had to be sufficiently different from any other statistics-related subject, and to complement any introductory statistics course and enrich and challenge students taking such a course. These requirements also meant that students could not be assumed to have a mathematical background. This type of 'breadth' subject also teaches students more generic skills including those relating to presentation, written communication, and teamwork.

The semester-long CTWD course was first taught at the University of Melbourne in 2008, and will run for the seventh time in 2013. Up to 150 students have taken the course in different years. This has included students from the Arts, Science, Engineering, Commerce, Environments, and Biomedicine. CTWD has been taught by different lecturers in different years; many of the examples we present in this chapter come from the course in 2011 when it was taught by the authors.

2   Dispositions and Statistical Thinking

What does it mean to be statistically literate? Gal (2002) described two related aspects of statistical literacy: '(a) people's ability to interpret and critically evaluate statistical information, data-related arguments, or stochastic phenomena, ... and when relevant (b) their ability to discuss or communicate their reactions to such statistical information, such as their understanding of the meaning of the information, or their concerns regarding the acceptability of given conclusions.' (pp 2–3).

Many authors have described what it means to 'think statistically'; here we adopt one framework for characterising what CTWD aims to teach. Pfannkuch and Wild's foundational work articulates the nature of applied statistical thinking (Pfannkuch and Wild 2000; Wild and Pfannkuch 1999). Their four-dimensional framework for describing statistical thinking in empirical enquiry was based on the literature and interviews with students and practising statisticians. This framework characterises the *dispositions* needed for applied statistical thinking. The dispositions reflect creative and critical thinking: scepticism, imagination, curiosity and awareness, openness to challenging ideas, and persistence in seeking explanations.

The fundamental types of statistical thinking were also identified (Pfannkuch and Wild 2000). These related to recognising the need for data, the understanding that statistical information can take different forms and is manipulated in different ways, which they term 'transnumeration', considering variation, reasoning with statistical models, and an understanding of the integration of statistical information into the real-world context.

Our goal was for students to develop these ways of statistical thinking to become statistically literate in Gal's (2002) sense.

We describe our approach to the statistical literacy course through Pfannkuch and Wild's framework by discussing the role of context, our approach to teaching the fundamentals and to encouraging statistical dispositions.

3   The Statistical Content of a Statistical Literacy Course

Statistical thinking requires statistical knowledge, context knowledge, and understanding of the information in data (Wild and Pfannkuch 1999). What should the statistical content of a statistical literacy course include? As a general education in statistics is not traditionally part of an Australian undergraduate curriculum, there are few examples to follow.

The content and structure of the statistical literacy course were informed by a number of perspectives including the literature, liberal arts approaches to statistics, interdisciplinary experience in teaching statistics, and statistical consulting. Fundamental topics identified in the literature include Utts (2003) seven topics for the statistically literate citizen.

Wild and Pfannkuch (1999) described a cycle of statistical enquiry that characterises 'the way one acts and what one thinks about during the course of a statistical investigation' (p 225) (see also MacKay and Oldfield 1994; Marriott et al. 2009). The cycle reflects stages in statistical consulting described, for example, by Kenett and Thyregod (2006). The cycle of Problem, Plan, Data, Analysis, and Conclusion (PPDAC) also proved useful in thinking about the content of a statistical literacy course. It emphasises the importance of focusing on the 'big picture' questions, such as 'where do the data come from?' Our aim was to cover all stages of the cycle, giving students some experience with aspects of the data analysis phase. This was limited, but included descriptive statistics and graphing, and the calculation of an approximate margin of error and confidence interval for a proportion. Emphasis was given to interpreting inferential statistics—P-values and confidence intervals for measures of risk, mean differences, and so on.

In course development, we also drew on our broad experience in statistical consulting across all stages of the PPDAC cycle to identify a range of topics from applications of statistics appearing in both research and everyday contexts. The statistical content covered four themes with 15 topics, as shown below:

Finding data as evidence

1. Data quality: Anecdotes, intuition, or evidence?
2. Context: Data—a number with social value
3. Variation: Embracing the way we vary
4. Sampling: Sampling matters
5. Designed experiments: Evidence—by design

   Examining evidence in data

6. Graphics: Good pictures paint a thousand words
7. Summaries: Understanding relationships
8. Observational data: Relationships can be deceptive
9. Statistical models: Bell curves and other interesting models

   Understanding uncertainty in data

10. Probability: Objective and subjective

11. Risk: Understanding risk
12. Psychological influences on probability

Drawing conclusions from evidence in data

13. Confidence intervals: How certain can we be?
14. *P*-values: How 'significant' is our evidence?
15. Meta-analysis: Resolving inconsistencies and accumulating knowledge

Topics proposed by Utts (2005) in *Seeing through statistics* include many of the topics in CTWD, and the themes are similarly organised. This represents a convergence of approaches and thinking, rather than an explicit decision to base the subject content on Utts' design.

The detail of each topic was developed by considering the specific learning outcomes that we wanted to achieve for that topic. This ensured that lecture material, tutorial activities, and assessment tasks were well tailored to the types of knowledge, skills, and dispositions we wished students to develop.

Consider the topic 'graphics'. The lecture content provides illustrations of good and bad graphical practices and discusses important features of good graphs and research on interpretation of graphs. Lectures give examples of media reports including graphical displays and model how to critique and improve graphical displays. Five principles for good graphics are presented, and a number of standard forms were recommended. The principles were developed from the work of Tufte (1983) and Cleveland (1994).

The learning outcomes required students to develop skills in producing graphs for small data sets, and importantly to understand:

Good graphical presentation of data rarely happens automatically.
Good graphs are simple in design.
Graphs should, above all else, show the data clearly.
Good graphs have transparent visual encoding.
Good graphs have accurate titles and well-labelled axes with measurement units
    defined.
Axes should generally maintain constant measurement scales.
Good graphs identify the source of data.
Pie charts and pictograms are poor choices for representing data.
Aligning data along a common horizontal axis facilitates accurate comparisons.
Standard forms have been developed for particular purposes.
Unusual values should be investigated carefully for information they may provide;
    they are not necessarily mistakes.

For the first few years of the course students were not expected to use software to produce graphs, although those wishing to do so were encouraged. In 2011, the use of *Gapminder* software was introduced, and students were expected to be able to produce graphs using *Gapminder*.

The content of CTWD differs from an introductory course in statistics or an allied discipline in a number of important ways. A number of the topics included reflected the breadth of the application of statistics and divergence from what might be considered standard content; these topics were context, designed experiments and observational studies, risk, psychological influences on probability, and meta-analysis. These topics might be touched on in introductory courses, but the treatment is more extensive in CTWD.

The content is presented in a strongly applied setting which allows an emphasis on practical interpretation and potential misconceptions. The topic P-values is a useful example, covered in two to three lectures as the second last topic in the course. This allows introduction of P-values in the context of many case studies that are already familiar to students. The formal concepts covered are the null hypothesis and the P-value, with the meaning illustrated graphically in the context of examples. Many examples of the parameters of interest are provided: differences of means, differences of proportions, risk ratios, correlations, and so on. The logic of testing a null hypothesis is discussed in the context of research expectations. Teaching stresses the P-value as a probability, not a binary decision criterion, as in some disciplines, although it is often used to indicate 'statistical significance'.

Formal calculation of a P-value is not included. Material about the use and interpretation of P-values discusses the errors of equating statistical significance and importance or practical significance, using the P-value as a measure of certainty, and making inferences that 'no effect' had been found based solely on a large P-value.

The content also includes the controversy about the use of P-values in many disciplines, and the historical precedence of P-values over confidence intervals, particularly in some disciplines. The takeaway messages for students are that P-values should be reported and interpreted with estimates and confidence intervals and that they should be suspicious of studies relying on P-values alone for interpretation. This depth of discussion, without recourse to detail about how P-values are calculated, is important in a statistical literacy course to enable critical reading of statistical reports in other disciplines. By the end, it is intended that students understand the rhetorical title of the topic: 'How "significant" is our evidence?'.

3.1   Content Delivery

CTWD is delivered as a 36 lecture series. Six of the lectures are presented by eminent guest lecturers—experts in an applied discipline. These lecturers each contribute two 'bookend' lectures, around a content section of CTWD. The first lecture is scene setting, raising substantive questions that can be answered in a statistical framework; the second lecture provides answers to those questions, drawing on the statistical content covered in the course. The guest lecturers provide a statistical orientation from different disciplines as well as engaging and realistic examples; anecdotal feedback from students about this aspect of the course is very positive. The disciplines involved over the past 6 years include epidemiology, actuarial science, and environmental science.

The lectures generally follow the sequence of topics described above and make extensive use of 'rich media'. Examples of video include George Bush dismissing the findings of the first Iraq survey, elephants reacting to bee sounds in Lucy King's study of the possible deterrent effect of bee sounds on elephants (King et al. 2007), Rosling (2013) demonstrating the Gapminder software showing several dimensions simultaneously, and several stakeholders commenting on the Toowong breast cancer cluster (Stewart 2007).

Our goal in lectures is to deliver rich statistical content while relying very little on mathematical notation. As we elaborate below, rich context and interesting case studies are used to motivate and illustrate the important lessons. There are many examples, rather than a few. Visual representations are used wherever possible— pictures, diagrams, graphics, and graphs. Simulations are presented using StatPlay (Cumming and Thomason 1998). In developing the lecture material, we tried to consider the many different ways an idea, concept, or principle might be represented. We attempted to model good statistical thinking and practice in the lectures—to exemplify the disposition of a statistical consultant.

Students attend weekly tutorials that are designed to supplement lectures without presenting new content. Tutorial activities give students the opportunity to look at data-based investigations in detail and to practise applying their developing statistical dispositions. We discuss the tutorial content in more detail below.

CTWD makes extensive use of the University of Melbourne's Learning Management System (LMS). Lectures are recorded with Lectopia, and access is provided via the LMS; course content is always available to students. CTWD does not have a set text. Students are asked to read accessible background material, such as Bland and Altman's British Medical Journal Statistics Notes, tailored to each topic. For example, on the topic of P-values, readings include Gardner and Altman's (1986) 'Confidence intervals rather than P values: estimation rather than hypothesis testing' and Cohen's (1994) 'The earth is round ($p < 0.05$)'.

The University of Melbourne's Digital Repository allows students to access reading materials and multimedia. The Digital Repository for CTWD currently contains about 150 items—newspaper articles, video clips, links to websites, and academic papers. There is material directly relevant to the statistical content as well as material related to the case studies. Only a small proportion of the material available on the Digital Repository is required reading. Background material on case studies and supplementary material on statistical content are available so that students can follow up on content that particularly interests or challenges them.

4   Statistical Literacy and Context

In the development of statistical education, it is now recognised that students' learning should be grounded in real-world applications to encourage the development of statistical thinking along with an understanding of the importance of data collection methods and study design (e.g. Cobb 2007; Wild and Pfannkuch 1999).

A common complaint about statistics majors is that they have not learned to solve real problems, for example: 'Statistics departments and industry report that graduates of statistical programs are not well oriented toward real problems' (Minton 1983); for a similar recent discussion, see Brown and Kass (2009). This is also a difficulty for students learning statistics in service courses (Singer and Willett 1990). The need to engage these students with genuine relevant data has been identified in many applied disciplines (e.g. Love and Hildebrand 2002; Paxton 2006; Singer and Willett 1990).

The statistical literacy course is replete with case studies and examples. The case studies are data-based investigations that had richness and some complexity and are often relevant to more than one topic. We describe three broad types of case studies.

There are excellent historical case studies, familiar to many in statistical education. These include stories such as John Snow and the Broad Street pump, the Challenger disaster, the very early randomised trials in England (streptomycin for TB, pertussis vaccine for whooping cough), the mice experiments for penicillin, the failure of the Literary Digest to predict the 1936 US Presidential election, and the Salk vaccine trials.

Contemporary examples are case studies with global significance if they are international or with a local media presence if they are Australian. A rich case example is the problem of estimating the civilian death toll due to the war in Iraq (Brownstein and Brownstein 2008; Zeger and Johnson 2007). In the context of this case study, the need for data rather than opinion is discussed, along with survey methodologies, measurement procedures, and methods of sampling; it also included an example of a poor graph.

The third type of case study challenged the stereotype of statistics as a dull science. These were chosen to communicate the wide effectiveness of statistical science and to engage students. These are novel examples, often with 'good news value' and they were identified through news sources—newspaper or radio. One example used Coren's (2006) study of dogs and earthquakes to illustrate outliers and issues in observational studies. For these case studies and the contemporary ones, materials both from the media and academia were sought, given the educational potential in the use of the media as a motivating source of material (Watson 1997).

The content includes many smaller examples chosen to illustrate particular concepts or points. All topics and concepts are introduced in the context of a real-world application. We illustrate the embedding of context in content for the topic of risk.

Risk is a popular term for media reports and the topic begins with a range of media examples, including a study of the risks of using the drug 'ice' (Bunting et al. 2007). The statistical content covers the definition of 'risk' as the probability of an adverse outcome, comparisons of ways of reporting risks (risk ratio, risk difference, and odds ratio), discussion of the choice of baseline risk, and concepts of test effectiveness: true positive, true negative, false positive, false negative.

The case study of the breast cancer cluster at the ABC studios in Toowong, Queensland (Stewart 2007) is an important part of the topic risk, as are some other medical examples including the trial for Torcetrapib for high cholesterol halted in

2006 (Barter et al. 2007) and Australia's largest outbreak of Legionnaires' disease in Melbourne's aquarium in 2000 (Greig et al. 2004). Other applications included the Minneapolis domestic violence experiment (Sherman and Berk 1984) and the effectiveness of roadside drug tests.

Common errors in the interpretation of risk measures are illustrated using media articles, and this is revisited in the topic on psychological influences on probability where research on judgments of risk and uncertainty is presented.

Wild and Pfannkuch (1999) describe a continual shuttling between the context sphere and the statistical sphere in all stages in the PPDAC cycle; statistical knowledge informs the understanding of the context and vice versa. The embedding of statistical learning in rich case studies and examples exemplified this synergy. Students are expected to have good knowledge of the content of a specified number of the case studies. The case studies are used in assessment tasks in various ways; weekly quizzes and exam questions make reference to the case studies and in the examination students are expected to be able to discuss the nature of the statistical content or to choose examples of statistical applications from these case studies.

5 Learning About Fundamentals

Here we describe how CTWD addresses Pfannkuch and Wild's (2000) fundamental types of statistical thinking. The statistical literacy course was shaped by the fundamental idea of data in context, as described above. In a broad sense, the need for data is illustrated in the richness of this context. The first topic, 'Data quality: Anecdotes, intuition or evidence?', deals with the need for data more directly. There are two important motivating examples. The first is the prosecution and fining of GlaxoSmithKline for false claims about the vitamin C content of Ribena following a school experiment by two 14-year-old girls in New Zealand (Eames 2007). The second example is the civilian cost of the Iraq war where world leaders dismissed empirical estimates in preference to anecdotal information. The need for data is revisited in the discussion of psychological influences on probability where the reliability of personal estimates of probabilities is contrasted with empirical data.

'Variation: Embracing the way we vary', the third topic, covered the pervasiveness of variation in most quantitative research, the need to interpret patterns in data against an understanding of the variation in a given context, the idea of 'background' or 'unexplained' variation, and how different types of distributions reflect variation in different ways. The presentation of these quite abstract and deep ideas is essentially non-mathematical and relies heavily on graphical presentation in the context of case studies. Time series data are ideal for illustrating the need to make comparisons in the context of all relevant variation. The topic begins with the Antarctic ice-core data (Barnola et al. 2003; Etheridge et al. 1998) illustrated in a sequence of graphs that expand the time interval considered. This is shown in Fig. 1.

Variation in samples from the same source and in groups from different sources is presented with graphs. Students are asked to make intuitive graphical comparisons

Carbon dioxide concentration in parts per million (ppm) over time
at two sites in Antarctica

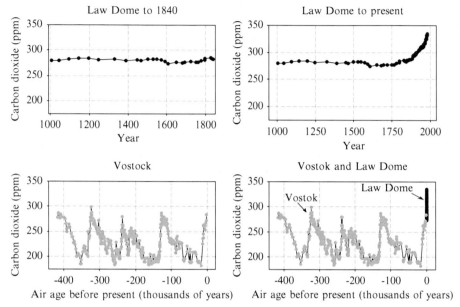

Fig. 1 Sequence of graphs from Antarctic ice core data

of groups and to suggest when the distributions might be sufficiently different to consider the differences to be 'real'. Assessment tasks include evaluation of evidence for a systematic difference between two groups represented as pairs of histograms; students were asked to choose between four pairs of histograms.

This approach does not entail mathematical formulae or formal testing. It is however like the processes of data exploration and hypothesis generation used by an experienced statistician (Pfannkuch and Wild 2000), and is a different type of 'analysis'.

Students are, for example, asked to engage in data exploration and hypothesis generation in a tutorial on that used the death rates from the sinking of the Titanic (Dawson 1995). The rates are provided broken down by four explanatory variables: age, gender and class, and eye colour (a fictitious addition). The event is not identified. Working collaboratively, students produce dotplots of the rates in terms of various combinations of explanatory variables to evaluate differences against background variation. They consider how the explanatory variables alone and in combination influenced the rates. The simple substantive question, 'From what event did these data come?', genuinely engages students. With the graphics displayed around the room, a range of possible explanations can be generated. Typically a number of incorrect events are proposed along with the right answer.

Pfannkuch and Wild (2004) suggest that 'even much simpler tools such as statistical graphs can be thought of as statistical models since they are statistical ways of representing and thinking about reality' (p 20). The introduction of the idea of reasoning with statistical models is challenging in the context of a statistical literacy course, but the Titanic example illustrates one way this can be done using graphs.

Statistical modelling is discussed in the topic on variation with reference to the Coren's (2006) study of dogs and earthquakes, but at a high level of abstraction. The model of an outcome, y, is a function of explanatory variables plus random error, expressed in a qualitative rather than mathematical form. Random error is characterised as arising from explanatory factors that are yet to be identified or understood, and as unavoidable variation. The idea of assessing the effects of explanatory variables by comparing their observed effects to the random background variation is introduced.

Graphical representation is a strong component of the lectures on statistical models. High-profile sporting examples introduce the problem of deciding when an observation is extreme or unusual in the context of a statistical model. Content includes Wild's (2006) idea of a distribution as a model and lens through which we can view data. The use of Normal, binomial, and Poisson distributions in addressing inferential questions goes as far as illustrating the calculation of probabilities.

The idea of 'transnumeration' is considered the most fundamental for students of statistics (Wild and Pfannkuch 1999). Transnumeration involves ways of obtaining, manipulating, and modelling data that support good understanding of the problem at hand. It encompasses the measurement process of quantifying observable characteristics, summarising, graphing, and finding models for data.

Tutorial activities give students some experience of transnumeration in designing survey questions and making responses, and in graphing, summarising and considering 'models' for data as illustrated by the puzzle of the Titanic. There are limitations to the amount of data manipulation that can be done in this statistical literacy course. One approach we use is to limit the size of the data set students use and to distribute tasks in class groups. Another approach that is adopted to facilitate the understanding of representation and manipulation of data is through activities and assessment tasks based around critical analysis of data presentations and reports. Were the graphs and summary numbers coherent? Did the summary numbers seem plausible, given the contextual information? By developing an enquiring disposition and a sense of number, the consistency between different mappings of data can be assessed and understood. Whenever possible the many different ways an idea, concept, or principle might be represented are presented. The use of both media and academic or background reports on the case studies and examples provide different kinds of representation of the same data-based investigation.

6   Developing Dispositions

A core skill of an applied statistician is asking good questions (Derr 2000; Kenett and Thyregod 2006). The statistical literacy course focused on developing dispositions of curiosity and scepticism, and a questioning approach in being persistent in finding explanations—dispositions that require attention to detail. Such dispositions can be taught although they may depend on how engaging a problem is to an individual (Wild and Pfannkuch 1998).

The rich embedding of statistical learning in context is one strategy used to try and engage students. The lectures made extensive use of newspaper reports, pictures, diagrams, graphics, graphs, and sound and video clips from news programmes, documentaries, and the Internet. Many case studies and examples, of the three types described earlier, are used, rather than a few. Case studies and examples are integrated into as many topics as possible and are taken from a very broad range of disciplines.

In lectures, case studies and examples are presented through the perspective of an enquiring statistician. Messy details are not sanitised or glossed over. Background material to media reports and examples proved to be an excellent source of enrichment. Importantly, the need to 'dig deep' and follow through to resolve uncertainties and understand how the story in representations of data fitted together is modelled in lectures. This kind of story telling is exemplified in Tufte (1997). Tutorial activities are structured to practice questioning and find coherence. Assessment tasks require follow through to explain representation, reports, and commentary of a statistical nature.

An example of modelling the interrogative approach is from a case study of a graphic showing greenhouse gas emissions by source in Australia, published in Royal Auto in March 2008 (Negus and Mikedis 2008); see below.

Source: Royal Auto, March 2008, p 12

This is an extraordinary example of poor graphics; the proportions represented do not preserve the order of the data in a table provided with the article, let alone represent the quantities accurately. Transnumeration from a table to a graph is incoherent.

The example is very useful even if only the Royal Auto article was considered. However, to better understand the categories represented, the original data source was obtained (Department of Climate Change and Australian Government 2008). This provides lessons about the importance of context given that the icons did not always appear to be appropriate representations of the categories intended, clarity in communication of the results of data-based investigations, and the need for attention to detail in reporting. Alternative representations of the data are presented and discussed. In this, and many other examples, an appropriate and coherent graphical representation was created if the original was inadequate.

As well as modelling an enquiring approach, explicit summaries of the main messages for each topic, lists of questions to ask about a study, and tips for making better judgements are provided. Topic summaries included tips on when to be suspicious about the meaning of a data-based claim or report. The 'questions to ask' provide a structure for scepticism, and are like the 'worry questions' of Wild and Pfannkuch (1998), featured in the University of Auckland subject 'Lies, Damned Lies and Statistics', and the seven critical components described by Utts (2005).

Tips for improving personal judgements provide a summary of the material on psychological influences on probability, as shown below.

Tips for improving your judgements

When given a probability estimate, make sure the reference class is clear.

When making personal estimates of probability, remember you might be being optimistic.

When your doctor gives you an estimate of risk, ask if this is his or her personal estimate or one based on long-term relative frequency. Doctors are not always well calibrated.

Think about reasons why your judgement might be wrong; this can reduce overconfidence.

When evaluating risks, try re-describing the problem in terms of frequencies; you might find the risk easier to calculate.

When judging sequences, check your assumptions. Do you think the events are independent or not? What role do you think chance has in the outcome?

When making estimates, does the framing of the information influence your answer?

When making estimates, was an anchor provided that might have influenced your answer?

Experienced statisticians can ask the right kind of 'trigger questions' to obtain important information about context (Wild and Pfannkuch 1999). A budding statistically literate citizen needs a structure of questions to ask throughout the PPDAC cycle (Wild and Pfannkuch 1999). This structure is provided, modelled, and practised.

7 Learning Activities

We have described the role of tutorial activities and assessment in the learning of the fundamentals of statistical thinking and the development of a statistical disposition. We now describe some of the learning activities designed for CTWD in more detail.

7.1 Tutorials

One-hour weekly tutorials provide opportunities to apply statistical thinking in practice. The table below illustrates the tutorial activities used in 2012, for example.

	Activities
1	Designing a class survey
	Interpreting a reading from Dicken's *Hard Times*
2	Further question design for a class survey
3	Solving the Titanic puzzle
4	Identifying issues in four real-world sampling problems: adolescent drug use, museum art works, counting unregistered vehicles, and lizard abundance
5	Reviewing early trials of streptomycin for tuberculosis
6	Reviewing a study of selective imitation in dogs
7	Applying principles of good graphics to examples
	Reviewing published graphs
8	Poster presentation preparation
9	Assessing claims in newspaper reports of empirical research
10	Poster presentations
11	Review of National Statistics Omnibus Survey (UK) of understanding of breast cancer risk
12	Review of a National In-Service Vehicle Emissions Study

The materials for the tutorials are chosen with a focus on the content covered in recent lectures. However the tutorial activities require students to apply what they have learnt from the course to date. The early activities involve designing a survey and participating in the data collection; results from the survey are presented as examples in lectures and are also used in assessment tasks. Although the students do not carry out analysis of the survey data, they are actively involved in the design, data collection, and interpretation of the results.

In the sixth tutorial in 2012, for example, students worked on 'Selective imitation in domestic dogs' (Range et al. 2007). This study was reported in a newspaper article 'Dogs prove that they really are right on the ball'. Students first consider the newspaper article and the information it provides about design, experimental control, and measurement; they check some of the reported proportions. Then the detail of the scientific report is considered and questions are revisited. Students construct a summary table from data in supplemental material provided with the scientific report. The overall quality of the study and the validity of the conclusions drawn are evaluated. This diversity of tasks, based on different parts of the PPDAC cycle, is used in tutorials wherever possible.

There are no laboratory classes in CTWD, and no formal training in statistical software (other than *Gapminder*) is provided. However, as the example above illustrates, students are expected to make common-sense checks of summary statistics, manipulate data, compute some summary measures, and construct simple displays and tables. As mentioned earlier, they calculate an approximate margin of error and confidence interval for a proportion. This is often practised as group work.

As the table above indicates, students work on a large assignment during tutorial time; one part of this large assignment is completed in a tutorial devoted to a poster display and presentation session. This is discussed in more detail below.

7.2 Assessment

The lecture content and tutorial activities have remained relatively constant in CTWD. A big challenge was developing appropriate assessment methods and choosing suitable materials for assessment. This has developed over time.

The case studies presented in lectures and tutorials are one source of material for assessment tasks. We needed to find additional studies, novel to the students, with a mix of good and bad features, and with the right degree of complexity to allow students to apply their critical enquiry skills. Simple flawed examples of data-based investigations can assess an important principle but have limited scope for assessing varying depths of students' understanding; examples that are too complex can include features that are not relevant to the assessment task but may mislead students to focus on them. Media reports proved to be a rich trove for this kind of material.

We describe the most recent assessment structure which includes quick quizzes, fortnightly tests, short assignments, a large assignment, and a final examination. There are ten weekly online quick quizzes of up to ten questions that can be attempted three times. These quizzes, worth 5 % of the total marks, are for revision of lecture content and concepts. Five fortnightly single-attempt online tests are time limited; these are worth 15 % in total. The fortnightly tests require students to read some preparatory material, such as a newspaper article and scientific paper, before starting the online test.

There are three short assignments worth 5 % each with a strict work limit of 200 words. A major assignment is worth 25 % and has two components—group work and an individual write-up. The final exam is worth 40 % and is modelled on the other assessment components. We discuss the various forms of assessment in more detail below.

CTWD includes topics such as variation and statistical modelling, but without formal mathematical treatment. Setting appropriate assessment tasks was challenging, requiring a tailored approach that would evaluate the application of generic skills and data-related thinking skills. Importantly, the marking of the assessment tasks has to reward good statistical thinking; this might sound self-evident, but this proved to be an important element in the design of suitable assessment tasks. We provide examples of our assessment tasks of various types below.

7.2.1 Weekly Quizzes

Students complete the weekly quick quizzes via the University's LMS. This allows a variety of different types of questions to be asked including multiple choice, multiple answers, exact numerical, matching, ordering, fill-in-the-blank, and 'hot spot'; all of these types of questions are used in CTWD. A 'hot spot' question gives a visual representation, and students need to click on the area of the representation that gives the answer to the question.

The use of visual representations is one approach used to present and assess concepts and principles traditionally given a formal mathematical treatment. For example, this hot spot question examined the idea of sampling variation:

Each histogram below shows a sample of 75 net weights dumped by a garbage company. Three of the histograms are from samples taken from the same population. Click on the histogram that is most likely to have come from a different population.

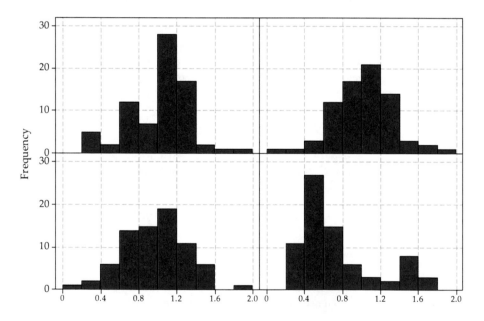

In the quick quizzes, the emphasis is on interpretation of representations of data and the application of knowledge appropriately to real-world problems. In a traditional course, students might be asked to produce suitable graphs to describe distributions of data. In CTWD we ask, for example:

Consider the graphs below. Match the histograms and boxplots.

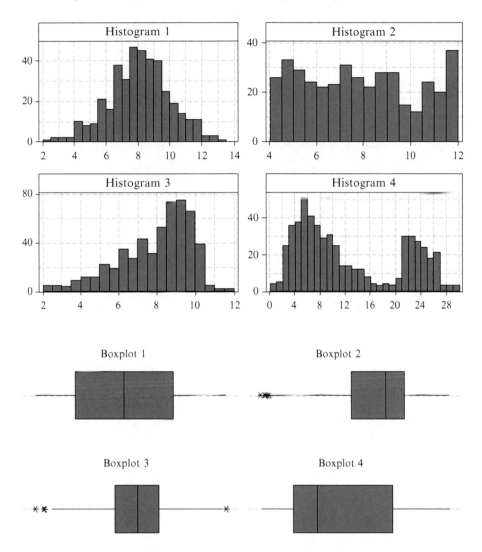

This question requires an understanding of the construction of histograms and boxplots and their relationship; skills in estimating the median and quartiles help with finding the correct answer. Quick quiz questions also test students' ability to produce simple summary numbers, such as proportions, risks, risk ratios, odds, and odds ratios.

7.2.2 Short Assignments and Fortnightly Tests

Broadly, the short assignments ask for a 200-word critical evaluation of a data-based argument, report, or representation. The short assignments assess critical thinking skills in an applied context and also require good reading comprehension and writing skills. The assignments give very clear guidance about the nature of the task, the content to consider, and the points to be addressed in the word limit.

For example:

This assignment is based on an article that was published in The Australian on November 22, 2010 entitled 'Cars to blame in most accidents involving cycles, says research calling for new road rules'.

There is a research report of the study alluded to in the article: 'Naturalistic cycling study: identifying risk factors for on-road commuter cyclists'. This report was written by researchers from the Monash University Accident Research Centre (MUARC) and is referred to here as the 'MUARC Report'.

You should start by critically reading the article in The Australian (referred to here as 'the article') and in particular thinking about the quantitative information, reasoning, and conclusions it contains. You should then read the MUARC Report.

Imagine that you are one of the researchers who conducted the study and wrote the MUARC Report.

You are approached by the ABC Four Corners programme, who are planning a comprehensive piece on the increase in commuter cycling in many cities in Australia and associated policy issues. They know about your study and have emailed you to set up a meeting. Before the meeting, a Four Corners researcher, Jennifer, writes to you and asks if you were happy with the article in The Australian. She writes: 'Did it do a good job of summarising the key features of the study? Did you have any problems with it?'

For Short Assignment 1, you need to prepare an email response to this enquiry. You should start the email message with the following, which does not count towards the 200 words.

Dear Jennifer,
Thank you for your interest in our study. I am happy to provide feedback about the article in The Australian.

In your 200 word limit, you should pay attention to the following.

- The words used in the article for the main message: are they appropriate? Is the main message of the article consistent with the MUARC Report?
- Any deficiencies in the reporting of the quantitative information in the article.
- Any problems with the descriptions of other relevant information about the study in the article.
- Matters identified as important in the MUARC Report that are not mentioned in the article.

Important: These issues are provided as a guide only. Your assignment should read naturally as an email to Jennifer, answering her enquiry.

An important issue for students to identify in the media report is the mislabelling of 'incidents' (neither collisions nor near collisions) as 'accidents'. Additional issues to be raised include the small number of cyclists involved, the ad hoc sampling method, the potential subjectivity of judgements about driver behaviour, and the generality of the claims being made. Marking of this assignment rewarded attention to these types of issues explained in context, rather than descriptive summaries of the MUARC report.

Another short assignment asked students to use the Gapminder software to produce a graphic to include in a short report describing the relationship between rates of HIV/AIDS infection and some other variables in the six worst-affected African countries: Botswana, Lesotho, Namibia, Swaziland, Zambia, and Zimbabwe. Students were asked to provide one graph and a report to address the changing relationship between life expectancy and HIV infection rates in the six countries of interest, between 1995 and 2006. The assignment specified a number of questions to be addressed, and marking rewarded a graphical representation that suitably supported commentary on those questions.

In the first years of CTWD, there were more short assignments and no fortnightly tests. Fortnightly tests were introduced as structured assessment tasks that modelled the type of analytic reasoning needed for the short assignments without the writing requirement.

An example of a fortnightly test used early in the semester was based on Bicycle Victoria's 'Super Tuesday' bicycle count. Students were instructed to prepare for the test by examining some particular pages on the relevant website. As this test was early in the course, the test focused on issues of measurement, data collection, and sampling.

7.2.3 Major Assignment

The major assignment involves a detailed review of a single research study. Students work on one of five case studies; each case study had a newspaper report and a published article.

The major project has two parts. Students work in groups on the first part on a (tutorial) poster display and presentation session. This session is modelled on a conference poster session, with a brief presentation followed by discussion with assessors and fellow students. The verbal presentation and the poster are assessed. One aim of this task is to help students understand the research study they are reviewing quickly and efficiently. We chose studies that are relatively straightforward in their design. As the complexity of the statistical analysis is quite varied, students are given help to understand the relevant details of the analysis where needed.

The second and individual part of the major project is a 1,200-word critical assessment of the reporting of the study in the news item and in the published article and of other aspects of the design, implementation, and analysis of the study. This critique requires description and explanation of both the strengths and weaknesses of the study. Students can comment on the analysis of some of the studies, for example, by critiquing the graphical presentation of the results.

Examples of the topics of the major assignment include 'circle sentencing' of Aboriginal offenders, heart failure and the 2004 World Cup, symptoms and medical conditions in Australian veterans of the 1991 Gulf War, directional tail wagging of dogs, and the mortality of rock stars. Again, the marking schemes for the piece of assessment are designed to reward appropriate context-based evaluation of the statistical content.

7.2.4 Examination

The examination includes quick quiz type questions, short answer questions, and longer questions. Here is an example of a short answer question:

Oscar the cat lives in an end-stage dementia unit in the USA. Staff in the dementia unit believe that Oscar is able to predict patients' deaths. Oscar is not friendly to people in general, but he has been observed to sleep on the beds of patients in the last hours of their lives. As of January 2010, it was claimed that Oscar had, in this way, accurately predicted approximately 50 patients' deaths.

(a) Consider assessing the claim that Oscar can predict patients' deaths by collecting data each time a patient dies. List three variables that you think would be important to measure.
(b) What is the primary design flaw in the study implied in (a)?
(c) Suggest a design that would overcome this flaw.

Here is an example of a longer question:

In April 2008, The Herald Sun published a series of articles over several days about a poll of Victorian police officers. The headline on the first day was 'POLL: POLICE FACE CRISIS'. There had been heightened tensions between the Police Commissioner, Christine Nixon, and the secretary of the Police Association, Senior Sergeant Paul Mullett.

The Herald Sun published the following information about the survey:

'The Herald Sun wrote to more than 11,000 Victoria Police officers and invited them to complete the Herald Sun survey—3,459 responded. That means 30 % of the force's sworn officers took part in the unprecedented poll.

The survey sample of 3,459 is much larger than those commonly used by pollsters. Galaxy Research vetted the questions to ensure they conformed to accepted standards of conducting research in Australia.

The Herald Sun provided each police officer with a unique password, which expired after it was used once. That ensured they could not submit multiple entries.

The completed surveys were analysed by an independent market research company, which provided the results to the Herald Sun for publication'.

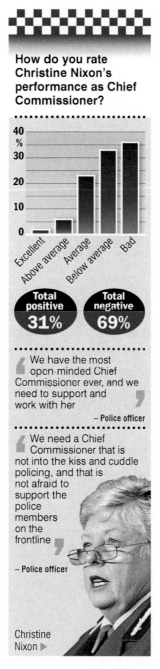

How do you rate Christine Nixon's performance as Chief Commissioner?

Total positive 31%   Total negative 69%

We have the most open-minded Chief Commissioner ever, and we need to support and work with her
— Police officer

We need a Chief Commissioner that is not into the kiss and cuddle policing, and that is not afraid to support the police members on the frontline
— Police officer

Christine Nixon ▶

Unless otherwise specified, use the extracted information from the Herald Sun article to answer the following questions.

(a) What was the sample frame for this survey?

(b) The day the survey was released, Chief Commissioner Christine Nixon was critical of it: 'I don't intend to go anywhere', Ms Nixon said this morning after a landmark survey raised a raft of serious issues about the state of the force. Speaking on radio 3AW this morning, Ms Nixon attacked the results because just 30 % of sworn officers—3,459—had responded to the survey. 'That is despite the sample being much larger than that commonly used by pollsters.' (Herald Sun) What was the basis of Christine Nixon's criticism?

(c) What argument is the Herald Sun using in its comment: 'That is despite the sample being much larger than that commonly used by pollsters.'? Provide an analysis of the argument.

(d) What strengths and weaknesses of the survey are apparent from the Herald Sun's description of the survey? Comment on any features you have not discussed in (b) and (c) above.
Parts (e) to (h) refer to the extract from the Herald Sun.

(e) What type of information is in the two quotes supplied at the bottom? What weight should be attached to these comments as evidence?

(f) Examine the bar chart at the top of the extract. What type of data is represented?

(g) Comment on the descriptions of the categories of response ('Excellent', 'Above average' … 'Bad') and the effect they could have had on the responses.

(h) How were the two simpler categories—'Total positive' and 'Total negative' constructed? Comment on this construction.

This relatively simple example allowed questions to be asked about many aspects of the processes involved in a data-based investigation as well as to assess examples of statistical reasoning. In the examination, questions about this type of material were quite structured given the time constraints within which students had to work.

8 Student Feedback

Given the diversity of the student cohort in CTWD, there is a diversity of feedback; for some students the subject is mathematically challenging, and for others, it is too light on mathematics. The end-of-course *Student experience survey* was completed by 73 of the 155 enrolled students in 2011. Eighty-four percent agreed or strongly agreed that they had learnt new ideas, approaches, and/or skills in CTWD, and 81 % agreed or strongly agreed that they had learnt to apply knowledge in practice. In the open-ended commentary, there were many positive comments about the content presented by the guest lecturers. More common negative comments referred to the amount and variety of assessment in the course, and to the uncertainty about 'what I need to know'. We have attempted to address this problem of 'what to know' by making aspects of the course structure more explicit and by providing end of topic summaries that can be linked back to the case studies as examples of the important messages. Continuous assessment and practice of critical thinking skills are an essential element in the design of CTWD; without this kind of practice, students may simply rote learn lists of critiquing principles and fail to appropriately identify real problems in context.

A follow-up survey of students in the first year of CTWD (2008) was carried out in 2010, as the students completed 3 years of undergraduate study, and in many cases, their degree. Around three-quarters of the 55 students responding to the survey agreed or strongly agreed that they could critically evaluate research reports, which include quantitative data, and media reports that include quantitative data. Students were asked an open-ended question: 'What advice would you give a student contemplating studying Critical thinking with data?'

Here are some responses:

'Critical thinking with data will change the way you think and analyse in general and not only in relation to data'.

'Make the most of Critical thinking with data, because the lessons you learn can be used throughout your degree, and potentially the rest of your life. Be willing to discuss/think about the information presented and really understand it. It's an interactive subject—lecture attendance is highly recommended'.

'I would say, Critical thinking with data is definitely one of the most useful subjects I've done in my 3 years at university. No matter what degree you are doing or what area you want to work in the future, we are constantly surrounded with information in the form of data, and having the skills to be able to critically analyse this information is very important in any successful career'.

In general terms, the feedback shows encouraging signs that the aims of the subject are being largely fulfilled, and in some individuals, deep insights have been established.

9 Some Lessons

Applied statisticians need to be able to think creatively and flexibly to find suitable solutions to sometimes novel real-world problems that can be understood by users. Developing a statistical literacy course suitable for all-comers amplifies the challenge to think creatively and flexibly about how to explain and represent the fundamental ideas, and how to engender the important dispositions. What lessons did we learn in addressing this challenge?

While the need for teaching statistics in context is widely recognised in statistics education, our experience in developing CTWD highlighted the deep embedding of statistical knowledge in a context. Content was chosen to reflect this deep embedding; case studies are presented as unfolding stories with rich detail to be probed and queried. CTWD provided the opportunity to work with material in this way given students are not burdened with learning software or details of analytic techniques. Storytelling from an applied statistician's perspective provides a way of modelling dispositions of curiosity, scepticism, and persistent enquiry. From the students' point of view, it might seem like there is a lot to learn in a short time. This learning is guided by the structured tutorial and assessment tasks that requires detailed and persistent enquiry and tips and questions for the sceptic.

Statisticians are encouraged to pursue excellence in their graphics (Gelman 2011; Gelman et al. 2002; Tufte 1983) as a means of communicating data. The need to teach the fundamentals in a non-formulaic way extended how we thought about the use of graphs, and they became a primary mechanism for presenting many concepts and fundamental ideas. The words of one statistician are apt here: 'I think graphical techniques are a much more fundamental part of statistics than the statistical tests and the mathematical statistical theory which we have that underlies all these things. I think that to a large extent what we are trying to do is put a sound basis to what are usually reasonably sound judgements, personal judgements about what's going on here. So I think a lot of statistical thinking really relies on good understanding of how to use graphics.' (Pfannkuch and Wild 2000, p 147)

The claim that context and graphics play an important role in an undergraduate statistical literacy may seem axiomatic to many statistical educators. The lessons we learnt in the development of CTWD were in how much we could exploit data stories and how often we could illustrate a concept or assess understanding with graphics. In CTWD, context is not just a cover story to give meaning to numbers to be analysed. It is the whole story of the data through all stages of the PPDAC cycle, with an important emphasis on enquiry. Context is the story uncovered by questioning. There is a great deal of work in finding and developing good case studies. However, the more the material is integrated across topics and in assessment tasks, the less likely it is that the case study is simply seen as an 'interesting story'. We hope that the examples we provide of the integration of stories and graphics into teaching and assessment will encourage other statistical educators to perceive the exciting potential in this approach for students to engage in deep statistical learning that is not within a formal mathematical paradigm and which is a crucial element of becoming a mature statistical thinker.

10   Note

Developed from papers presented at the Sixth Australian Conference on Teaching Statistics, July 2008, Melbourne, Australia.

This chapter is refereed.

Acknowledgements The authors thank Associate Professor Robyn Pierce and Dr. Robert Maillardet for providing some of the results of a survey of the 2008 cohort, carried out in 2010. We also acknowledge the collaboration of Professor Chris Wild, Dr. Maillardet, and other members of the advisory committee in course development.

References

Barnola, J. M., Raynaud, D., Lorius, C., & Barkov, N. I. (2003). Historical CO_2 record from the Vostok ice core. In *Trends: A compendium of data on global change*. Oak Ridge, TN: Carbon Dioxide Information Analysis Center, Oak Ridge National Laboratory, U.S. Department of Energy.

Barter, P. J., Caulfield, M., Eriksson, M., Grundy, S. M., Kastelein, J. J., Komajda, M., et al. (2007). Effects of torcetrapib in patients at high risk for coronary events. *New England Journal of Medicine, 357*, 2109–2122.

Brown, E. N., & Kass, R. E. (2009). What is statistics? (with discussion). *American Statistician, 63*, 105–123.

Brownstein, C. A., & Brownstein, J. S. (2008). Estimating excess mortality in post-invasion Iraq. *New England Journal of Medicine, 358*, 445–447.

Bunting, P. J., Fulde, G. W., & Forster, S. L. (2007). Comparison of crystalline methamphetamine (ice) user and other patients with toxicology-related problems presenting to a hospital emergency department. *Medical Journal of Australia, 187*(10), 564–566.

Cleveland, W. S. (1994). *The elements of graphing data*. Monterey Park, CA: Wadsworth.

Cobb, G. (2007). One possible frame for thinking about experiential learning. *International Statistical Review, 75*(3), 336–347.

Cohen, J. (1994). The earth is round (p < .05). *American Psychologist, 49*, 997–1003.

Coren, S. (2006, November). *Can dogs predict earthquakes? A possible auditory answer.* Paper delivered at the Psychonomic Society meetings, Houston, TX.

Cumming, G., & Thomason, N. (1998). StatPlay: Multimedia for statistical understanding. In L. Pereira-Mendoza, L. Kea, T. Kee, & W-K. Wong (Eds.), *Statistical education: Expanding the network. Proceedings of the Fifth International Conference on Teaching of Statistics* (pp. 947–952). Voorburg, The Netherlands: International Statistical Institute.

Dawson, R. J. M. (1995). The "unusual episode" revisited. *Journal of Statistics Education, 3*(3). www.amstat.org/publications/jse/v3n3/datasets.dawson.html

Department of Climate Change, Australian Government. (2008). Australian National Greenhouse Accounts, National Inventory Report (revised) 2005.

Derr, J. (2000). *Statistical consulting: A guide to effective communication*. Pacific Grove, CA: Duxbury Thomson Learning.

Eames, D. (2007, March 24). Schoolgirls' study nabs food giant. *New Zealand Herald.*

Etheridge, D. M., Steele, L. P., Langenfelds, R. L., Francey, R. J., Barnola, J. -M., & Morgan, V. I. (1998). Historical CO_2 records from the Law Dome DE08, DE08-2, and DSS ice cores. In *Trends: A compendium of data on global change*. Oak Ridge, TN: Carbon Dioxide Information Analysis Center, Oak Ridge National Laboratory, U.S. Department of Energy.

Gal, I. (2002). Adults' statistical literacy: Meanings, components, and responsibilities. *International Statistical Review, 70*(1), 1–25.

Gardner, M. J., & Altman, D. G. (1986). Confidence intervals rather than *P* values: Estimation rather than hypothesis testing. *British Medical Journal, 292*, 746–750.

Gelman, A. (2011). Why tables are really much better than graphs. *Journal of Computational and Graphical Statistics, 20*(1), 3–7.

Gelman, A., Pasarica, C., & Dodhia, R. (2002). Let's practice what we preach: Turning tables into graphs. *American Statistician, 56*, 121–130.

Greig, J. E., Carnie, J. A., Tallis, G. F., Zwolak, B., Hart, W. G., Guest, C. S., et al. (2004). An outbreak of Legionnaires' disease at the Melbourne Aquarium, April 2000: Investigation and case-control studies. *Medical Journal of Australia, 180*, 566–572.

Kenett, R., & Thyregod, P. (2006). Aspects of statistical consulting not taught by academia. *Statistica Neerlandica, 60*(3), 396–411.

King, L. E., Douglas-Hamilton, I., & Vollrath, F. (2007). African elephants run from the sound of disturbed bees. *Current Biology, 17*, R832–R833.

Love, T. E., & Hildebrand, D. K. (2002). Statistics education and the "making statistics more effective in schools of business" conferences. *The American Statistician, 56*(2), 107–112.

MacKay, R. J., & Oldfield, W. (1994). *Stat 231 course notes fall 1994*. Waterloo, Ontario, Canada: University of Waterloo.

Marriott, J., Davies, N., & Gibson, L. (2009). Teaching, learning and assessing statistical problem solving. *Journal of Statistics Education, 17*(1). Online. http://www.amstat.org/publications/jse/v17n1/marriott.html

Minton, P. D. (1983). The visibility of statistics as a discipline. *The American Statistician, 37,* 284–289.

Moore, D. S., & McCabe, G. (1999). *Introduction to the practice of statistics.* New York: W.H. Freeman.

Negus, B., & Mikedis, S. (2008, March). The road to greener motoring. *Royal Auto,* pp. 12–13.

Paxton, P. (2006). Dollars and sense: Convincing students that they can learn and want to learn statistics. *Teaching Sociology, 34*(1), 65–70.

Pfannkuch, M., & Wild, C. (2000). Statistical thinking and statistical practice: Themes gleaned from professional statisticians. *Statistical Science, 15*(2), 132–152.

Pfannkuch, M., & Wild, C. (2004). Towards an understanding of statistical thinking. In D. Ben-Zvi & J. Garfield (Eds.), *The challenge of developing statistical literacy, reasoning and thinking* (pp. 17–46). Dordrecht, the Netherlands: Springer.

Range, F., Viranyi, Z., & Huber, L. (2007). Selective imitation in domestic dogs. *Current Biology, 17,* 868–872.

Rosling, H. (2013). *Gapminder.* Retrieved August, 2013, from http://www.gapminder.org/

Sherman, L. W., & Berk, R. A. (1984). The specific deterrent effects of arrest for domestic assault. *American Sociological Review, 49,* 261–272.

Singer, J. D., & Willett, J. B. (1990). Improving the teaching of applied statistics. Putting the data back into data analysis. *The American Statistician, 44*(3), 223–230.

Stewart, B. W. (2007). "There will be no more!": The legacy of the Toowong breast cancer cluster. *Medical Journal of Australia, 187*(3), 178–180.

Taplin, R. (2003). Teaching statistical consulting before statistical methodology. *Australian and New Zealand Journal of Statistics, 45*(2), 141–152.

Tufte, E. (1983). *The visual display of quantitative information.* Cheshire, CT: Graphics Press.

Tufte, E. (1997). *Visual explanations.* Cheshire, CT: Graphics Press.

Utts, J. (2003). What educated citizens should know about probability and statistics. *American Statistician, 57,* 74–79.

Utts, J. (2005). *Seeing through statistics* (3rd ed.). Belmont, CA: Thomson Brooks/Cole.

Watson, J. M. (1997). Assessing statistical thinking using the media. In I. Gal & J. Garfield (Eds.), *The assessment challenge in statistics education* (pp. 107–121). Amsterdam: IOS Press.

Wells, H. G. (1903). *Mankind in the making.* London: Chapman & Hall Ltd.

Wild, C. (2006). The concept of distribution. *Statistics Education Research Journal, 5,* 10–25.

Wild, C., & Pfannkuch, M. (1998, June 21–26). What is statistical thinking? In L. Pereira-Mendoza, L. Kea, T. Kee, & W. Wong (Eds.), *Proceedings of the Fifth International Conference on Teaching Statistics* (Vol. 1, pp. 335–341). Voorburg, Netherlands: ISI.

Wild, C., & Pfannkuch, M. (1999). Statistical thinking in empirical enquiry (with discussion). *International Statistical Review, 67*(3), 223–265.

Zeger, S., & Johnson, E. (2007). Estimating excess deaths in Iraq since the US—British-led invasion. *Significance, 4*(2), 54–59.

Spreadsheets and Simulation for Teaching a Range of Statistical Concepts

Graham Barr and Leanne Scott

Abstract Fundamental statistical concepts remain elusive for many students in their introductory course on Statistics. The authors focus on the teaching of Statistics within a spreadsheet environment, wherein the students are required to master the basics of Microsoft Excel (Excel) to perform statistical calculations. This approach has the advantages of developing the students' ability to work with data whilst also building an understanding of the algebraic relationships between elements embedded in the formulae which they use. The use of a classroom experiment aimed at exploring the distributions of a number of randomly generated pieces of information is demonstrated. Teaching sessions are built around a suite of Excel-based simulations based on this experiment that attempt to demonstrate the concept of random variation and show how statistical tools can be used to understand the different underlying distributions. The methodology is then extended to a further, more advanced, bivariate teaching example which investigates how this spreadsheet methodology can be used to demonstrate the effect of expected value and variability on the statistical measurement of covariance and correlation. The authors reflect on their experience with the spreadsheet-based simulation approach to teaching statistical concepts and how best to evaluate the approach and assess levels of student understanding.

Keywords Simulation • Teaching • Statistics • Spreadsheets • Excel • Random Sampling

G. Barr (✉) • L. Scott
Department of Statistical Sciences, University of Cape Town, Cape Town, South Africa
e-mail: gdi@iafrica.com; leanne.scott@uct.ac.za

H. MacGillivray et al. (eds.), *Topics from Australian Conferences on Teaching Statistics:*   99
OZCOTS 2008-2012, Springer Proceedings in Mathematics & Statistics 81,
DOI 10.1007/978-1-4939-0603-1_7, © Springer Science+Business Media New York 2014

1 Introduction

Teaching mathematical and statistical principles through a spreadsheet platform offers significant advantages. Problem structuring through the medium of a spreadsheet develops in the student a general algebraic way of thinking as the process requires skills in expressing numerical relationships using algebraic notation. In the field of Statistics, the advantages of spreadsheets for teaching purposes are particularly marked as spreadsheets can simultaneously present an easily navigable yet extensive vista of numeric information stored in multiple rows and columns, along with the formulaic links between them. In addition, spreadsheets are able to simultaneously give a rich graphical depiction of the same data. However, over and above the tractability of spreadsheets for straightforward data analysis, the richest feature that a spreadsheet offers to the teacher of Statistics is its ability to show how one can mimic the process of repeated statistical experiments. It is this through this feature of spreadsheets that we demonstrate those concepts of statistics which are particularly hard to convey at an introductory level. This "repeated sampling" feature of Excel can be captured and exploited through simulation, an analytical tool which we believe is extremely useful to expose students to, even at an introductory level. By simulating statistical sampling, one can reveal a range of subtle and often misunderstood ideas which are central to basic statistical knowledge, such as those of randomness and statistical distributions. Moreover, Excel, the most often used spreadsheet in academia and commerce, has a powerful built-in programming language, Visual Basic for Applications (VBA). This allows teachers and students to enhance and leverage basic Excel power and functionality to a new level of flexibility and sophistication with click button automation and slickness.

In South Africa, education is increasingly seen as a key to overcoming multiple socio-economic problems. At the tertiary level, an overwhelming number of students come from disadvantaged backgrounds with associated deficits in foundational education and supportive resources (including a lack of access to computers and the benefits of technology-rich environments). Microsoft (MS) Partners in Learning program have made MS software easily accessible to educators and learners at schools and universities, so Excel is a particularly attractive and low-cost medium for education in the field of Statistics. In addition, Excel currently dominates the spreadsheet market making it the package that students are mostly likely to encounter in the workplace, and thus an obvious vehicle for teaching.

However, the spreadsheet model, either with programming enhancements or not, is not the only potential vehicle for teaching Statistics through a simulation approach; other statistical resources have been developed which have parallel simulation features. The most common of these are web-based tools written as Java applets; a good example being that of the Rice Virtual Lab in Statistics (RVLS) developed by David Lane in the early 2000s. This site has over 20 Java applets that simulate various statistical concepts (http://onlinestatbook.com/rvls.html). We believe the most important reason for using Excel as against web-based resources at the university level remains the fact that spreadsheets give students far greater control over the material being presented. The row/column formula-based format of Excel with associated graphical support moreover provides a good, accessible base for algebraic learning as against the prescriptive, and often inflexible, web-based material.

In this chapter we will relate how our experience with teaching first year Statistics courses at the University of Cape Town (UCT) has shown that the key concepts of randomness and distribution can be most effectively taught through simulation in an Excel-based spreadsheet environment using a 2-stage approach, first using a formula-based spreadsheet, and subsequently with the enhancement of VBA programs. The first teaching case study in this chapter uses a carefully crafted example which demonstrates to students how the random sampling of a set of distinct attributes associated with a set of individuals may reveal completely different distributions of each attribute. A key component of this teaching example is a set of associated customized Excel spreadsheets, firstly without and then with VBA enhancements, which elucidate these ideas and have been shown to be very didactically effective. We then go on to showcase a more advanced example which demonstrates the extent to which the measured covariance and correlation between two random variables are related to the first two moments of the two distributions.

2 Spreadsheet Learning in the Mathematical and Statistical Sciences

As early as 1985, 2 years after the launch of Lotus 1-2-3, the spreadsheet had been recognized as a force in Statistics education (Soper and Lee 1985). It is now universally recognized that the two-dimensional structure of spreadsheets, along with their associated graphical components, can facilitate the comprehension of a wide range of mathematical and statistical concepts by providing a supportive platform for conceptual reasoning. Baker and Sugden (2003), for example, give a comprehensive review of the application of spreadsheets in teaching across the mathematical, physical, and economic sciences. Black (1999) was one of the first authors to recognize the usefulness of spreadsheets in a simulation context for teaching complex statistical concepts.

The idea of using simulation, especially within VBA, was found by Barr and Scott (2008) to be particularly useful and effective for the teaching of first year Statistics to large classes and they confirm the sentiments of Jones (2005) that statistical concepts and procedures taught within the context of a spreadsheet tend to be transparent to pupils, allowing them to look inside the "black box" of statistical techniques. A comprehensive survey of the use of simulation methods for teaching statistical ideas has been done by Mills (2002). We argue that the literature supports the notion that spreadsheets are an effective teaching tool, with Excel as the de facto spreadsheet standard.

3   Teaching the Notion of Random Variation with Excel

The core part of this chapter is to showcase the teaching of the two foundational statistical concepts of *randomness* and underlying *distribution* through simulation, using both simple spreadsheet functions and more sophisticated VBA programs. Our experience has led us to believe that VBA-structured spreadsheets by themselves provide a difficulty for a large cohort of students; a leap into the dark to some extent. However, when properly scaffolded by first teaching the students how to construct a standard spreadsheet with appropriate formulae, it becomes a more effective learning tool. By themselves, VBA simulation programs or simulation programs written in Java on the web are neat and impressive but constitute too much of a "black box" for students. Leading students through the process of first developing their own spreadsheets to explore statistical problems effectively provides appropriate building blocks to support subsequent exposure to VBA simulation programs.

3.1   How Well Is the Notion of Random Variation Understood?

One of the fundamental concepts of Statistics is that of random variation. It is a notion that we as Statistics educators frequently assume people have an intuitive understanding. It is, however, a subtle notion that apparently random and unpredictable events have underlying patterns that can be uncovered through (*inter alia*) long-term observation.

An open-ended invitation to describe their understanding of "randomness" and how it affects our day-to-day lives was extended to a group of 20 adult learners, all of whom were tertiary educators in non-quantitative disciplines themselves. A brief discussion on the perceived need for an understanding of Statistics preceded this, touching on the fact that very little of what happens in life can be predicted with certainty, and indicating that Statistics provides a mechanism to manage uncertainty associated with random variation. A variety of notions of randomness were articulated, from which some unexpected themes emerged, in particular a pervasive view of randomness as being a "victimizing" force or a tool of malevolent authorities, associated with poor planning and discipline. In many cases, randomness was associated with chaos. All of the descriptions volunteered by the students were devoid of any notion of underlying pattern or distribution. Subsequent discussions confirmed that they believed the existence of an underlying pattern was, in fact, contradictory to the very idea of random variation. From an educational point of view, it could be suggested that the consequence of (these) students' views of the nature of random variation is that there is an inflated view of the power of Statistics to impose order on randomness or a jaded view of the discipline as a tool to disguise chaos and unfairness.

It is suggested that beginning the Statistics journey with a description of the world as containing innate patterns and order which are hidden from us through the random and unpredictable way in which individual outcomes are free to vary, may open up the power and interest of the discipline in a way that the traditional approach of teaching "theory followed by its application" fails to do. We will show below, using an appropriate experiment, that the spreadsheet environment is an ideal canvas on which to sketch and unveil the ideas around random variation and underlying patterns.

3.2 So How Does One Convey the Concept of Random Variation? The Class Experiment: Random Selection; Different Patterns!

We begin the class experiment with a discussion with the students about different types of numbers, reflected both by the different measurement scales we choose to assign to them, and by the process that generates them. We ask them to consider the following experiment in which each student in the class will contribute four pieces of information, viz: (1) their first name; (2) their height (in cm); (3) an integer randomly selected within the range 1–50; and, finally, (4) their personal results of a (to be explained) experiment involving mice! Once we have generated this data we will be collecting it from everyone and constructing four separate graphs of each of the four information types. As part of this experiment we will be constructing ways to generate data for (3) by using, and exploring, the Excel random number generator. Data for (4) involves a mouse training experiment which tests the ability of 5 (simulated) randomly selected mice to navigate a simple maze, recording the number of successful mice. Students will be able to run the Excel models to record their own data for the class experiment. The focus of this class exercise is for students to answer the key question: *What shapes do we anticipate for these graphs?* We proceed by considering the randomly generated number.

3.3 The Random Numbers

The students are led through the first exercise by a process that typically has the following format:

Suppose we are interested in mimicking the National Lottery and (repeatedly) generating a random number which lies between 1 and 50. Each draw can be likened to drawing a number from a hat with the numbers 1–50 in it. If we want to keep drawing a number from this hat in such a way that all numbers are equally likely, we would have to also suppose that we have a very large hat that can hold such a large (and equal) quantity of each of the numbers that it doesn't limit our thinking about the situation;: What would a histogram of these numbers look like? Some discussion would probably lead us to conclude that we would expect all of the bars in the histogram to be of equal height. What sort of patterns do we get when we randomly take numbers out of the hat? If we just pick one number it could pop up at

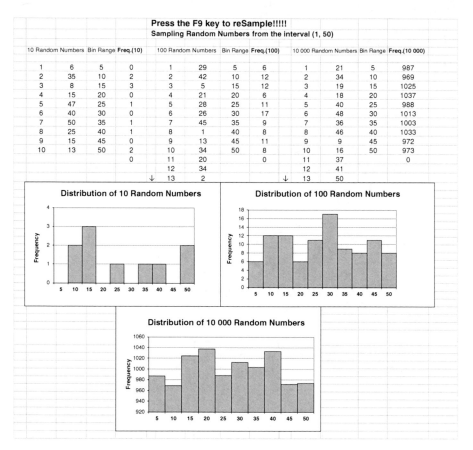

Press the F9 key to reSample!!!!!
Sampling Random Numbers from the interval (1, 50)

10 Random Numbers		Bin Range	Freq.(10)	100 Random Numbers		Bin Range	Freq.(100)	10 000 Random Numbers		Bin Range	Freq.(10 000)
1	6	5	0	1	29	5	6	1	21	5	987
2	35	10	2	2	42	10	12	2	34	10	969
3	8	15	3	3	5	15	12	3	19	15	1025
4	15	20	0	4	21	20	6	4	18	20	1037
5	47	25	1	5	28	25	11	5	40	25	988
6	40	30	0	6	26	30	17	6	48	30	1013
7	50	35	1	7	45	35	9	7	36	35	1003
8	25	40	1	8	1	40	8	8	46	40	1033
9	15	45	0	9	13	45	11	9	9	45	972
10	13	50	2	10	34	50	8	10	16	50	973
			0	11	20		0	11	37		0
				12	34			12	41		
				↓ 13	2			↓ 13	50		

Distribution of 10 Random Numbers

Distribution of 100 Random Numbers

Distribution of 10 000 Random Numbers

Fig. 1 Simulating the uniform distribution. Samples of size 10, 100, and 10,000 demonstrate the empirical distribution

any point in the specified interval (1:50). The fact that the number is randomly selected means there is no way of telling (from the preceding numbers or, in fact, any other source of information) exactly which number is going to pop up next.

In order to see a pattern of numbers we need to observe more than one randomly drawn number. Let's see what happens when we generate 10 random numbers. Perhaps the pattern looks a little obscure... sometimes it looks very different from what we might have expected. What happens when we draw 100 numbers; or 1,000? (see Fig. 1). It seems that as the pool of numbers that we are drawing grows, so the pattern of the numbers in the hat gets revealed. A small pool can give quite a mis-leading picture of the histogram of the numbers in the hat! However, a big pool is less likely to do so.

So how big a pool of numbers do we need to have access to in order to get a reliable picture of the numbers in the hat? Imagine that we hadn't known the shape of the histogram of the numbers in the hat. The numbers might, for example, have had a different (other than flat/rectangular) pattern of distribution. Let's use some spreadsheet commands to model our thinking of the above. We can easily simulate the drawing from our hat of a number (where all the numbers in the hat are integers that lie between 1 and 50 inclusive) by typing the formula: =(RANDBETWEEN(1,50)) into our spreadsheet. This computes (and rounds to the nearest integer) a single random number in the interval (1, 50). We can then resample this number by pressing the F9 key. We can also display this visually by plotting the number in a simple histogram. We should first set up some bin intervals, the simplest is the set of 10 intervals between 1 and 50 with interval width 5. We then use an array formula to compute the frequencies: ={FREQUENCY(data_range, bin_range)} which can then be plotted in a histogram.

We replicate this procedure for a sequence of sample sizes from 1 (single random number), through to 10 (10 random numbers), then 100 and finally 10,000. By repeatedly pressing F9 (which simply recalculates the formulae and effectively re-samples) we get to replay the (random) selection of different sized pools of numbers from the hat. Each time, we get a different selection of numbers. One feature becomes apparent. The pattern (of the numbers in the hat) becomes increasingly clearly revealed as we observe larger and larger pools of numbers from the hat. Although we cannot at any stage predict what the next number will be, by observing randomly selected pieces of information from the hat we can begin to piece together what the pattern of numbers in the hat must look like. In real life this is likely to mean we have to observe the pattern (of randomly revealed pieces of information) over time, before we can begin to understand something about the nature (or distribution!) of the numbers in the hat.

3.4   The Mouse Experiment: Generating Data from a Binomial Distribution

The Uniform case above constitutes a starting point and necessary platform to consider more complex distributions. In particular, as it allows us to transparently allocate a binary outcome probabilistically, it leads naturally onto the Binomial Distribution. We consider a fixed number of trials, where at each trial we have assigned a fixed probability of success or failure (i.e. a Uniformly generated random variable on (0, 1) indicates a success if U is between 0 and p and failure otherwise). In order to generate the fourth piece of information for our class experiment we ask the students to consider the following scenario:

We, as statisticians, have been approached to mediate on a claim that an animal trainer has managed to train a group of ten mice to turn left at the end of a tunnel. As evidence of this feat he has cited the fact that in an observed demonstration, out of 10 trained mice, 9 of them turned left! In an attempt to pronounce on these results we try and build a model that mimics the behavior of the mice.

We start with the sceptical view of the situation and assume that the mice have NOT, in fact, been trained. Thus each mouse makes an arbitrary choice of left or right at the end of the tunnel which means that the probability of them turning left is 0.5.

Table 1 The outcomes of 10 experiments, where $p = 0.5$

Trial	1	2	3	4	5	6	7	8	9	10
Fail = 0, success = 1	0	1	0	0	0	0	1	0	0	1

We are interested now in what different bunches of 10 hypothetical mice might do in terms of "left turning" behavior. Let's assume we have a large number of mice at our disposal, which are very similar and which we put through the identical training regime. To run the experiment we select 10 mice, put them through the tunnel and record the number of successful mice, i.e. the number out of the 10 who turned left. We could then select another 10 randomly and perform the experiment again; in fact, we could repeat this as many times as we like. Our selection assumes the mice are always "fresh", referred to as sampling "without replacement" (i.e. our mice don't get tired or complacent or difficult!).

We use the spreadsheet to help us produce some simulated results for batches of 10 mice. The mechanics of this are as follows: We select a random number between 0 and 1. If the number is less than or equal to 0.5 we assume the mouse went left, if not we assume it went right. Then we do this for 10 mice, labelling the results Trial 1, through to Trial 10. This comprises one experiment. We can see how easy this is on the spreadsheet. In cell C3 we put the value of p, in this case 0.5. The following formula: =IF(RAND() < C3,1,0) results in a 1 if the random number calculated is less than 1 and a 0 if it is not (that means it is greater than 1). We then copy this formula down a further nine cells to get the model results for one experiment. In this case, Trials 2, 7, and 10 were a "success" (mouse turned left). Trials 1, 3 to 6 and 8 and 9 were failures (turned right) (Table 1).

We could repeat this experiment under the same theoretical conditions (i.e. the same ineffective training program which yields p=0.5 and with the mice being selected randomly). Excel allows us to repeat the experiment just by pressing the F9 (recalculate) key. Next time we might get 6 successes for the same value of p. If we keep pressing F9 we see we can generate a series of results, sometimes the same, sometimes different.

These simulated results seem to form some type of pattern for a particular p. For example out of 10 mice we often get 4, 5, or 6 who are successful. This makes us curious to examine the pattern further. Let's repeat this experiment 50 times and keep the results. We could write down the results but it's a lot easier to copy the set of formulae across a number of different columns so that we can store the results of many random experiments. The results are shown below. We can use the spreadsheet to calculate and display the pattern of results over the 50 experiments, each of which comprised 10 trials. That is, we have repeated the 10-trial experiment 50 times. Note the interesting pattern (Table 2) which we have plotted in a histogram (Fig. 2). There are relatively high frequencies for 4, 5, and 6 successes, a number for 2, 3, 7, and 8 successes, BUT none for 0, 9, and 10 successes. The histogram gives an interesting picture. In fact the histogram is beginning to make us think that it was pretty unusual that nine out of the animal trainer's ten mice successfully turned left IF they hadn't been trained! Is it even possible that nine out of ten turned

Table 2 The number of successes over 50 experiments for 10 mice with $p = 0.5$

# Successes	0	1	2	3	4	5	6	7	8	9	10
Frequency (50 experiments)	0	1	3	4	11	12	9	7	3	0	0

Fig. 2 Histogram of data in Table 2

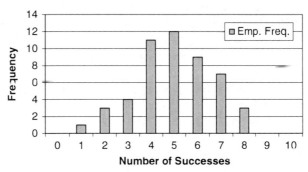

left purely by chance on this occasion? We could keep repeating the experiment and see if we ever get an occasion when we get 9 out of 10 successes. However, perhaps we should consider that the training was effective!

How would we model effectively trained mice? Well, it's unlikely to have been 100% effective (after all most educational programs have less than 100% pass rates) so perhaps it is effective to the tune of say 80%. This would equate to a p parameter of 0.8. We don't know what the true p value is but it does seem that the observed data are not consistent with a p of 0.5.

3.5 Putting all This Together: Making Sense of the Co-existence of Pattern and Randomness

The teaching script for the final consolidation exercise typically runs as follows:

The interesting feature of the results of our "mouse training" model is the pattern of results it revealed. Each time we repeated the 50 experiments, each consisting of 10 trials, we observed a different set of numbers but they appeared to keep a number of common features. This pattern became clearer the more times we repeated the experiment. The pattern was different from that of the numbers we drew out of the hat. What does this tell us about randomness? What causes the patterns? Remember we said the hat was our mechanism to ensure random selection, in other words to mimic the way data might present itself to us in an unpredictable way. All the numbers were mixed up in the hat and we drew them out in a way that meant no particular numbers were favored or prejudiced. What if we put different pieces of information (as distinct from random numbers) into the hat, shuffled them and drew samples of them out of the hat? Will the patterns related to different pieces of information all be the same? We are now in a position to conduct our class experiment and find out; we collect from each member in the class the piece of paper bearing the four pieces of information we specified earlier.

We construct histograms of the samples of paper as we draw them from our hat. We might not be surprised to observe that the random numbers have a flat, rectangular distribution. We also see that the "successful mice observed" have the same shaped distribution as the one we saw repeatedly with our electronic mice running model. The heights of students may show one bell shaped histogram, or may have a hint of two humps of data, with the males being taller than the females. The "names" may well show a few modes, depending on popular names and prevalence of language groups.

Our "randomizing" hat has had the effect of giving us the data in random and unpredictable order, but the distinct patterns associated with each different piece of information have been preserved, and are revealed as we have access to more and more data. In fact, our reconstruction of the data into histograms reinforces two facets of random variation. On the one hand, it is reflected in the unpredictable way we frequently encounter (information in) life (stocks vary on the stock exchange, increments of growth of children, number of cars on highway at a particular time, etc.). But, perhaps paradoxically, randomisation also provides the best mechanism to uncover the true pattern of an unknown measurable (e.g. household income). Selecting data (sampling) randomly ensures we have the best chance to see as broad a spectrum of the unknown pattern of data as quickly (efficiently) as possible!

The classroom tutorials and accompanying Excel spreadsheets referred to in the section above are freely available on the UCT OpenContent website at http://opencontent.uct.ac.za/Science/Teaching-Fundamental-Concepts-in-Statistical-Science.

4 Challenging the Students with a More Advanced (Bivariate) Example

The overall variability of two variables working in tandem represents a useful extension of these ideas and lends itself well to spreadsheet exposition. In this example we use the simulation approach to demonstrate the association between the estimate of correlation and the underlying variability of the component random variables. Through simulation, students can track the variation in the estimates of covariance and correlation stemming from simulated pairs of the random variables with the same fixed underlying parameters (true standard deviations and true correlation). In our second year course, we take these concepts further and demonstrate the degree of variance reduction when two random variables are combined; this principle of variability reduction has many applications, most notably in constructing a portfolio of shares to demonstrate risk (proxied as standard deviation) reduction. This example and the associated active learning exercise were written for the second/third year commerce students who do courses in Statistics as part of their curriculum and complete allied courses in Finance.

4.1 Demonstrating the Effect of Correlation Between Random Variables on Variability

Our first didactic step is to demonstrate the nature of correlation between variables on a spreadsheet and how changing the correlation coefficient changes the nature of the relationship between X and Y in a way that is graphically demonstrable and statistically measurable.

We can demonstrate these ideas neatly on a spreadsheet by simulating a variable X (with given E(X) and Var(X)) n times along with a variable Y (also with given E(Y) and Var(Y)) *such that Y has some given prespecified correlation ρ with X.*

In order to sample from this bivariate distribution (X, Y) we use the Cholesky decomposition method (see, for example, Horn and Johnson 1999).

For the simple 2-variable case where the correlation matrix $\Sigma = \begin{pmatrix} 1 & \rho \\ \rho & 1 \end{pmatrix}$ we will have $L = \begin{pmatrix} 1 & 0 \\ \rho & \sqrt{(1-\rho^2)} \end{pmatrix}$.

Then $\begin{pmatrix} X \\ Y \end{pmatrix} = \begin{pmatrix} 1 & 0 \\ \rho & \sqrt{1-\rho^2} \end{pmatrix} \begin{pmatrix} X \\ X^* \end{pmatrix} = \begin{pmatrix} X \\ \rho X + \sqrt{1-\rho^2} X^* \end{pmatrix}$ where, say, X is drawn from some distribution and X^* is independently drawn from the same distribution. Then the generated Y will be from the same distribution and have an expected correlation of ρ with X. This Cholesky decomposition allows us to generate a bivariate distribution $\begin{pmatrix} X \\ Y \end{pmatrix}$ with correlation matrix Σ. Note that Y and X can then be suitably scaled and mean shifted so as to have individually any mean and variability (variance or standard deviation) required, allowing us to generate a bivariate distribution $\begin{pmatrix} X \\ Y \end{pmatrix}$ with any given parameters μ_X, μ_Y, σ_X, σ_Y and ρ.

In a spreadsheet exposition, we could let cell C2$=\mu_X$, cell C3$=\sigma_X$; E2$=\mu_Y$, cell E3$=\sigma_Y$ and cell G2$=\rho$

Then the scaled and mean shifted Cholesky decomposition for a normally distributed bivariate distribution translates on the spreadsheet to:

X [cell C7]$=\$C\$2+$NORMINV(RAND(),0,1)$*\$C\$3$, and

Y[cell D7]$=\$E\$2+\$E\$3*(\$G\$2*(C7-\$C\$2)/\$C\$3+$SQRT$((1-\$G\$2^2))*$(NORMINV(RAND(),0,1)))

where NORMINV(RAND(), 0, 1) is the Excel formula which generates independent drawings from a $N(0, 1)$ distribution.

The first step for teaching this module through the medium of a spreadsheet is to demonstrate how the scatter plot between X and Y varies in shape as the mean and variance of X and Y varies and critically as the correlation between X and Y varies. The scatter plot we use on the spreadsheet is for 50 points and also shows the regression line of the ordinary least squares (OLS) regression between X and Y. The slope of this line has a link to the estimated correlation since it will equal the correlation coefficient if the data is pre-standardized (have a mean of zero and variance of 1).

Table 3 Table of parameters
and estimates for case with
$\rho = -0.5$ and Std.Dev(Y) = 1.5

	True parameters	Estimates (3 dec. places)
E(X)	1.0	0.941
Std. Dev.(X)	1.0	0.897
E(Y)	1.5	1.531
Std. Dev.(Y)	1.5	1.667
Cov.(X, Y)	−0.75	−0.681
Corr. (X, Y)	−0.5	−0.455
α	2.25	2.327
β	−0.75	−0.846

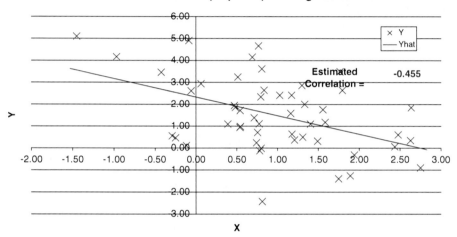

Fig. 3 Scatter plot of simulated data to demonstrate variability and correlation ($\rho = -0.5$) with regression line imposed

We may demonstrate repeated sampling of these 50 points through pressing the F9 key and hence capture the variability in the estimated correlation as well as the variability in the estimated regression line parameters. We have also used the Excel function LINEST to estimate the regression line.

In the example above, using the parameters in Table 3 and depicted in Fig. 3, the estimates are drawn from a bivariate sample of 50 (X, Y) points with a correlation of −0.5 (in this case the estimated correlation for the sample is −0.455) and a (theoretical) standard deviation in Y of 1.5.

In the second example above, using the parameters in Table 4 and depicted in Fig. 4, the estimates are also drawn from a bivariate sample of 50 (X, Y) points, but with a correlation of 0.5 (in this case the estimated correlation for the sample is 0.390) and a (theoretical) standard deviation in Y of 0.5. Note that in this second example, apart from the positive correlation and positively sloped regression line, the Y variability is much less than in the example above.

Table 4 Table of parameters and estimates for case with $\rho=0.5$ and Std.Dev$(Y)=0.5$

	True parameters	Estimates (3 dec. places)
E(X)	1.0	1.070
Std. Dev.(X)	1.0	0.893
E(Y)	1.5	1.452
Std. Dev.(Y)	0.5	0.385
Cov.(X, Y)	0.25	0.134
Corr. (X, Y)	0.5	0.390
Alpha	1.25	1.272
Beta	0.25	0.168

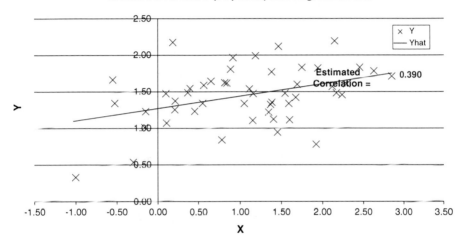

Scatter Plot X and Y (50 points) and Regression line

Fig. 4 Scatter plot of simulated data to demonstrate variability and correlation ($\rho=0.5$) with regression line imposed

There are two interesting ideas which we can demonstrate at this stage to the class as an aside from the main exercise:

1. We can use the opportunity to discuss with the class the "dimensionless" nature of correlation (and we may discuss the relationship between covariance and correlation). We note that the sample correlations in the two examples are quite close to the true values as we would expect them to be, and stress the difference between the true and estimated value for correlation. In both cases we may show how the sample estimate varies as we take repeated samples (F9 key). Similarly we can track the estimated values $\hat{\alpha}$ and $\hat{\beta}$ for the regression line and the way these compare to the true alpha and beta values.

2. We can explore whether the variability of the estimated correlation is dependent on the difference between the means of X and Y; on the variability of the underlying X and Y populations and/or on the size of the sample we use to estimate the correlation. We find (by trial and error) that for some particular ρ, it is only dependent on the size of the sample! We could ask the class how we could

demonstrate this finding more rigorously (by, for example, plotting the observed variance of our estimate of correlation as a function of (1) sample size, (2) difference in population means and (3) $\sigma_{(X)}$ or $\sigma_{(Y)}$)

4.2   Student Responses to the Bivariate Example

When interrogated, the students found the visual demonstration of variability particularly enlightening. An important didactic feature of this example is the effect of random variation on estimation. This begins, in this example, with the fact that correlated variables with some fixed prior correlation may present, assuming some given sampled distributions, with different looking scatter plots. This observation shifts the idea of correlation (in a regression context) from that of a fixed deterministic entity to that of the notion of correlation as a random variable with a distribution. Some students comfortably made the connection that the "observed correlation measure" from a particular dataset was a random variable, likely to change from sample to sample; however, for many, this remained a difficult concept to grasp.

5   Evaluating Students' Understanding of the New Teaching Approach

5.1   Questionnaire to Evaluate Efficacy of This Approach

It is clearly difficult to assess the value of a new teaching approach in the absence of an unambiguous baseline (or control). The approach outlined was developed in the context of a teaching environment that is both multi-lingual and in which students generally had poor groundings in mathematics. The questionnaire we used to probe the students' experiences and the efficacy of this teaching approach focused on the students' ability to discriminate between empirical and theoretical distributions as our preliminary investigations indicated that these are likely to be new concepts for all students. In a given experimental setting (e.g. a Poisson process), students were asked to distinguish between distributions which are either (1) empirical or which (2) reflect the underlying theoretical model. Between 70 and 95 % (taken across a number of questions) of around 1,000 first year Statistics students who participated in the survey answered these questions adequately. Anecdotal evidence has suggested that a lower proportion of advanced Statistics students (i.e. of those exposed to traditional teaching approaches) make this distinction correctly, despite these students having passed much higher levels of statistical theory.

The table below (Table 5) shows the assessed level of understanding (by a first year Statistics lecturer using a subjective scale from 0 to 5) based on students' descriptive responses to the indicated questions. Students were directed to use their own words ("parrot" definitions from text books were excluded from the analysis).

Table 5 Assessed level of Student Understanding

Graded response	What is a pdf? (%)	What is an empirical pdf? (%)	When will the empirical dbn. of data look like the (underlying) theoretical dbn? (%)
0 (no understanding)	6	10	12
1	47	50	18
2	17	19	8
3	15	14	13
4	11	5	22
5 (perfect understanding)	4	2	27

Understanding of the concept of probability distributions was generally poor, as was expected to be the case at this introductory level. However, the notion of empirical data was found to be relatively well appreciated. The authors assert that this discrepancy can (tentatively) be attributed to the explicit way in which the notion of a theoretical model is linked to empirical data in the spreadsheet approach. In terms of qualitative assessment, both teachers and tutors who worked through these Excel-based lectures and tutorials with the students, reported excitement at various levels of breakthrough. One tutor's experience was particularly noteworthy; this tutor's group "discovered and articulated" the Central Limit Theorem themselves in a way which the tutor had sceptically thought was beyond students' scope at this stage of the subject.

5.2   Evaluation in an Excel Environment

This new approach has also been formally tested through an "online" test based on a spreadsheet environment. The test is set up to probe the understanding of statistical concepts (such as statistical significance and the distributions of sums and multiples of random variables) and also to test the basic Excel competence. The test, which has been run by the authors over a period of four semester-long courses, has the following format:

Students are given a dataset comprising a number (m) of samples of data (each with n observations) generated from some (known) distribution with known mean (μ). The students are tasked with testing that the true mean of the sample data is μ, for each of the m samples of data. The observed proportion of false rejections is then compared to the stated level of significance. The online test also requires students to compare theoretical and empirical standard deviations and standard errors. In general, the test probes students' understanding of the notion of empirical as opposed to theoretical results as well as some fundamentals of statistical inference.

The authors have found the process of preparing students for the test to be a very valuable learning exercise. Through having to generate statistical tests and other results which demonstrate the principles of statistical inference, and comment on these, the students have been found to express new levels of insight into these topics.

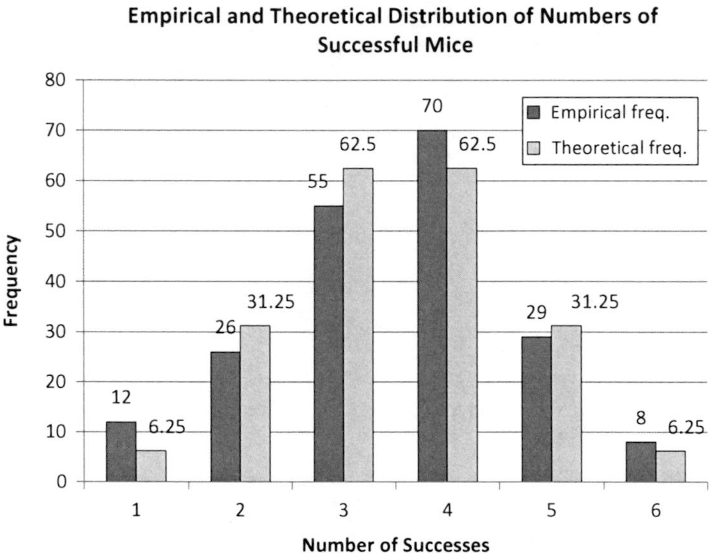

Empirical and Theoretical Distribution of Numbers of Successful Mice

Fig. 5 Bar chart of empirical and theoretical distribution of numbers of successful mice

Also, teachers have expressed renewed interest in their teaching through finding a new medium through which to teach "old" concepts. The Excel test comes in the last week of the course and the authors found that it provided an opportunity to synthesize the theory and practical aspects of the work covered throughout the course. Many students indicated achieving breakthroughs in their understanding of statistical concepts during the process of preparing for this evaluation.

5.3 Formal Written Evaluations

Some questions, such as those in Fig. 5 above, which test the understanding of the concepts taught using this approach, are also included in the more traditional assessments (written exams and tests)

Question 1: A total of 1,134 students wrote class test 1 this year. The frequency distribution of the results for this test is given below in Table 6.

(a) Calculate the empirical probability of passing the class test.
(b) The average mark for the class test was 61.4 % with a standard deviation of 17.2 %. Assume the test marks follow a Normal distribution with this mean and standard deviation and compute the theoretical probability of passing the class test.
(c) Comment on the difference/similarity between the empirical and theoretical probabilities computed in parts (a) and (b)
(d) Why might the Normal distribution be appropriate for modelling test results?

Mark (%)	Frequency
0–9	18
10–19	6
20–29	18
30–39	59
40–49	122
50–59	253
60–69	281
70–79	211
80–89	131
90–100	35

Table 6 Frequency distribution of marks for test 1

Question 2: An experiment is conducted to determine whether mice can read. A simple maze is constructed with only two passages: one leads to a piece of cheese and the other to a mouse trap. A written sign is placed at the beginning of the maze directing the mice to the cheese. Five mice are placed in the maze, one at a time, and allowed to run through the maze until they reach either the cheese or the trap. The number of mice who successfully reach the cheese is recorded. You may assume that the mice cannot smell the cheese and that successive mice cannot follow the scent of preceding mice. The above experiment was repeated 200 times. The empirical results are summarized in the bar chart below in Fig. 5, together with the theoretical results that may be expected if mice cannot read.

(a) Write down the sample space, the sample size, and the number of trials in this experiment.
(b) If mice cannot read, what is the probability that at least two mice correctly navigate the maze and find the cheese?
(c) The theoretical frequency of three successes is displayed on the bar chart as 62.50. Show how this frequency is derived.
(d) Using the theoretical and empirical frequencies, perform a goodness-of-fit test to determine whether mice can read. State your hypotheses, test statistic, critical value (from tables), and an appropriate conclusion at the 5 % significance level.

Hint: The empirical frequencies are what have been "observed" in the experiment, whilst the theoretical frequencies are what you would "expect" if mice cannot read.

(e) In light of your conclusion in part (d), why do the theoretical and empirical frequencies differ? Will they ever be the same? Discuss.

5.4 Authors' Experiences with Implementing the New Teaching Approach

The spreadsheet-based approach to teaching introductory Statistics was phased in over a period of about 5 years. Formerly, this course had been taught using traditional "chalk and talk" methods, supplemented by the use of menu-driven applications in software packages which perform statistical calculations. The enrolment for this course is well in excess of 1,000 students, many of whom will not be learning in their mother tongue. The complication of language difficulties, along with increasingly poor mathematical preparedness of the students, was a further impetus for adopting spreadsheet-based simulation tools for teaching. These tools have the strength of being visual in nature and of encouraging and reinforcing the basic rules and structure of algebra.

This large course with its multiple challenges provided a fertile terrain for teaching innovation. However, the sheer size of the course, involving a teaching team of five lecturers and some twenty tutors, made the transition from one teaching orientation to another particularly difficult; and the process of change management ultimately became the biggest challenge of the undertaking. Change processes involving large groups are slow and require methodical "bedding down" of new practices. The process has been faced with the contradictory imperatives of promoting innovation versus encouraging consistency and clarity. However the authors believe that a coherent educational package is emerging!

6   Conclusion

Our experience has led us to believe that the spreadsheet has many powerful didactic facets. It provides a flowing, dynamic model which links cells in a transparent way. At a basic level it provides a two-dimensional, visible, matrix-like calculating machine where at the press of a button the whole matrix may be recalculated. This feature can be used to simulate samples with different underlying distributions. Randomly sampled bits of information have statistical distributions which reflect how these bits of information are generated and what attributes of life they reflect. While randomly selected numbers will reflect a Uniform distribution, heights of people will reflect a Normal distribution, the results of a simple binary experiment, a Binomial distribution and names across different cultural communities often a multi-modal distribution. A key insight for learners is that although each attribute is selected from the hat of our experiment *randomly* the *patterns* or *distributions* of the attributes can be quite different. This experiment thus constitutes a very powerful mechanism for learners to differentiate between the concepts of randomness and the underlying pattern of the attribute itself.

Our experience with the extensive use of Excel for first year Statistics courses has led us to conclude that a pure VBA approach may be somewhat intimidating for learners and teachers alike so we have often adopted a 2-stage approach, first showing students how to build simple spreadsheets for themselves to demonstrate basic statistical principles and thereafter introducing the automated VBA-based point-and-click programs. These latter VBA-based spreadsheets allow more flexibility and sophistication and in particular allow automated comparisons showing students how empirical distributions converge on theoretical distributions.

In addition, this approach has been shown to be effective beyond the introductory level where the exercise on the distribution of the estimated correlation coefficient has been successful at second year level. Overall, we believe this approach is a step forward in the teaching of foundational statistical concepts, such as randomness and distribution, which are often poorly understood, even by Statistics graduates.

7   Note

Developed from a paper presented at Seventh Australian Conference on Teaching Statistics, July 2010, Perth, Australia.

This chapter is refereed.

References

Baker, J. E., & Sugden, S. J. (2003). Spreadsheets in education. The first 25 years. *eJournal of Spreadsheets in Education, 1*(1), 18–43. Retrieved August 25, 2010, from http://www.sie.bond.edu.au/

Barr, G. D. I., & Scott, L. S. (2008). A new approach to teaching fundamental statistical concepts and an evaluation of its application at UCT. *South Africa Statistical Journal, 42*, 143–170.

Black, T. R. (1999). Simulations on spreadsheets for complex concepts: Teaching statistical power as an example. *International Journal of Mathematical Education in Science and Technology, 30*(4), 473–81.

Horn, R. A., & Johnson, C. A. (1999). *Topics in matrix analysis.* Cambridge: Cambridge University Press.

Jones, K. (2005). Using spreadsheets in the teaching and learning of mathematics: A research bibliography. *MicroMath, 21*(1), 30–31.

Mills, J. D. (2002). Using computer simulation methods to teach statistics: A review of the literature. *Journal of Statistics Education, 10*(1). Retrieved October 26, 2010, from http://www.amstat.org/publications/jse/v10n1/mills.html

Soper, J. B., & Lee, M. P. (1985). Spreadsheets in teaching statistics. *The Statistician, 34*, 317–321.

Navigating in a New Pedagogical Landscape with an Introductory Course in Applied Statistics

Martin Gellerstedt, Lars Svensson, and Christian Östlund

Abstract During the last few decades, a great deal of effort has been put into improving statistical education, focusing on how students learn statistics and how we as teachers can find effective ways to help them. At the same time the use of computers, the Internet, and learning management systems has grown rapidly, and offers new educational possibilities. In this chapter, we will discuss how these changes in the pedagogical landscape have affected our introductory course in applied statistics. The course and teaching context are presented in relation to guidelines for assessment and instruction in statistics and to seven principles for effective teaching. Teaching strategies, course content, and examples of course material are included. Furthermore, results from evaluations are discussed, especially focusing on diversity in student characteristics. These results indicate a variation in learning styles both between and within groups. Finally, we present some of our ideas for future development including strategies for individualization and the use of educational mining.

Keywords Teaching strategies • Learning style • Evaluation • Monitoring and adaption

1 Introduction

During the last few decades, teachers of statistics have faced a variety of thrilling challenges. Firstly, there has been a statistical education movement focusing on how students learn statistics and how we as teachers can find effective ways of helping them. An important step in this development was the introduction of exploratory

M. Gellerstedt (✉) • L. Svensson • C. Östlund
School of Business, Economics and IT, University West, 461 86 Trollhättan, Sweden
e-mail: martin.gellerstedt@hv.se; lars.svensson@hv.se; christian.ostlund@hv.se

H. MacGillivray et al. (eds.), *Topics from Australian Conferences on Teaching Statistics: OZCOTS 2008-2012*, Springer Proceedings in Mathematics & Statistics 81, DOI 10.1007/978-1-4939-0603-1_8, © Springer Science+Business Media New York 2014

data analyses (EDA), introduced by Tukey in the late 1960s—early 1970s. Tukey's ideas really showed that statistics is the art of transforming data into knowledge, and his box-plot illustrates well that "a picture paints a thousand words," a saying that could have a statistical version, namely "a graph could say more than 1,000 numbers." Another milestone was roughly 20 years ago when a report (Cobb 1992) called for a change in statistical education addressing a greater focus on statistical thinking in terms of: understanding the need for data and importance of data production, the omnipresence of variability, and its possible quantification and explanation. The Cobb report had a great impact and generated several changes in statistical education regarding *content* (more focus on data analyses, less on probability theory), *pedagogy* (more active learning, less passive instruction), and *technology* (for example, using statistical software), (Moore 1997). In order to summarize the experiences obtained during the reformation of statistical education, the American Statistical Association (ASA), funded by a member initiative grant, developed the following guidelines for assessment and instruction in statistics: GAISE in the K–12 curriculum and for the introductory college statistics course (ASA 2005). The guideline for a college course put forth the following six essential recommendations:

1. Emphasize statistical literacy and develop statistical thinking
2. Use real data
3. Stress conceptual understanding, rather than mere knowledge of procedures
4. Foster active learning in the classroom
5. Use technology for developing conceptual understanding and analyzing data
6. Use assessments to improve and evaluate student learning

In parallel to the movement of reforming statistical education, there has been a general reformation of education due to the technical developments that offer new possibilities for designing, producing, distributing, and evaluating education. The use of computers, the Internet, and learning management systems (LMS) has expanded from an attractive fad for enthusiastic pioneers to now being a natural ingredient in more or less all of higher education. The fact that new educational technologies call for the rethinking of educational strategies and adaptation of principles for effective teaching has long been known (Chickering and Gamson 1987). Nevertheless, in the age of online and blended education, it remains a central challenge in this new pedagogical landscape to develop principles, which increase learning (for example, Graham et al. 2001; Trentin 2002; Garrison and Vaughan 2007; Barczyk et al. 2010). Research studies evaluating online teaching point to numerous success stories (for example, Volery and Lord 2000; Moore and Kearsley 2011), but there is also research that points to a gap between promises and reality (for example, Wilner and Lee 2002; Hartnett 2012).

The literature on educational technologies highlights some of the concerns that are potentially problematic with the use of educational technologies for example, the difficulties of mediating social presence (Gunawardena 1995), sense of community (Svensson 2002a, b), student frustration with technical problems, ambiguous instructions, and a lack of prompt feedback (Hara and Kling 1999). A problem of significant importance for educators is student dropout rates, which may be partly explained by factors such as lacking a sense of community and student interaction (Rovai 2002, 2003). At present, online education and the use of digital technology seem to be moving from pilot testing and small-scale initiatives into a more mature phase where large-scale programs are rigorously evaluated (Svensson and Gellerstedt 2009), and there are a number of suggested guidelines and principles, which summarize lessons learned that are possible to use in the design and evaluation of an online course (for example, Graham et al. 2001; Buzzetto-Moren and Pinhey 2006; Herrington et al. 2001; Wood and George 2003). The work by Graham et al. includes the following seven principles, which could be regarded as a practical lens useful for developing and evaluating online courses: "Good practice"...

i.   Encourages student–faculty contact
ii. Encourages cooperation among students
iii. Encourages active learning
iv. Gives prompt feedback
v.   Emphasizes time on task
vi. Communicates high expectations
vii. Respects diverse talents and ways of learning

We have also seen that new educational technologies used in distance education for example, the use of LMS, pre-recorded lectures, and discussion boards have also become an ingredient in standard campus-based education (Svensson and Nilsson 2010), and we believe that insights from campus and online teaching should be used in synergy for optimizing this blended learning.

Along with the technical development described above and changes in teaching strategies, the availability of online courses may attract students who normally would not attend a traditional campus course (Betts et al. 2009). Naturally, it is interesting to study the characteristics of this new student population. Are there any demographic differences or differences in learning style or study process? Do these students differ in psychological variables related to student achievement for example, self-efficacy (Zimmerman 2000)? Are there other important variables beyond course context, for academic achievement online, for instance locus of control, that is, the extent to which individuals believe that they can control events that affect them, which may be relevant for completion of a course (Dille and Mezack 1991; Pugliese 1994; Parker 2003)?

Hence the last decades have certainly been challenging for a teacher of statistics, with statistical education reform, the digital era, and pedagogical issues both online and on campus, and possibly also enrollment of new types of student cohorts. Since the late 80s, we have worked to tackle these challenges in introductory courses in statistics. We have made changes, evaluated them, and accordingly refined aims, pedagogical strategies, assessments, etc. many times in this period. In 2008 we decided to use a holistic approach to consolidate our experiences in designing modern courses in applied statistics.

When working with development and evaluations, we have found that the 3p-model: presage, process, and product is also valuable and supports a process view (Biggs 1989). This model describes the interaction between the following three components: student and teaching context (presage), student study strategies/orientation (process), and outcome—results (product).

In this chapter we describe what we have learned over the years and principles that we have developed through thought, experience, and from literature. We discuss our efforts with reference to GAISE and the seven principles of Graham et al (2001), which correspond well to our consolidated experiences, and have helped us to put our work in a wider context and facilitate further development. We also discuss the course and teaching context and present some results from evaluations structured by the 3p-model. Finally, we discuss a number of ideas on how courses like this could be modernized even further.

2 Applied Statistics at University West, Sweden

The University West, Sweden, was a pioneer institution, designing the first online B.Sc. program in Scandinavia, and has now been offering online courses in a wide range of educational disciplines for 15 years. Enthusiastic colleagues have taken part in practical work and research over the years. Also an LMS (DisCo) was developed in-house and is now standard at the University for both campus and distance courses (Svensson and Nilsson 2010). The university has work-integrated learning as a profile, which should permeate all our courses.

The courses in Statistics have changed and adopted new technologies and the online era has demanded new teaching strategies and course context. Furthermore, the pedagogical strategy has gradually changed towards a profile similar to the GAISE recommendations. We have increased focus on statistical literacy, statistical thinking, interpretation of results, and using computers for analyses rather than spending time on "mathematical masochism." We believe that this approach enhances skills useful for working life and that it increases the conceptual understanding of statistics. We also believe that the attitudes towards statistics, using this approach, are changing for the better (see Wore and Chastain 1989 for a similar conclusion).

2.1 The Program

In 2008 we decided to consolidate our experiences from the previous two decades to design modern attractive courses available on campus and online as well. The educational program in applied statistics, corresponding to a "minor in applied statistics" henceforth referred to as "the program," includes the following four courses (main content within brackets)

- "Applied statistics—to collect and summarize data" (survey methodology, descriptive stats, and regression)

- "Applied statistics—to draw conclusions from data" (probability, inference, classical parametric and nonparametric analyses)
- "Applied statistics—predictive modeling" (one/two-way ANOVA, multiple linear and logistic regression)
- "Applied statistics—independent analytical project" (factor analyses, reliability analyses, and producing a quantitative project)

All courses cover material corresponding to 5 weeks full-time studies. The first three courses are given full time on campus, but half-time pace online, i.e., 10 weeks duration. The project course is given at a slower pace—for a duration of 20 weeks both on campus and online. This chapter will focus on the first course: "Applied statistics—to collect and summarize data."

2.2   The Introductory Course

The course: "applied statistics—to collect and summarize data" is an introductory course covering: survey methodology, descriptive statistics, regression, introduction to hypothesis testing, and the chi-square-test. As a standard, it is offered on campus every autumn as a compulsory course for business administration students at the undergraduate level. It is delivered at a full-time pace during 5 weeks, including approximately 6–9 h of lectures per week, 4–8 h of scheduled supervision time in computer rooms, and roughly 4 "open-door-hours" for addressing various questions. The number of students typically varies from 60 to 80. Occasionally the course is offered on campus an additional time during a year for undergraduate students from various fields. Beyond the on-campus delivery of the course, the course is also offered online, starting at the same time as the on-campus course but at a half-time pace, i.e., 10 weeks duration. The online version is also offered at the beginning of each year at the same pace. The online courses are usually offered to a maximum of 60 or 80 students (dependent on available teaching capacity), which allows for an expected drop-out-rate of 10–20 students. The online course is delivered completely via the Internet—no physical meetings on campus at all. The online course recruits students from all over Sweden and usually a handful of the students are located outside of Sweden. The online course is an elective course and attracts participants from various backgrounds, business, sociology, psychology, engineering, and occasionally also statisticians who want to improve their ways of communicating statistics.

The course is divided into three modules with the following main topics: descriptive statistics, regression, and finally survey methodology and hypothesis testing. Each module is completed with an assessment. The two first modules are assessed by computer labs (using IBM-SPSS) including around 50 "true-or-false-statements" to be answered in a template document including a standardized table for responding to the statements (see also Sect. 3.1). The last module includes a fairly comprehensive survey project. In the campus version of the course, all three assessments may be undertaken in a group of no more than three students, and an individual oral examination is given to complement these three assessment pieces. The online version excludes the oral examination, but we demand that all assignments are undertaken individually, even though students can, and are encouraged to, discuss the assignments via the course homepage. The computer labs and oral examination are all graded as fail or pass. The survey project also uses the grade, pass with honor. The students have to pass all assessments in order to pass the course. For receiving the grade pass with honor, the students must pass the survey project with honor. The online version is designed and adapted for participants who work full time. The online course has a flexible design, with no physical meetings at all and no fixed time for lectures, but there are three deadlines for submission of assessments. The course is supported by pre-recorded lectures (see, for example, http://bit.ly/gellerstedt_video_mreg)

3 Course and Teaching Context

In this section we discuss our course context with reference to both guidelines for statistical education (GAISE 1–6 presented above) and the seven principles for effective teaching on campus and online (i–vii presented above). We limit our discussion to our most central ideas and keep the discourse relatively brief. To complement the discussion, some examples of teaching material are available for downloading at: http://bit.ly/gellerstedt_statistics .

3.1 Emphasize Statistical Literacy and Statistical Thinking

Literacy can be defined as: the ability to read, to write coherently, and to think critically about the written word. Statistical literacy could be defined as an extension of this definition by simply adding "and statistics" to the end of the definition. This demands knowledge of vocabulary, the ability to read summarized statistical measures, tables, and graphs and a basic understanding of statistical ideas. We have tried to implement statistical literacy through students learning to:

Produce and consume statistics—read and write stats. Using computer labs with real data examples, where students produce data and have to choose adequate statistical measures that focus on the interpretation of statistical results (for an example of a lab, see document: lab1 in http://bit.ly/gellerstedt_statistics).

Speak statistics. Students on campus are required in the oral examination to be able to explain statistical terms and concepts with their own words. We simply want to know that students are capable of generalizing a concept and putting it in a context other than the textbook.

Make yourself understandable. This also emphasizes the "ability to write statistics." In the final survey report a criterion for pass is that the chosen statistics are adequate and that they are presented in a pedagogical and understandable way even for non-statisticians. That is, we emphasize the importance of well-conceived tables and graphs with variables placed in a reader-friendly position, adequate statistical measures, no overkill in decimal places, etc. We also demand that statistical conclusions such as "reject null-hypothesis" are phrased in a conclusion in terms of the actual application.

Avoiding naked data, i.e., numbers without a context. It makes it much more interesting to learn to "write and read" statistics if there is a story behind it.

Statistical thinking is described in GAISE as "the type of thinking that statisticians use when approaching or solving statistical problems." In GAISE there is also an analogue to carpentry, meaning that a course should not only teach how to use tools like: planes, saws, and hammers (corresponding to topics such as descriptive stats, graphs, and confidence intervals) without also discussing how to build a table (corresponding to teach students to use statistics in order to answer a question). If we scrutinize statistical education, we see that there are still many statistical text books focusing very much on tools but very little on the "statistical handicraft." In implementing statistical thinking in a number of ways, some of the most important ideas we have employed are:

Teach and discuss the handicraft. We have developed our own descriptions of "statistical handicraft" partly based on process models found in data mining literature. For instance, the model CRISP-DM includes the following steps: business understanding (question to be addressed), data understanding, data preparation, modeling, evaluation, and deployment. These are illustrated with a wheel symbolizing the fact that going through the process sometimes has to be done several times and that the process also often generates new questions. Another similar workflow description is SEMMA developed by the SAS institute, (Azevedo and Santos 2008).

Go for the whole journey. Give the students the opportunity to carry out the whole process from addressing questions to analyses and writing a report. This is implemented in the survey project, which is the most comprehensive assessment item in the course. An example of how such a project is described to students is available at http://bit.ly/gellerstedt_statistics in the file: survey_project_happiness. The project starts with asking students to suggest interesting themes to be addressed in the survey. The theme should preferably be based on currently discussed questions in the news and current affairs. Examples from recent courses include: sleep disorder and financial crises, personal finances, and happiness. In the second step, teachers construct a poor questionnaire and students are required to suggest at least three improvements (rephrasing questions/alternative answers, adding crucial questions, etc.). Based on these suggestions, a somewhat better questionnaire is constructed (see questionnaire_happiness in http://bit.ly/gellerstedt_statistics) and each student is asked to collect 10 completed surveys and enter the data in a ready-made coded template in IBM-SPSS, which is pooled into one single database by the teachers (takes roughly 1 h). During the first survey project, we noticed that several students lost the "statistical focus" since they were unaccustomed to writing a quantitative report.

We received a lot of questions regarding structural issues, number of pages, headings, font size, etc. In order to get students back into statistical focus, we decided to support students with a suggested report skeleton in the survey project, (see structure_report_happiness in http://bit.ly/gellerstedt_statistics). Usually we try to include an established constructed index in the questionnaire, which gives us the possibility to discuss index-construction, reliability, and validity.

Don't take data for granted. We try to discuss "operationalization" both before and as variables are identified. For instance, during lectures students may receive a very open and rather vague question such as, "Is it a good idea to play music for a herd of cows?" Students must clarify the question (for example, to increase production, improve quality of milk, or support harmonic healthy cows), discuss potential variables (students are asked to find both quantitative and qualitative variables), and discuss potential designs (observational study, cross-over study, randomized trial, etc.). Another example of a task for students is: "Assume that you gain access to Facebook log files—what kind of variables would you find and what kind of questions could be addressed?" In this specific case, we could address ethical issues as well.

Have fun. It is easier to learn something if it is interesting and fun. Students are encouraged to occasionally focus on interesting and fun data sets, even if the application in itself may fall outside the usual range of specific discipline contexts. We mix our business cases with statistical applications and data from various contexts such as wine, music, love, and happiness. If at least a part of our passion for statistics can be spread across as many contexts as possible, it would help improve the prospects for effective learning. Using jokes can also strengthen the relationship between students and teachers, decrease stress, and sometimes statistical jokes actually include pedagogical aspects (Friedman et al. 2002).

3.2   Use Real Data

In consolidating our teaching experiences, we agreed that the reason we find statistics interesting is because it helps us to understand reality. The magic happens when the statistical patterns are translated into the studied phenomena and answer addressed questions that may generate new interesting ideas. It may be hard to convince students that for example, a telling graph describing a complex real-life phenomenon could be equally esthetically appealing as a telling painting, photograph, or a piece of music which captures a part of our reality. However, in order to show at least a part of such beauty, we urge that, in contrary to the Swedish impressionist Zorn's objects, statistical objects (data) should not be naked—data must be dressed properly, that is, with an interesting context. Some examples of our efforts to work with real data are:

- Each year students help us collect answers to a number of questions using a survey project, usually with around 500–1,000 respondents per project. These data sets typically include interesting topics discussed in the latest news and include various types of variables which allow descriptive statistics, inferential statistics, and sometimes also regression modeling. Students suggest the theme of the survey project, which increases the likelihood that students find the data interesting and engaging.

- We use the Internet for obtaining data. For instance, we used a site with cars for sale and downloaded information about cars such as price, age, and mileage for a specific brand. This is also a data set that we are using in several courses.
- We are using our own research data in courses.
- We also use data that we have gathered during statistical consulting.
- We have used the Internet resources for accessing secondary data sets.
- We have bought a data set. For example, a small fee is required for a data set including all real estate sold during 1 year in a municipality close to the university. The small fee (around 200 euro) was well worth the data set, which we are using in our first three courses for descriptive stats, inference, and multiple regression analyses. Even when the data start to get old, our experience is that it still leads to interesting questions to discuss with students, i.e., possibilities to extrapolate results, if patterns still hold, but on another price level.
- Sometimes, we need data for illustrating a specific point, and in such a situation we usually choose to simulate data, but it is still based on a real-life situations, i.e., realistic data, rather than real data.

It is also important to consider what we actually mean by real data? We must realize that the "real data" students first and more frequently face is data outside the academic world (Gould 2010). For instance, we guess that the most commonly used data these days are "thumb up or thumb down," "like"/"dislike" choices familiar from social media sites. To discuss the potential use of analyzing a person's "thumbs-up/down-profile" and what that could reveal about that person's personality and interests is a good example of modern real data—a real context example.

Beyond using real data, we also believe that it is important to use real data in a genuine way, for instance, to discuss the potential use of the statistical analyses, deployment in an organization, the use of statistics for making decisions, or ethical considerations. As pointed out in the previous section, it is important that students are trained in converting statistical results and conclusions into a discussion and conclusion in terms of the actual application. We want teachers of our courses also to be active outside the academic world, working with statistical consultations and applied research. We believe that such experiences make it easier to illustrate that statistics are interesting and fun. In short, real data becomes more interesting if it is used in a real-life context.

3.3   Focus on Conceptual Understanding and use Technology

Over the past two decades, we have gradually limited the use of mathematical formulae to a minimum and instead focused on using statistical software. As a standard we use IBM-SPSS, which is commonly used software for applied statistics in Sweden.

However, we still have some formulae included in our lectures, since we believe that the mathematical formulae sometimes illustrate the statistical idea in a very comprehensible way. For instance, the standard form of calculating the standard deviation illustrates that variability can be measured by comparing each single observation to the mean. In fact, we also encourage the students to make this calculation manually, at least once, to get a hands-on feeling, but thereafter we recommend the use of software and focus on understanding the result and how it could be used.

Regarding descriptive statistics, we always start with a lecture aiming at discussing descriptive statistics in general terms, i.e., statistics aiming at describing location, spread, and shape. We emphasize the meaning and use of the different statistical characteristics of a data set, and the importance of not only focusing on location. At the next lecture, we continue by discussing commonly used corresponding measures.

When introducing hypothesis tests, we have chosen to start with the chi-square-test. In the first course, when introducing a hypothesis test, the chi-square-test is actually the only test that we use, allowing us to focus on the philosophy of hypothesis testing in general (we also discuss standard error for an average and a proportion, but not formalized as tests). We do not demand that students should be able to calculate the test function manually, but we discuss the formula and emphasize the idea of comparing what we have observed in relation to what we expected, given the null-hypothesis. The formula actually captures this key idea, just as the formula for standard deviation illustrates that the distance between each observation and the mean could be used for measuring spread.

For the on-campus version of the course, we use oral examination as one mandatory assessment. In the beginning of the course, students receive roughly 30 questions, which cover essential parts of the course. We declare that on the oral examination students are expected to be able to answer any of these 30 questions. We demand that students find their own examples for illustrating different concepts, such as "give your own example which describes the difference between a randomized trial and an observational study," "explain with your own example how standard deviation could be interpreted and what it could be used for." We do not accept examples from the textbook or examples found after some "googling"—we recommend students find examples based on their own interests.

3.4 Active Learning

When discussing learning styles, *active* learning is often related to "learning by doing" and includes working in a group. An alternative learning style is *reflective* characterized by learning by thinking things through, preferably working alone or with a colleague or friend. Since the teachers in the course have a background in mathematical statistics and have preferences towards reflective learning (and, frankly, disliked group work as students), the ideas of active learning have been discussed intensively. However, there may be a natural difference in teaching a class for future mathematical statisticians compared to a class of business administration students who must be able to apply statistics. We agreed that active learning, as suggested by both GAISE (recommendation 4) and "the seven principles" (iv) was a good choice for our course. We also believe that the principles i–ii, communication student–student and students–faculty, are important for active learning.

We also agree that learning is not a spectator sport. Active learning means essentially that students, in some sense, must let learning be a part of them and their daily life. We want our students to write about statistics, to talk about statistics, and we actually want students to "live with statistics," i.e., relate statistics to their experiences, interests, and daily life. The question is how? We struggled, and are still struggling, with the question of whether we should let the students work in a group or not, to solve assignments, and deliver assessments in a group or not (see the section about our evaluations for further discussion). We do not believe that working in a group is a necessity for active learning. We also believe that it is a question of when group work is initiated. For some students, it would be appropriate to start group discussions directly from the beginning and together discuss and learn during the course, while for some students, those with preferences to reflective learning, a group discussion is perhaps valuable after a period of individual studies where a student can think things through. As described in the previous section, students on our campus course work together and deliver assessments together, while our online students are not placed in groups; they could still help each other, for example via the discussion board, but they must deliver assessment items individually.

The most important strategies we use for encouraging active learning are:

Write, read, and speak statistics, make yourself understandable, practice as a handicraft, and go for the whole journey. The strategies used for statistical literacy and statistical thinking described above actually go hand in hand with the ideas of active learning. Our assessments: IBM-SPSS-labs, an oral examination and a survey project are essential components for active learning and we have tried to design the assignments in a way that supports the idea of "read, write, and speak statistics."

Continuity. In order to support the idea that students should "live with statistics," we believe in continuity. In order to avoid intensive studies at the end of the course, we divided all our courses into modules. Each module is described in study guidelines and is accompanied by a corresponding assessment. Regarding the first two labs, the students have to answer roughly 50 "true/false statements." Since they deliver their answers in a standardized format, correcting the assessments takes little time, and the results together with a presentation of the correct answers with accompanying explanations is usually presented the day after the submission deadline of the assessment. In that way students receive prompt feedback and can compare and discuss their answers.

Communication. We try to make the discussion board on our course homepage as active as possible and we fervently encourage its use. Normally, we ask students who email us questions to address the question on the course home page instead: "Hi, that was a really good question, which I believe is of common interest, could you please ask that question on the discussion board." We have the policy to visit the discussion board on a daily basis. On campus we offer specific hours as "open-door" when we are available at the office for discussions. We also have booked computer labs, roughly 4–8 h a week, for our classes to use. It is not mandatory, and many of our students work with IBM-SPSS at home (it is free to download for our students), but approximately one third of our students use these lab hours. Two of these hours are supervised by teachers in the room, but we also usually take a stroll in the computer rooms during the other hours—"surprise supervision."

WRIB. Beyond the discussion board, we have introduced something we call a work-related issue board (WRIB). On our course homepage, this discussion board is available for addressing questions, which may be out of the scope of the course. On this board students are welcome to start discussions and ask questions regarding statistical problems in their profession or related to statistical results presented in real life, newspapers, and TV. We welcome all kinds of problems, even if the questions are beyond the scope of the course. As an example, in the latest course we had questions regarding: construction of items in a survey, how to make quality acceptance control, and how to handle dropouts in a clinical trial. Our experiences so far are that it is really exciting for the teachers to discuss real-life problems, that it is positive that other students actually see that statistics are used in practice and in various fields, and that other participants actually face these problems. The WRIB and discussion board have proven to be important tools to create a sense of community among teachers and participants.

3.5 The Use of Assessments

We regard our assessments mostly as learning objects. We also use our assessments for maintaining a good study pace and for engaging students to work continuously throughout the course. Our two IBM-SPSS labs focus on interpretation of results and the ability to choose appropriate statistical measures. The labs include self-guided instructions on how to use IBM-SPSS procedures, and students have to discuss the results and consider a total of 50 true-false statements, which are divided in 10 groups with 5 statements per group. If a group of 5 statements is answered correctly, a student receives 1 point; if one of the five statements is answered incorrectly a student receives 0.5 points, otherwise 0 points. Thus, the assessment can have a maximum of 10 points and at least 5 points are needed to pass. The correct answers with detailed explanations are presented on the course homepage the day after the deadline for delivering the answers. The students are encouraged to compare the suggested solution to their own answers, and we stress that this is an important learning activity. In this way students get immediate feedback. Even if a student fails the assessment, studying the solution may help a student increase understanding and continue with the course. A lab reassessment is offered at the end of the course, after all deadlines for other assessments. There is also usually a discussion about the labs during the labs and also after the deadline on the discussion board of the course homepage.

The course ends with the survey project, as described in Sect. 3.1. In this project there are no available IBM-SPSS instructions, so students are left on their own. However, if they have technical questions regarding how to produce specific output such as graphs or tables, we offer guidance because we want the students to focus on statistical content and not spend too much time on technical issues.

3.6   Explicit Instructions and Standardized Material

One important experience from one and a half decades of distance education is the importance of explicit guidelines and structure. The highly standardized concept we have used in online settings is also used on campus. Some examples of our structural ideas:

Modules: For supporting a continuous, uniform study pace, we have divided all our courses into three to five modules. In each module a study guide describes: theoretical content, keywords, mandatory assignments, study advices, links to pre-recorded lectures, and deadlines.

Standardization and time limits. All our produced documents have the same layout in all of our courses. We also provide templates for students to use when answering assignments. We have explicit deadlines, which are not negotiable, and if an assignment is delivered late, it will not be corrected. At the beginning of the course, a general document including explicit course instructions that explain all details and "game rules" is made available. This kind of standardization has helped us optimize the budget of the course and ensures that we use as much time as possible for teaching and as little time as possible for administration.

Standardized pre-recorded lectures: We try to implement standardization in all types of productions. The standard of our pre-recorded lectures has increased over the years. However, when designing the new courses in applied statistics, we decided to adopt a more standardized concept. Each pre-recorded lecture starts with a 1-min introduction—presenting an overview of the lecture. In all the videos the teacher also welcomes the students and toasts them with a cup of coffee saying "cup a cup" as a gimmick for supporting our informal teaching style that is recognized in each video. The idea is to create the feeling of a personal meeting by "taking a cup of coffee with the teacher and learning some stats." At the end of the video, the content is wrapped up and once again the cup of coffee is used. If you want to be welcomed by "cup a cup," please see http://bit.ly/gellerstedt_video_mreg.

Time aspects. In the course information, we explicitly declare our time policy and what students should expect, for example, "respect deadlines, late assignments will not be considered," "our aim is to respond to e-mails within 24-h."

Expectations: We try to communicate in the study guidelines exactly what we expect from students. As previously stated, we also use templates as a report skeleton, as the recommended way of reporting the survey. Furthermore, we usually publish a report produced from previous courses, a report graded as pass with honor, as one good exemplar, corresponding to principle vi.

Dynamic communication: To get students to interact with each other is often pointed out as a crucial success factor in distance education (Conrad and Donaldson 2011). For supporting communication and active learning, we want to have communication that is as dynamic as possible. We attempt to visit the discussion board on a daily basis, even on weekends, such as Sunday afternoon, since we know that most of our online students are working and do much of their studies during the weekend. According to our experiences frequent visits and quick responses are essential for a lively and dynamic discussion. This also goes hand in hand with four (i–iv) of the seven principles for effective teaching. We have even noticed that it is important to start the discussion. Usually, some really active students guarantee this, but if not, we try to get the discussion rolling by starting discussion threads ourselves. We also frequently ask questions to address existing questions on the discussion board, instead of students sending us an email. Regarding the "WRIB" presented in Sect. 3.4, we are planning to use some of the really interesting questions as a "start" in all our courses.

3.7 Respect Diverse Talents and Ways of Learning

It is rather common in pedagogical research to recommend key success factors, which are important for supporting "students." For instance, as mentioned in the previous section, in distance education, student interaction is pointed out as crucial, and both GAISE and the seven principles urge the use of active learning. Working in a group may be one way of supporting both student interaction and active learning. Nonetheless, we do not believe that this approach improves learning for all students. For instance, there may be a subpopulation of students who prefer to think things through on their own. In our online courses we know that a majority of the participants are working, some even full time, and maybe a collegial discussion at work would be more fruitful than working in a group with other students? Among the seven principles, the third recommendation is to "encourage active learning," which we support as described above. However, we believe that we must also respect that some students may want to study in a more reflective manner, and on their own, at least to start with before interacting with others. Ignoring such differences would indirectly imply contradictions between principle iii (encourage active learning) and principle vii (respect diversity). As it is now, we have chosen to encourage solving assessments in a group (maximum three students) in our campus version of the course, but on the online course we demand individual submissions, even if the students are allowed to cooperate. This choice was originally done for practical reasons, but according to our evaluations this may actually also have a pedagogical rationale as discussed in the next section.

Beyond the discussion about active vs. reflective learning, students may also differ in learning style from many other perspectives. Some students like sequential presentations, taking one step at a time and in the end understanding the whole picture, while other students prefer to get the whole picture first and thereafter work with the details. Some students may like and learn best by doing practical explanations, while some like conceptual models and abstract thinking better and others like visual presentations more than verbal and vice versa. In our production of pre-recorded lectures, we have tried to mix general lectures providing an overview with lectures that go into details. An idea that we have discussed to a great extent is the possibility to produce various study guides, meaning that a student could, for instance, choose between a sequential presentation or a presentation that starts with an overview and then continues with the details.

Another aspect of respecting diversity is related to different interests. As discussed previously, it is essential to use interesting cases and preferably real data for supporting learning. However, naturally all students do not share the exact same interests. In a course with business administration students, we guess that a large proportion of the students is, or should be, interested in business cases. Yet, there is certainly a mixture of other interests as well. Likewise, in an online course with elective students from various fields, the mixture of interests will be even greater. We try to satisfy different interest by:

- Using real data from various fields.
- Using the "WRIB," i.e., the discussion board, which is open to all kinds of statistical questions—regardless of application.
- Students are asked to suggest themes for the survey project. It is rather common that students reach an agreement on an interesting theme via the discussion board. In order to support various interests, we sometimes add some questions that may seem to be very far from the theme. In the survey project we usually address three "research-questions" that should be analyzed. However, for supporting various interests we allow students to change two of the three questions and address other ones based on their interests, and these questions may include the questions distant from the theme.

4   Student and Learning Context: Evaluations

We have continuously evaluated the program developments by using various sources of information, and some of the experiences found in these evaluations are presented previously (Gellerstedt and Svensson 2010). In this section we will discuss some preliminary results from a survey delivered to students in our introductory course in applied statistics the autumn 2012, both on campus and online. The study included two questionnaires, the first *pre-course-questionnaire* was delivered at the beginning of the course, during the first week, and the second *post-course-questionnaire* was delivered after the course. The results are structured by using the 3p-model that focuses on baseline characteristics (student presage), studying approaches and learning styles (process), and outcome (product).

The on-campus group consists of students in a Bachelor program in business administration, referred to as "campus students" henceforth. For campus students, the course is mandatory and given at a full-time pace (duration of 5 weeks). The questionnaire was answered by 54 among 66 registered campus students (82 %). The second group, referred to as "online students" studied the course as an elective course given online, at part-time speed (10 weeks). In this group the response rate was 44 of 61 students (72 %). More details about the course are described in Sect. 2.2.

The questionnaire was quite comprehensive and was designed to measure: demographical background, study intention, work experience, working situation, computer literacy, and attitude to the subject. The questionnaire also included psychological variables (self-efficacy, procrastination, and locus of control), learning style, and study process. These variables are explained more in detail in the appendix.

The *post-course-questionnaire* was much shorter and focused on: attitudes to the subject, how the practical work was organized, interaction between theory and practical assignments, and pros and cons of working in a group. Finally, this questionnaire asked students if they found the course: difficult, interesting, useful, and enjoyable and if they "learned something" (each attribute was judged using the 3-point scale: less/worse than expected, as expected, better/more than expected). 36 of the 66 registered campus students and 26 of the 61 registered online students (43 %) responded to the post-course-questionnaire (54 % response rate).

4.1   Different Populations of Students: Baseline

Teaching context should not just reflect a teacher's style; it should preferably be designed for all kinds of students. Yet, individual differences in background, learning goals, and learning styles are often ignored (Ford and Chen 2001). There may also be essential differences between different populations of students. We found some important differences when we compared the on-campus and online groups of students, see Table 1.

We found no difference in either gender or academic parents (academic mother and/or father), but online students were on average 11 years older, and as expected, they had longer working experience. A majority of online students were employed (20 % working part-time and 57 % working full time) which is consistent with a previous smaller study (20 % part-time 62 % full time, see Gellerstedt and Svensson 2010). Among campus students no one worked full time, but 65 % reported part-time work.

Among campus students 98 % thought that earning the credits for the course was important. Among online students only 77 % thought it was important to receive the credits for the course, even though there was no significant difference in intention to do the examination. Among online students 82 % reported that they wanted to work with statistics and analyses in a future occupation, while the majority of the campus students responded "do not know" to this question. Among the online students working full time, 84 % reported that the course was useful for their present occupation.

Table 1 Baseline characteristics by group, p-values from student's t-test or chi-square test

Variable	Statistic	Campus ($n = 54$)	Online ($n = 44$)	p-value
Gender	% Men/women	43/57	46/54	>0.20
Age	Mean (years)	23	34	<0.001
Work experience	Mean (years)	3.3	9.8	<0.001
Working now?	% No/part/full	35/65/0	23/20/57	<0.001
Academic parent(s)	% Yes/no	38/62	55/45	0.098
Credits important?	% Yes/no	98/2	77/23	0.001
Examination intention?	% Yes/no	98/2	98/2	>0.20
Working with stats in future?	% Yes/no/don't know	21/21/58	82/9/9	<0.001
Experience writing report	% Yes/no	75/25	95/5	0.007
Index for "math easy, fun"	Mean (sd)	4.1 (0.7)	3.8 (0.9)	>0.20
Stat is exciting and useful	Mean (sd)	4.2 (0.6)	4.6 (0.7)	0.004

We have, over the years, noticed that the written quantitative report (a final examination task) is frequently of higher standard among online students, which may be explained by experience. 95 % of the online students reported experience with writing a report while the corresponding figure among campus students was 75 %.

We also asked some questions regarding attitudes to statistics, for example, if students believe that statistics is exciting (mean of 4.1 and 4.3 for campus and online, respectively, on a scale from 1 to 5, t-test: $p > 0.2$) and useful (campus mean 4.3 and online mean 4.7 on a scale from 1 to 5, t-test: $p = 0.024$), which to our enjoyment, showed high averages in both groups, but highest among online students.

In short, our online students were older, more experienced, not equally interested in the credits, and had to a high degree realized that statistics is exciting, useful, and a desirable ingredient in a future occupation. There were no significant differences between student populations regarding the psychological variables: self-efficacy, locus of control, and procrastination.

4.2 Different Ways of Studying and Learning

One of our main interests in conducting this study was the learning style. As pointed out above, learning context should not only reflect a teacher's style. The way that courses and teaching context have changed over the years, and the way the program was designed, as described above, is different to the learning preferences of the teachers. The program is mainly developed by statisticians with a background in mathematical statistics, which is a discipline in which abstraction, thinking through, and individual reflection tend to be more common than learning by working in groups.

We cannot say that the pedagogical strategies commonly used in undergraduate studies in mathematical statistics are erroneous, but we strongly believe that there is a major difference in teaching students whose main subject is in another area,

Table 2 Learning styles by group, p-values based on chi-square test

Learning style dimension	Statistic	Campus ($n=54$)	Online ($n=44$)	p-value
Active vs. reflective	% Act/Ref	68/32	34/66	0.001
Sensitive vs. intuitive	% Sen/Int	87/13	80/20	>0.20
Visual vs. verbal	% Vis/Verb	74/26	61/39	0.178
Sequential vs. global	% Seq/Glob	67/33	55/45	>0.20

Table 3 Study approach by group, p-values corresponding to student's t-test

Study approach	Statistic	Campus ($n=54$)	Online ($n=44$)	p-value
Deep approach—index	Mean (sd)	34.0 (7.0)	40.5 (5.5)	<0.001
Surface approach—index	Mean (sd)	32.6 (6.4)	28.2 (6.8)	0.001

such as business administration, and who are supposed to be able to interpret and apply statistics in a future occupation, such as in a business situation, compared to students who intend to actually become statisticians. As mentioned previously, our courses have shifted from a high degree of mathematical abstraction and individual reflection to a more learning by doing in-context. Based on course evaluations and discussions with students, we have seen that our design concept is appreciated, but we had not formally investigated learning styles before this survey. As seen in Table 2, there are no significant differences between campus and online students except for the first dimension.

It is interesting to note that campus students prefer an active learning style, while online students prefer reflective. Even if a majority of online students are reflective, there is around one third of these students who have a preference towards an active learning style, just as around one third of the on-campus students are reflective while the majority have preferences towards an active learning style. On-campus students are encouraged to work in a group, but are allowed to work on their own as well. Among the on-campus students with a preference towards active learning only 8 % choose to work on their own, compared to on-campus students with a preference towards reflective learning of which 29 % choose to work on their own ($p < 0.004$, chi-square test).

Beyond the questionnaire delivered among students, four teachers (who are currently teaching or have been teaching in the course) were also asked to fill in the learning style questionnaire. All four teachers preferred reflective, intuitive, verbal, and global learning styles. Interestingly enough, this is completely contrary to student preferences, with the only exception that both teachers and online students are reflective rather than active (first dimension).

We are trying to support a deep and holistic approach for learning. We focus assignments on interpretation, concepts and the ability to put the statistical results in an applied context. Deep learning is partly driven by examination tasks (Svensson and Magnusson 2003; Ramsden 1992). However, we wanted to measure the approaches at the beginning of the course, before examination tasks had any major impact. And, interestingly, there was a difference between campus students and online students as seen in Table 3. This is not surprising, since campus students study statistics as a mandatory course included in their Bachelor's program, while online students have applied for the course voluntarily.

In summary, a majority of online students (two out of three) generally prefer a reflective learning style while the situation is the other way around on-campus, where roughly two out of three students prefer more active learning. The dimension "sensitive vs. intuitive" is the dimension with the lowest degree of diversity, but still more than 14 % favor the least preferred style (intuitive). There is clearly diversity in a preferred learning style. The differences in study approach were expected since the campus course is delivered as a mandatory course while the online course is for volunteer students. We did some preliminary explorative analysis, which indicates that attitude towards statistics is related to study approach: students who find the topic interesting and useful use a deeper approach than students with a less enthusiastic view of the topic.

4.3   Attitudes and Outcomes

An evaluation of a course must be based on the learning objectives of the course. The learning objectives in our course are to achieve statistical literacy within the statistical content, to obtain basic skills in the statistical handicraft, and to learn to use statistical software (IBM-IBM-SPSS). Since our way of assessment as described in the previous sections are constructed to correspond to these learning objectives, we take the final grade (not passing the course, passing, and passing with honor) as a reasonable measure of fulfilling the learning objects.

This is consistent with several interesting studies on how a student's success can be predicted, for instance by psychological factors or by using log files and data mining (for example, DeTure 2004; Minaei-Bidgoli et al. 2003). In such studies, successes are frequently measured by using the student's final grade of the course.

However, it may be important to also consider other possible manifestations of success than just the final grade. In general, desirable outcomes of a course may include deep learning, independent learning, critical thinking, and lifelong learning attributes (Biggs 1989). It may be important to consider such attributes and how they correspond to learning objectives outlined in the course syllabus.

We have also discussed our underlying objectives with the design of our course (teaching and course context, described in Sect. 3). These underlying objectives include that students will find studying statistics interesting, useful, and fun! We also hope that our course design encourages active learning and deep learning strategies, which are not explicitly mentioned in the course syllabus, but which may be crucial for fulfilling the learning objectives.

Furthermore, we want our students to use statistics in real life and we want them to continue to learn more statistics over time. We believe that positive attitudes towards statistics at least increase the chances to succeed with life-long attributes. In that perspective, it may be even more important than grades to measure attitudes towards the subject (and as described in the previous section learning style and

Table 4 Attitudes after course, post-course-questionnaire

Proportions (%): less, worse than expected/as expected/more, better than expected			
Question	Campus	Online	p-value (chi-square test)
Interesting?	3/47/50	4/15/81	0.033
Useful?	0/36/64	4/27/69	>0.20
Difficult?	6/64/31	27/54/19	0.056
Fun?	0/47/53	4/27/69	0.159
Learned something?	0/39/61	0/23/77	0.189

study approach). In our post-course questionnaire, we asked students if they found the course: interesting, useful, difficult, and enjoyable and if they "learned something" (each attribute was judged using the 3-point scale: less/worse than expected, as expected, better/more than expected) (Table 4).

Generally, it is pleasing to see that the proportions of "less/worse than expected" are small.

However, the subject is regarded as more difficult than expected by nearly 30 % of the on-campus students and roughly one fifth of the online students. The findings from the post-course-questionnaire must be interpreted carefully due to a low response rate, but the positive attitudes are supported by verbal (we have a verbal discussion with all on-campus students) and written (we ask all students, on-campus and online, to beyond standard course evaluations, send us feedback and suggestions for improvements), where we can find comments like: "Not as difficult as expected and it was even fun!", "Surprisingly interesting and important", "A true relief for me who is afraid of formulas". Regarding our intentions with statistical literacy, active learning, and that we want our students to "live with statistics," it is truly satisfying to frequently see comments similar to: "I cannot read a single newspaper or watch the news without starting to raise statistical questions and hesitations regarding numerical statements and studies."

We have done some preliminary exploratory analyses regarding the relationship between baseline conceptions and attitudes in the post-course-questionnaire. These analyses indicate a positive correlation, for example, students who at the start of the course had positive attitudes to statistics as being interesting, useful and fun to a high degree responded "more/better than expected" in the after course questionnaire. Eleven students responded that it was unimportant for them to earn the credits for the course. These were students with generally positive attitudes towards the subject and who studied simply because statistics was interesting.

The on-campus course had a high completion rate with only 8 % not passing the course; see Table 5 for details. In the online group we can see that 44 % of the students did not complete the course. Around half of these students did not event start the course at all. The third column in Table 3, illustrates the distribution of outcome if those who never started are excluded, and in this group we can see that the proportion who did not pass was 29 %.

Table 5 Outcome and final grade by study group

	On-campus N (%)	Online N (%)	Online starters N (%)
Registered students	66 (100)	61 (100)	48 (100)
Never started the course at all	0 (0)	13 (21)	
Started the course but didn't complete	5 (8)	14 (23)	14 (29)
Passed the course after some reassessments	11 (17)	11 (18)	11 (23)
Passed the course without any reassessments	48 (73)	18 (30)	18 (38)
Passed the course with honor	2 (3)	5 (8)	5 (10)

Based on Table 5, and the relatively high proportion of non-completers among online students, it is interesting to discuss if online teaching is a less effective way of delivering a course? To investigate this, requires reasons for non-completion. We did a follow-up study and via telephone interviews we interviewed 10 of the nonstarters. Out of these ten nonstarters, one of them did not understand the written instructions sent by ordinary mail, that is, did not understand how to register on the course homepage and download software, and one student had taken a similar course at another university. Among the remaining eight students all had an explanation based on external factors: moved to a new house, health problems, a new job, etc. We also interviewed 10 of the starters who did not complete the course. The explanations were similar, frequently referring to a lack of time due to unexpected external factors. In this latter group there may be an interview bias, but most of them wanted to register for the course again and complete it, indicating that external factors are reasonable explanations as opposed to reasons related to the course context.

Furthermore, we know from the questionnaire that the 14 online students who did not complete the course actually declared from the very beginning that earning the credits for the course was unimportant. Also, based on our questionnaire data we can see that among students who did not care about the credits, 64 % did not complete the course. As pointed out above, these participants are to a high degree studying for pleasure. They just want to follow the course without spending time on formal assessments. Naturally, it is nice to have such interested students enrolled, but since the financial budget we receive from the government is partly dependent on the completion rate, this is financially troublesome. However, if these 7 students are subtracted from the 14 non-completers, the figures start to be similar to the on-campus group. Furthermore, if we also observe that the proportion of "pass with honor" is greater among online students, we cannot say that there is any evidence regarding differences in teaching effectiveness according to our data. We have observed the same pattern over several years (see Gellerstedt and Svensson 2010), and learned how to include a drop-out rate in our budget. Finally, another important factor for explaining completion is employment. Among the students who work full time 40 % did not complete the course. The corresponding figure among the remaining students is 4 %.

4.4   *Evaluations Raise Questions*

To summarize some of the lessons, we have learned from the questionnaire study, we have considered the following questions as important in our future development:

- "Students" should not be regarded as one homogenous population, but the question is how to cope with diversity, including different learning styles?
- How can we increase deep leaning strategies, especially on mandatory courses?
- Success in terms of completion rate is high, but the proportion of pass with honor is low. How can we increase this proportion? We can also focus on improving the attitudes towards statistics.
- There are still rather high proportions of students who find statistics more difficult than expected—who are these students and how can we support them?
- We have tried to design a course, which participants who work full time should also be able to follow. Nevertheless, around 40 % of the participants who work full time do not complete. What can be done?

Some of these issues are related to our ideas concerning future development, discussed in the next section.

5   Future Development

In this section, we will discuss some ideas and issues in our planning for further progress.

5.1   *Using Log Data*

A conventional teaching environment allows face-to-face interaction with students and provides the possibility to receive feedback and to study student progress. In online education, this informal face-to-face interaction must be replaced with digital interaction, which may not be equally rewarding socially, but has the advantage that it automatically generates data in log files. Every e-mail, every click on the course home page, all downloads, submitted assignments, starting or responding to a thread on a discussion board, and so on, leave a digital footprint. This data explosion provides many opportunities for analyses of all kinds. During the last decades, data mining techniques have become popular for examining log file data (Romero and Ventura 2007). Within the education field, new frameworks called learning analytics and academic analytics have developed (Long and Siemens 2011). Learning analytics include methods for analyzing social networks, context analyses, adaptive systems, predicting success/failure, etc.: analyses of interest for learners and faculty and hence useful for Biggs classical 3-p model (Biggs 1989). This means that learning analytics could be used for measuring and visualizing student (and teachers) activity. Examples include automatically sending a push mail for nonactive students, generating instructions and suggested activities based on previous activities, predicting results, and perhaps modeling learning style and learning progress. As an example, among our distance students, 42 % took part in the discussion board at least once. Among these students 21 % did not complete the course as compared to 32 % among students who never took part in the discussion (but possibly read it). Among students who were active on the discussion board 74 % preferred a sequential learning style, while the corresponding figure was 40 % among the other students.

The field of educational mining and learning analytics has shown many promising results. However, a drawback is that data mining tools are rather complex and demand more effort than a teacher might want to give, and future tools need to be more user-friendly with automatic analyses and visualizations (Romera et al. 2008).

This is consistent with our experiences that handling log files needs data management and they are not easily analyzed. We have saved log files, but not yet related these files to the results from our questionnaire studies. Perhaps the next challenge for researchers is to develop systems that are easier to use, and can find the most vital analyses among all possible analyses with a mountain of data. We also suggest developing indexes, such as a validated index for student activity and teachers' activity on the LMS, which includes more informative variables than just the time spent on the LMS or number of visits, which are rather common measures used today.

Even if analysts and data mining researchers are optimistic regarding the potential with learning analytics, there is still a need for proofs of its pedagogical value. It might be questioned if abstract phenomena such as learning can be adequately measured by a small set of simple variables. Such apprehension may be justified, but we believe that log files could be used jointly together with questionnaires, self-assessments, and interviews.

5.2 Triggering the Student and Teacher by Monitoring

Most of the learning analytics and data mining projects so far consider analyses made for and presented for educators, but it is suggested that analyses should be available also for the users (students) (Zorrilla et al. 2005). We plan to design some kind of barometers to be seen on the LMS. We know from evaluations that some students find it difficult to grasp how active they are supposed to be, if they are doing enough and whether they are doing the right thing. One way would be to monitor all students' activity with a student-activity index, and comment on each student's index, for example "you are one of the 25 % most active students."

Such an index must include relevant variables related to performance. The idea is that such an index should be displayed in real time and could act as a trigger for the students. A different example of a "monitoring trigger" situation is the monitoring of physical training activities such as on http://www.funbeat.net/ . On this site a person can add all of their workouts and compare themselves with friends (and even strangers if desired), and hopefully be motivated to at least avoid being the laziest one. Maybe monitoring learning in real time could have the same energizing effect. Indeed, monitoring student activity is considered as one of the major predictors for effective teaching (Cotton 1998).

A student activity index is of course also valuable information for teachers, especially concerning online courses where dropouts are common. There are also other potential ways of measuring online achievement beyond log files. For instance, Bernard et al. (2007) suggested a 38-item questionnaire for predicting online learning achievement, which showed some predicting capacity, even if previous overall achievements according to the university records were a better predictor. There are also self-assessment tests available on the web for potential online students, aiming at measuring if online studies fit the student's circumstances, life-style, and educational needs: http://www.ion.uillinois.edu/resources/tutorials/pedagogy/selfEval.asp.

It would be an advantage if a monitoring system could be generic, that is, possible to implement regardless of which LMS is being used. A very interesting and promising study built a "watch dog" called Moodog implemented in the Moodle LMS (Zhang and Almeroth 2010). Moodog keeps track of students' online activities and shows results with simple graphs. Furthermore, Moodog is able to automatically send feedback and reminders to students based on their activities. This kind of "humble barking" from Moodog may influence both teachers and students to more effective learning, but that is a hypothesis which needs to be investigated. Also, the importance of teachers' activity could be addressed. Our experience has demonstrated that the results in a course correlate to the amount of teacher activity. Certainly, increased student activity leads to increased teacher activity, and student activity is related to results.

5.3   Adaptation and a "WRIB"

In this study we have confirmed some differences between campus and online students, for example that online students are more experienced and have more reflective learning styles. However, this is only a generalization based on averages. Even if campus students on average are active, we noticed that roughly one third of the students had preferences towards reflective learning. Similarly, regarding online students who are on average reflective according to this study, around one out of three is active and perhaps would benefit by working in a group. Several researchers in the field of online education claim that interaction between students is the key to an effective online course (Conrad and Donaldson 2011). We do believe in high interactivity and it is stimulating when there are lively discussions on the course home page. However, we have some doubts regarding the frequently common advice that students should work in groups. Among our online students, around three out of four are working and may have logistical problems with working in a group, especially if experience and workload varies among the members of the group. Consequently, the cooperation overload may not counterbalance the potential pedagogical benefit. In the post-course-questionnaire, we asked the participants about their opinion regarding working in a group. The response alternatives were: "The increased learning outweighs the cooperative logistics", "Increases learning, but does not outweigh the cooperative logistics", "Would decrease learning", and roughly one-third of the students chose each alternative.

As discussed here, traditional classroom teaching strategies have been refined and adjusted for online students according to the seven principles, but we believe that these principles may be adjusted somewhat for students who are also working. Presumably, working in a group would be beneficial if the assignment could be applied at participants' workplaces. We are planning to develop strategies for such work-integrated learning. Two years ago, we started to use a specific online discussion forum called "WRIB"—work-related issues board as presented in Sect. 3.4. Using the WRIB and discussion board are important tools to create a sense of community among teachers and participants. This might also be an opening for work-integrated learning in groups. In short, we believe that whether there is a pedagogical gain with working in a group or not depends on learning style, experience, and working situation.

We hope to find ways of making it possible to make individual adaptations, and we are planning to offer an optional self-assessment test at the beginning of the course. If a campus student has an active learning style, the recommendation will be to work in a group. If an online student has an active learning style, we will as an option, offer contact with one or two other active students in order to start a discussion group together, and if preferable, deliver assignments as a group. Nearly 80 % of our online students are working and 84 % of them reported that they have explicit use of statistics in their current profession. It would be beneficial for learning to explore the possibility to find efficient ways of taking advantage of this situation and find efficient ways of work-integrated learning. For instance, we are discussing the possibility of implementing collegial recruitment where a person who applies for the course also recruits one or several colleagues, thereby creating the possibility to study in a group and possibly also to use work-related projects as examination tasks.

Beyond adaptation based on active vs. reflective learning style, we have plans of offering different study guides based on learning style preferences. For instance, if a person is more sequential than global, a study guide would recommend a linear stepwise study plan in order to watch detailed pre-recorded lectures before the general global content videos. For a student with a more global learning style, the study guide would recommend the opposite. We believe that a future challenge is to find ways of individualizing teaching context, including study guidelines, pre-recorded lectures, examples, and Internet resources. Such individualization could be based on student characteristics such as demographical background, self-test assessment regarding learning style, theoretical background, and interest.

At present much research focuses on using electronic media for adaptation, personalizing course context (Franzoni and Assar 2009). In addition, it would be interesting to design a study, which makes it possible to evidence base the potential gain of individualization and adaptation as discussed above. A potential criticism to the idea of adaptation based on student preferences is that it could hamper the development of student characteristics. For instance, it is an advantage if you can handle all learning styles, when it is needed. So, if active students always receive study context adapted for active students, this learning style may be strengthened, but there will be no practice in being reflective. Thus, there is a need for evidence-based research.

6   Some Final Remarks

As pointed out in this chapter, the pedagogical landscape for statistical education has changed dramatically during the last few decades. Currently available guidelines and principles for effective teaching are useful and appreciated. Nevertheless, we call for more evaluations. To some extent, empirical evidence may confirm ideas already known due to experience. However, we believe that confirmation is also important and could be clarifying, and experience shows that even intuitively appealing ideas may turn out to be wrong. For example, as discussed in this chapter, a general recommendation such as encouraging active learning may be a valid recommendation for a large proportion of the students, but possibly not for all. It depends how the active learning is supported. Thus, to observe student diversity is important in future research. There are many studies performed, but not many randomized trials, and there appears to be a lack of meta-analyses confirming guidelines and principles. Just as "evidence-based medicine" is applied in health care or "business intelligence" in business, we believe that it is time for evidence-based education. As discussed, there may also be great potential with learning analytics and using online data and evaluations in real time. We are certainly in an era when educators have access to an ocean of data, and we believe that surfing the waves to navigate this vast information can be stimulating, gainful, and important for improving education. Working with pedagogical improvements is stimulating, and evaluations are a valuable tool, not only for confirming ideas, but for generating new insights and raising new questions as well. To summarize our experiences with pedagogical development and evaluations, we would like to use the old expression and state that we are "still confused but on a higher level."

7   Note

Developed from a paper presented at Seventh Australian Conference on Teaching Statistics, July 2010, Perth, Australia.

This chapter is refereed.

Appendix

Description of variables used in questionnaire.
  Psychological variables

- *Self-efficacy*: A 10-item (each item: 4-point scale) psychometric scale designed to assess optimistic self-beliefs (Schwarzer and Jerusalem 1995) and gives a score from 10 (low self-efficacy) to maximum 40.
- *Academic Locus of control (LOC)*: A 28-true/false-item scale, resulting score from 0 to 28. Low scores indicating external orientation, high scores internal orientation. This scale is designed to assess beliefs in personal control regarding academic outcomes. An illustrative example of a question is: "I sometimes feel that there is nothing I could do to improve my situation"—a yes answer would indicate external control whereas a no would indicate internal control. The scale has been shown to be valid and reliable (Trice 1985).
- *Procrastination*: A 20-item (5 pointed scale for each item), resulting in a score from 20 (low procrastination) to 100 (maximum procrastination). The scale is shown to have high reliability alpha of 0.87 (Lay 1986)

Learning style

- *Learning styles*: 48 dichotomous questions. The Felder–Silverman model (Felder and Silverman 1988; Felder and Soloman 2012; Felder and Spurlin 2005) classifies students as having preference to either of two categories in each of the following four dimensions):

  - *Active* (learn by doing, like working in groups) or *reflective* (learn by thinking things through, prefer working alone or with acquainted partner)
  - *Sensing* (concrete, practical, like facts/procedures) or *intuitive* (conceptual, innovative, like theories/models)
  - *Visual* (like visualization pictures/graphs) or *verbal* (like written/spoken explanations)
  - *Sequential* (learning in small steps, linear thinking process) or *global* (holistic, learning in large leaps)

Study process

- *Study process*: 22 items (5-point scale). The questionnaire results in: deep approach index and surface approach index (score from 11 to 55, higher score indicates high degree of approach). Both index could be divided into sub-indexes measuring motive and strategy for deep and surface approach, respectively (Biggs et al 2001).

References

American Statistical Association. (2005). *Guidelines for assessment and instruction in statistics education*. Retrieved from http://www.amstat.org/education/gaise/

Azevedo, A., & Santos, M. F. (2008). KDD, SEMMA and CRISP-DM: A parallel overview. In *Proceedings of the IADIS European Conference on Data Mining 2008* (pp. 182–185). Retrieved from http://www.iadis.net/dl/final_uploads/200812P033.pdf

Barczyk, C., Buckenmeyer, J., & Feldman, L. (2010). Mentoring professors: A model for developing quality online instructors and courses in higher education. *International Journal on E-Learning, 9*(1), 7–26.

Bernard, R. M., Brauer, A., Abrami, P., & Surkes, M. (2007). The development of a questionnaire for predicting online learning achievement. *Distance Education, 25*(1), 31–47.

Betts, K., Lewis, M., Dressler, A., & Svensson, L. (2009). Optimizing learning simulation to support a quinary career development model. *Asia Pacific Journal of Cooperative Education, 10*(2), 99–119.

Biggs, J. B. (1989). Approaches to the enhancement of tertiary teaching. *Higher Education Research and Development, 8*, 7–25.

Biggs, J., Kember, D., & Leung, D. Y. P. (2001). The revised two factor Study Process Questionnaire: R-SPQ-2F. *British Journal of Educational Psychology, 71*, 133–149.

Buzzetto-Moren, N., & Pinhey, K. (2006). Guidelines and standards for the development of fully online learning objects. *Interdisciplinary Journal of Knowledge and Learning Objects, 2*, 95–104.

Chickering, A., & Gamson, Z. (1987). Seven principles of good practice in undergraduate education. *AAHE Bulletin, 39*, 3–7.

Cobb, G. (1992). *Heeding the call for change: Suggestions for curricular action (MAA Notes No. 22), chapter teaching statistics* (pp. 3–43). Washington, DC: The Mathematical Association of America.

Conrad, R.-M., & Donaldson, J. A. (2011). *Engaging the online learner*. San Francisco: Wiley.

Cotton, K. (1998). Monitoring student learning in the classroom, in School Improvement Research Series Close-Up #4, May 1988. Retrieved from http://www.nwrel.org/scpd/sirs/2/cu4.html

DeTure, M. (2004). Cognitive style and self-efficacy: Predicting student success in online distance education. *American Journal of Distance Education, 18*(1), 21–38.

Dille, B., & Mezack, M. (1991). Identifying predictors of high risk among community college telecourse students. *The American Journal of Distance Education, 5*(1), 24–35.

Felder, R., & Silverman, L. (1988). Learning and teaching styles in engineering education. *Engineering Education, 78*(7), 674–681.

Felder, R., & Soloman, B. (2012). *Index of learning styles*. Retrieved September, 2012 from http://www.ncsu.edu/felder-public/ILSpage.html

Felder, M. R., & Spurlin, J. (2005). Applications, reliability, and validity of the index of learning styles. *International Journal of Engineering Education, 21*(1), 103–112. Retrieved from http://www.engr.ncsu.edu/learningstyles/ilsweb.html

Ford, N., & Chen, S. (2001). Matching/mismatching revisited: An empirical study of learning and teaching styles. *British Journal of Educational Technology, 32*(1), 5–22.

Franzoni, A. L., & Assar, S. (2009). Student learning styles adaptation method based on teaching strategies and electronic media. *Educational Technology and Society, 12*(4), 15–29.

Friedman, H., Friedman, L. W., & Ammo, T. (2002). Using humor in the introductory statistics course. *Journal of Statistical Education*. Retrieved from http://www.amstat.org/publications/jse/v10n3/friedman.html

Garrison, D. R., & Vaughan, N. D. (2007). *Front matter, in Blended learning in higher education: Framework, principles, and guidelines*. San Francisco, CA: Jossey-Bass.

Gellerstedt, M., & Svensson, L. (2010). *WWW Means Win Win Win in education—Some experiences from online courses in applied statistics*. OZCOTS conference, Fremantle, Australia 2010. Retrieved from http://www.stat.auckland.ac.nz/~iase/publications/ANZCOTS/OZCOTS_2010_Proceedings.pdf

Gould, R. (2010). *Statistics and the modern student*. Retrieved from http://escholarship.org/uc/item/9p97w3zf

Graham, C., Cagiltay, K., Lim, B.-R., Craner, J., & Duffy, T. M. (2001). Seven principles of effective teaching: A practical lens for evaluating online courses. *The Technology Source*, March/April. Retrieved from http://technologysource.org/article/seven_principles_of_effective_teaching/

Gunawardena, C. N. (1995). Social presence theory and implications for interaction and collaborative learning in computer conferences. *International Journal of Telecommunications, 1*(2/3), 147–166. Retrieved from http://www.editlib.org/p/15156

Hara, N., & Kling, R. (1999). Students' Frustration with a web based distance education course. First Monday, Peer-reviewed Journal on the Internet. Retrieved from http://firstmonday.org

Hartnett, M. (2012). Relationships between online motivation, participation, and achievement: More complex than you might think. *Journal of Open, Flexible and Distance Learning, 16*(1), 28–41.

Herrington, A., Herrington, J., Oliver, R., Stoney, S., & Willis, J. (2001). Quality guidelines for online courses: The development of an instrument to audit online units. In G. Kennedy, M. Keppell, C. McNaught, & T. Petrovic (Eds.), *Meeting at the crossroads: Proceedings of ASCILITE 2001* (pp. 263–270). Melbourne: The University of Melbourne.

Lay, C. (1986). At last, my research article on procrastination. *Journal of Research in Personality, 20*, 474–495.

Long, P., & Siemens, G. (2011). Penetrating the fog—Analytics in learning and education. *EDUCAUSE Review, 46*(5). Retrieved from http://www.educause.edu/ero/article/penetrating-fog-analytics-learning-and-education

Minaei-Bidgoli, B., Kashy, D. A., Kortemeyer, G., & Punch, W. F. (2003). Predicting student performance: An application of data mining methods with the educational system LON-CAPA. In *33rd ASEE/IEEE Frontiers in Education Conference*. Retrieved from http://lon-capa.org/papers/v5-FIE-paper.pdf

Moore, D. (1997). New pedagogy and new content: The case of Statistics. *International Statistical Review, 65*(123–165), 1997.

Moore, M. G., & Kearsley, G. (2011). *Distance education: A systems view of online learning* (3rd ed.). Belmont, CA: Wadsworth, Cengage Learning.

Parker, A. (2003). Identifying predictors of academic persistence in distance education. *USDLA Journal—A Refereed Journal of the United States Distance Learning Association, 17*(1). Retrieved from http://www.usdla.org/html/journal/JAN03_Issue/article06.html

Pugliese, R. R. (1994). Telecourse persistence and psychological variables. *American Journal of Distance Education, 8*(3), 22–39.

Ramsden, P. (1992). *Learning to teach in higher education*. London: Routledge.

Romera, C., Ventura, S., & Garcìa, E. (2008). Data mining in course management systems: Moodle case study and tutorial. *Computers and Education, 51*(1), 368–385.

Romero, C., & Ventura, S. (2007). Educational data mining: A survey from 1995 to 2005. *Expert Systems with Applications, 33*(1), 135–146.

Rovai, A. P. (2002). Building a sense of community at a distance. *International Review of Research In Open and Distance Learning*, ISSN:1492-3831.

Rovai, A. P. (2003). In search of higher persistence rates in distance education online programs. *The Internet and Higher Education, 6*(1), 1–16.

Schwarzer, R., & Jerusalem, M. (1995). Generalized self-efficacy scale. In J. Weinman, S. Wright, & M. Johnston (Eds.), *Measures in health psychology: A user's portfolio. Causal and control beliefs* (pp. 35–37). Windsor, England: NFER-NELSON. (Translated to Swedish by Koskinen-Hagman 1999 Retrieved from http://userpage.fu-berlin.de/~health/swedish.htm).

Svensson, L. (2002a). Interaction repertoire in a learning community. In *Proceedings of Computer Support for Collaborative Learning (CSCL 2002), CD-ROM*. Univ. Colorado Boulder.

Svensson, L. (2002b). Discursive evaluation in a distributed learning community. *Australian Journal of Educational Technology, 18*(3), 308–322.

Svensson, L., & Gellerstedt, M. (2009). A reality check for work-integrated e-learning. In *Proceedings of e-Learn Asia Conference*. Seoul, South Korea.

Svensson, L., & Magnusson, M. (2003). Crowds, crews, teams and peers: A study of collaborative work in learning-centre based distance education. *E-Journal of Instructional Science and Technology, 6*, 3.

Svensson, L., & Nilsson, S. (2010). Learning models in online education: On the transformation from tutors to producers. In *Proceedings of Global Learn Conference*. Penang Island, Malaysia.

Trentin, G. (2002). From distance education to virtual communities of practice: The wide range of possibilities for using the internet in continuous education and training. *International Journal on E-Learning, 1*(1), 55–66.

Trice, A. D. (1985). An academic locus of control for college students. *Perceptual and Motor Skills, 61*(3), 1043–1046.

Volery, T., & Lord, D. (2000). Critical success factors in online education. *The International Journal of Educational Management, 14*(5), 216–223.

Wilner, A., & Lee, J. (2002). *The promise and the reality of distance education* (Vol. 8, pp. 3). Update. Washington, DC: National Education Association, Office of Higher Education.

Wood, D., & George, R. (2003). Quality standards in online teaching and learning: A tool for authors and developers. In G. Crisp, D. Thiele, I. Scholten, S. Barker, & J. Baron (Eds.), *Interact, Integrate, Impact: Proceedings of the 20th Annual Conference of the Australasian Society for Computers in Learning in Tertiary Education*. Adelaide, 7–10 December 2003.

Wore, M., & Chastain, J. (1989). Computer-assisted statistical analysis: A teaching innovation? *Teaching of Psychology, 16*(4), 222–227.

Zhang, H., & Almeroth, K. (2010). Moodog: Tracking student activity in online course management systems. *Journal of Interactive Learning Research, 21*(3), 407–429.

Zimmerman, B. J. (2000). Self-efficacy: An essential motive to learn. *Contemporary Educational Psychology, 25*, 82–91.

Zorrilla, M. E., Menasalvas, E., Marin, D., Mora, E., & Segovia, J. (2005). Web usage mining project for improving web-based learning sites. In *Web mining workshop* (pp. 1–22). Cataluna.

The *Golden Arches*: An Approach to Teaching Statistics in a First-Year University Service Course

Małgorzata Wiktoria Korolkiewicz and Belinda Ann Chiera

Abstract The realities of large first year service courses add substantially to the challenges of creating an environment conducive to learning. Given the increased understanding of the importance of context in Statistics education, discipline relevance is a key consideration in designing effective and engaging curriculum. However, students enter university with increasingly diverse levels of competency in quantitative subjects, and a survey conducted in the first week of teaching typically reveals negative perceptions of Mathematics, and by extension of Statistics, tempered with anxiety about quantitative subjects in general. We present strategies to overcome some of these challenges in relation to a quantitative methods course for first year Business students. Analysis of a follow-up survey at the end of the course reveals a positive shift in students' attitudes and improvement in student success in the course.

Keywords Maths anxiety • Large first year classes • Service teaching

1 Introduction

Life Sciences and Business programs typically include some form of Statistics among first year courses to equip students with the quantitative skills they need. These courses are often part of a service teaching arrangement with content delivered by an academic from a discipline outside that of the students. Such an arrangement can introduce tensions between teaching Statistics as a discipline in its own right, or as 'methodology serving some other field' Moore (2005). From our own experience,

M.W. Korolkiewicz (✉) • B.A. Chiera
School of Information Technology and Mathematical Sciences, University of South Australia,
City West Campus, Adelaide, SA 5001, Australia
e-mail: malgorzata.korolkiewicz@unisa.edu.au; Belinda.Chiera@unisa.edu.au

H. MacGillivray et al. (eds.), *Topics from Australian Conferences on Teaching Statistics:*
OZCOTS 2008-2012, Springer Proceedings in Mathematics & Statistics 81,
DOI 10.1007/978-1-4939-0603-1_9, © Springer Science+Business Media New York 2014

there is also a tendency to fill service courses with everything students 'should know' whether they are ready for it or not. Larger than average classes tend to be taught in traditional lecture theatres and include diverse groups of students, many of whom perceive the subject matter as 'boring' and 'not used outside of the classroom'. The majority of students undertake these courses only because they are compulsory and will at best be 'occasional users' of Statistics Nicholls (2001).

Another important aspect is the diversity in students' backgrounds and life experiences, coupled with varied levels of competency and ability in quantitative subjects. Students at our institution are likely to study part-time and have work and family commitments, which provide a significant challenge to succeeding at university Scutter et al. (2011). Among the school leavers, there are students with advanced Mathematics subjects in Year 12 (final year of High School), but also many with no Year 12, or even Year 11 Mathematics. Even if students enter university with Year 12 Mathematics, they often report not having enjoyed the experience or having achieved particularly good results. On the other hand, mature-age students, who may have been away from study for many years, worry they will not be able to keep up. Consequently, first year service courses in Mathematics and Statistics carry the stigma of high fail rates, poor student engagement and poor student evaluations of teaching. Under these circumstances, creating a learning environment for students, in which they can engage with and reinterpret Statistics as personally meaningful knowledge, and a tool for use in professional and individual life, is one of the key challenges Gordon (2004).

Discipline relevance and diversity of students' backgrounds are indeed an issue, but they often mask a deeper problem of students' maths anxiety Preis and Biggs (2001). A recent study by Lyons and Beilock (2012) suggests that when anticipating a mathematical task, for example receiving a maths textbook or realising a certain number of maths classes will have to be taken to meet the requirements for graduation, people with higher levels of maths anxiety experience brain activity in regions associated with threats and pain. The study relied on modern magnetic resonance imaging and the subjects had previously been identified as either 'high maths anxiety' or 'low maths anxiety'. However, the authors noted that this effect was absent when their study subjects were actually performing a maths task. Whether or not the findings have cross-cultural validity, given that the experiment was conducted at a North American university, the study nonetheless suggests another explanation for the attitudes towards Mathematics and Statistics observed in our students. Confronted with having to take a maths subject, many students appear to want to postpone it for as long as their degree structure will allow, or otherwise enter the classroom with at least some level of apprehension. This effect is particularly noticeable in service courses. What can we then, as university teachers, do to get our students over the invisible barrier between anticipating an unpleasant 'maths' experience and actually doing the tasks?

In this chapter we present an approach we call *The Golden Arches of Teaching and Learning* which aims to overcome some of the challenges of service teaching in the context of a first-year quantitative methods course that includes a selection of introductory Statistics topics. We follow with a prior perception analysis based on an anonymous survey conducted in the first week of teaching. The survey is used to explore students' perceptions of Mathematics, and by extension, Statistics.

A recurring theme arising from successive administration of the survey centres on negative perceptions of Mathematics and Statistics, tempered with feelings of inadequacy and a lack of confidence in quantitative subjects in general. We close by presenting results of an end-of-semester analysis where comments from student evaluations are used to gauge the merits of our approach. The efficacy of *The Golden Arches of Teaching and Learning* is supported with results showing a greater satisfaction and a positive shift in attitudes towards Mathematics and Statistics and its role in both everyday and professional life.

2   The *Golden Arches*: Teaching Strategies for Learning

Students expect service courses, particularly those of a quantitative nature, to offer concepts and number-crunching tasks 'foreign' to them and 'irrelevant' to their degree and future career. Revising these assumptions is often difficult. Thus how the subject is taught is an important component and often requires teaching staff to adjust their own perceptions of service teaching as well. Beyond relevance, student evaluation surveys identify the main reasons underlying student dissatisfaction and anxiety about quantitative service courses to be course organisation, clarity and consistency of guidelines and feedback, and availability of support resources for students as learners (Nankervis 2008). Addressing these issues is particularly challenging in larger courses, which in our case translates into enrolment sizes between 400 and 600 students per semester.

The Golden Arches of Teaching and Learning (Fig. 1) have been developed to embody our perspectives on delivering a first-year quantitative service course and achieving 'success'. When viewed from a teaching perspective, we define success as student satisfaction and higher pass rates. From the student perspective however, success could simply represent passing the course with an added although possibly unexpected bonus of an enjoyable and relevant learning experience. The use of the phrase *Golden Arches* is in honour of a real-life example of the McDonald's

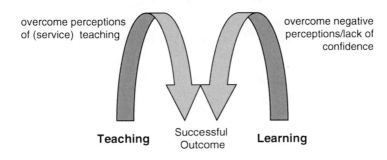

Fig. 1 *The Golden Arches of Teaching and Learning*

corporation used in lecture material, in one of many attempts to address student requests for topical relevance.

More specifically, *The Golden Arches of Teaching and Learning* are used to express the gap in perceptions of both teacher and learner, when aimed at a common focus. That is, the teacher often forgets the perspective of the student, usually for the reason they are now well-practised in the subject, whereas the student is often grappling with issues of comprehension, anxiety towards quantitative subjects and a lack of confidence in their own capabilities. By the same token, the student sees the subject as uninteresting yet challenging, delivered by a potentially unapproachable teacher who is perhaps incorrectly seen as a fount of effortless wisdom. In addition, when the student is not supported in their learning, such distorted comparisons are reinforced and may result in the student preventing themselves from achieving their own successful outcome.

An aid to bridging the gap between the two arches is a deeper understanding of the nature of maths anxiety and, by extension, reasons for student disinterest in the subject. There appear to be two clear barriers to retaining student engagement Thompson et al. (2012): procedural understanding, that is the fundamental analysis skills required to navigate through quantitative content; and contextual understanding, when a student needs to communicate the results of a quantitative analysis in non-specialist terms.

From the teaching perspective, the realisation and, most importantly, acceptance that a first-year service course is distinct from a first-year course within the teacher's own discipline, is a first step towards achieving a successful outcome as embodied in Fig. 1. In practical terms, we advocate:

- Creating and/or building upon a foundation of good course design that reflect the challenges of service teaching.
- Providing a solid structure of all aspects of the course.
- Reviewing the 'big picture' weekly to anchor the current topic within the overall course structure.
- Introducing real-life/realistic case studies to provide topic context.
- Recognising lack of confidence and general anxiety as likely reasons for negative perceptions students have of quantitative subjects.
- Removing the sense of remoteness between teacher and student by showing a genuine interest in the students' progress.

From the learning perspective, the emphasis is on gaining student trust, facilitating engagement, consolidation of knowledge and providing ongoing support. More specifically, we focus on:

- Creating, through energy and enthusiasm, a relaxed atmosphere conducive to learning.
- Having a 'conversation' with students in a lecture as opposed to lecturing at them.
- Being an easily accessible source of trust and confidence both in-person and online.

- Relating subject material to students' current life-stage (e.g. Smarties as data, analysing review distributions on iTunes and Amazon.com).
- Using discussion-provoking relevant real-life examples in audio–visual format.
- In-class exercises requiring student input and participation.
- Regular friendly reminders of opportunities for online and in-person help.

The discussion that follows is informed by an analysis of student questionnaire data collected after the above principles were implemented in a first-year quantitative methods course for Business students. In many respects, *The Golden Arches of Teaching and Learning* represents generally accepted tenets of good teaching. However, compared to a 'traditional' approach for designing an introductory course with Statistics content, our implementation of the *Golden Arches* is based on a subtle shift in emphasis which ultimately impacts positively on course delivery. Specifically, there is a deliberate shift away from Statistics and Mathematics content as perceived by discipline experts. Instead, more emphasis is placed on Business as well as other real-life situations in which Statistics and Mathematics can provide means of analysis or help identify effective solutions. In other words, the course becomes explicitly focused on the use of quantitative tools to answer questions of interest to students both as people and as future business professionals. We consciously move away from presenting material that appears to be about 'mathematical' tasks with no direct connection to 'real life', which we find to be a source of anxiety for many students. As will be shown next, this change in emphasis, coupled with varied and easily accessible learning support provided by dedicated teaching staff, has led to outcomes for students that were more positive than they would have otherwise expected.

3   Prior Perception Analysis

To gauge student attitudes, expectations and perceived abilities prior to the commencement of study, an anonymous and voluntary survey—*The Maths Anxiety Survey*—is administered in the first week of the course. We call it a 'Maths' survey, as we believe that students generally enter university without much awareness or understanding of distinction between Mathematics and modern Statistics. The survey, administered to three different student cohorts over three semesters, originally consisted of nine 5-point Likert-scaled items for quantitative analysis ranging from *Strongly Disagree* to *Strongly Agree*, interspersed with open-ended questions for qualitative analysis which will be discussed first.

A summative-based content analysis of responses to the first open-ended question was performed which involved the two authors independently identifying, coding and categorising themes from student responses. Interpretations were compared to reach a consensus on the set of codes to use. When responses included several different aspects, the responses were subdivided into distinct 'explanatory units', which were coded according to themes derived from author agreed-upon interpretations.

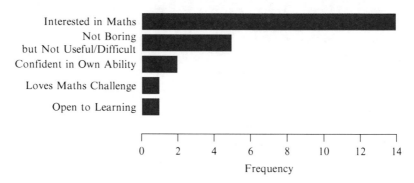

Fig. 2 Categories of prior perceptions (positive). The total sample size for all coded units is 23

The first of the open-ended questions within the survey, *If you tend to zone out in maths classes why is it? What could be done to help you stay motivated?* was asked to understand in the students' own words, the issues concerning lack of student engagement in maths. The responses fell broadly into two distinct categories: students who felt positive or ambivalent towards taking a maths subject versus those who voiced concerns. For those students who expressed a generally positive attitude towards maths, the following categories were identified, as shown in Fig. 2.

These results indicate that although it may be generally assumed Mathematics is not of interest to students as it is outside of their own discipline; there are still a number of students who look forward to taking a maths subject. Even so, it is not necessarily clear to the students that Mathematics will be of use to their chosen program of study.

In contrast, when examining the so-called negative responses (Fig. 3) lack of understanding and the perceived dryness of Mathematics are principal reasons behind 'zoning out' in class. It is interesting that in this case there is a marked difference between cohort responses—Cohorts 1 and 3 are predominantly similar in their responses however Cohort 2 stands apart in their perception of why they zone out (Fig. 3). Further investigation of the individual cohorts revealed that students who were enrolled in programs which obviously aligned to Mathematics (e.g. Commerce, Applied Finance) formed the bulk of Cohorts 1 and 3, whereas students in programs with tenuous connections to Mathematics (e.g. Tourism and Event Management, Human Resource Management) as well as respondents repeating the subject, tend to form the core of Cohort 2.

When asked what might help them stay motivated (Fig. 4), there was again a distinction between cohorts. Given the background of the students enrolled in Cohort 2, it is understandable why they believe it would be beneficial to have topics clearly explained in a step-by-step format (20 out 114 coded units), more examples and non-typical stimuli such as Audio and/or Visual teaching aids (11 out of 114 coded units). Interestingly, all cohorts noted the need to make the subject relevant to 'real life' (22 out of 114 coded units) and opportunities to interact with teachers and fellow students in class (20 out of 114 coded units) as best and most commonly

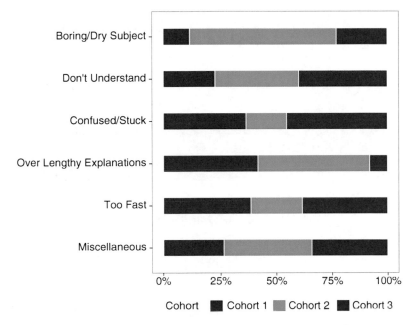

Fig. 3 Categories of prior perceptions (negative) regarding the behaviour of 'zoning out'. The total sample size for all coded units is 140

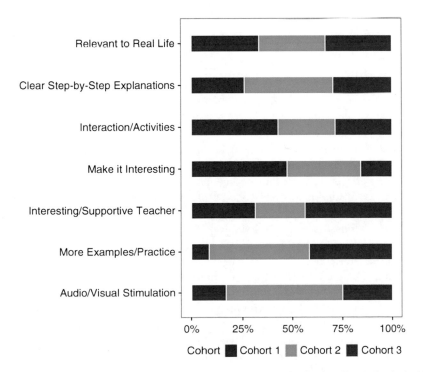

Fig. 4 Student suggestions to avoid 'zoning out'. The total sample size for all coded units is 114

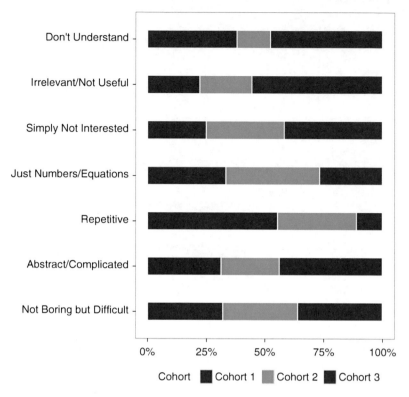

Fig. 5 Categories of prior perceptions for finding maths boring. The total sample size for all coded units is 116

cited solutions. Therefore, it seems that based on students' past experiences with maths, content and how it is taught is what matters most.

Since *I just find it boring* is a recurring sentiment, we asked a second open-ended question in the survey: *If you find maths boring why do you think that is? What could be done to help you stay motivated?* A lack of understanding seems to be the main reason behind student disengagement (28 % of total number of coded units) although an absence of relevance or general disinterest is also commonly cited (17 and 13 % of total number of coded units, respectively). As one student put it, maths is *just numbers and formulas and rules to follow*, which, according to another student, are *getting repeated over and over*. From the responses per cohort (Fig. 5), some differences between the groups are revealed.

In particular, Cohort 3 feels the most strongly about the irrelevance of Mathematics, tempered with a lack of understanding of the topic which is seen as abstract and/or complicated. Cohorts 1 and 2 provide largely similar responses, with the exception of Cohort 1 who more often cite repetitiveness whilst Cohort 2 is the least concerned with understanding the subject matter. Both of these groups are not overly concerned with the relevance of the course to their program of study. Finally, all cohorts agree equally that the subject is *Not Boring but Difficult*.

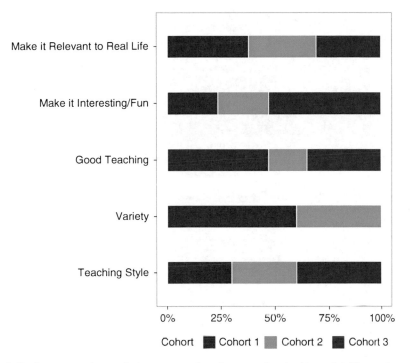

Fig. 6 Student suggestions to help overcome boredom associated with maths. The total sample size for all coded units is 78

Overall, the results seem to indicate many students come to university with a largely procedural knowledge of Mathematics without much conceptual understanding.

Interestingly, while some students may cite repetitiveness as the reason they 'zone out' or find maths boring, for others clear step-by-step instructions and practice exercises are means of becoming motivated (Fig. 6). This suggests a lack of understanding as the main reason behind maths anxiety, with repetitiveness as a way to cope or master the content that is perceived purely as a collection of number rules and formulae without much use outside of the classroom. Nonetheless, relevance to 'real life' and making the subject matter interesting are most commonly quoted as ways to promote interest (24 out of 78 coded units in both cases).

To complement the results of the content analysis, we analysed the quantitative responses to the 5-point Likert-scaled questions. Originally, the quantitative component of the survey consisted of nine questions (items); however, one of these questions was modified during the latter end of the study. For this reason, in what follows we will analyse only the eight questions that were consistently asked throughout the study.

The Maths Anxiety Survey results, broken down by survey question and cohort, are summarised in Fig. 7. The percentages of respondents who agree with each statement are shown to the right of the zero line; the percentages of respondents who

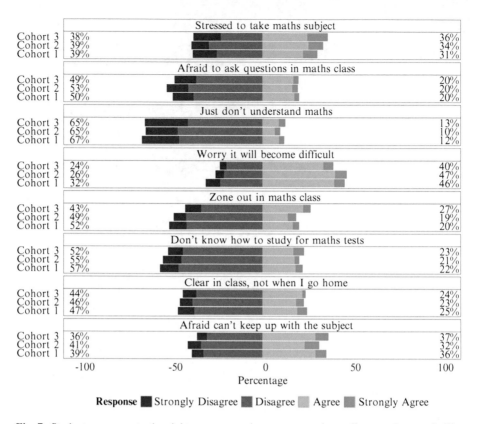

Fig. 7 Student responses to the eight survey questions represented as a divergent bar graph. The responses to the 'Neutral' category have been suppressed

disagree are shown to the left. The percentages for respondents who neither agreed nor disagreed have been suppressed.

For all three cohorts, the proportion of respondents who agreed with the statements is, with one exception, much lower than the proportion of respondents who disagreed. This indicates relatively low levels of anxiety with regard to maths overall, at least in relation to the dimensions explored by the current survey. The highest level of disagreement corresponds to the statement *I just don't understand maths*, suggesting reasonably high levels of confidence with maths, at least as it has been experienced by students thus far.

In contrast, the strongest agreement in Fig. 7 corresponds to the statement *I understand it now, but worry it will become difficult later*, indicating a degree of apprehension about the upcoming tasks and perhaps also some feelings of inadequacy when it comes to maths. For the two statements *I am stressed about taking a maths subject* and *I am afraid I will not be able to keep up with this subject*, there is an approximately equal degree of agreement and disagreement. Comparing responses by cohort reveals that Cohort 3 students were most likely to admit to

Table 1 Summary of rudimentary factor analysis statistics

Cohort	n	Cronbach alpha	Correlation matrix determinant	Bartlett's test of sphericity
Cohort 1	312	0.73	0.257	$\chi^2 = 418.1$, P-value $= 0$
Cohort 2	231	0.83	0.089	$\chi^2 = 548.1$, P-value $= 0$
Cohort 3	219	0.86	0.040	$\chi^2 = 692.4$, P-value $= 0$

zoning out and to have chosen 'Neutral' in response to *I understand maths now, but worry it will become difficult later*. Overall, however, there appears to be little difference in responses given by the three student cohorts.

Chi-square tests were performed in Minitab for each question to formally establish whether there was a significant association between student responses and student cohort. For all but one survey question, the hypothesis of independence could not be rejected at 1, 5 or 10 % confidence levels. The only statistically significant result was obtained for the statement *I understand maths now, but worry it will become difficult later* (chi-square statistic $\chi^2 = 17.046$, df $= 8$, P-value $= 0.03$). The largest contribution to the chi-square statistic came from the greater than expected number of 'Neutral' responses by Cohort 3, thus confirming the observation made earlier based on Fig. 7.

We have also investigated each student cohort using Exploratory Factor Analysis to obtain further insights into students' attitudes towards studying maths. An advantage of analysing the data per cohort is that we are in one sense validating the robustness of not only the factor solution, but also the applicability of the *Golden Arches* methodology to each group of students. Note that no transformations were applied to the data as the Skewness and Kurtosis statistics fell comfortably within the recommended range of +2 to −2 in each case, indicating no severe departures from Normality Hair et al. (1998).

Rudimentary statistics of the responses broken down by cohort are given in Table 1, including the values obtained for the Cronbach alpha test and the correlation matrix determinant test. Note that list-wise deletion was used in the case of missing values and that the sample sizes (n) reported in Table 1 reflect the number of complete cases per cohort.

We can see from Table 1 that all sample sizes are sufficiently large for factor analysis, in that each passes the dimension test requiring at least 20 observations per item Field (2009). The Cronbach alpha statistics range from 0.73 to 0.86, suggesting good internal consistency in the student responses in each round of the survey. It should also be noted that all inter-item correlations fell between 0.3 and 0.8 (not shown here, for space considerations) indicating that all items measured the same underlying characteristic Nunnally and Bernstein (1994).

Since Exploratory Factor Analysis generally relies on estimating factors from a correlation matrix of the data, we checked the correlation matrices for each cohort firstly for multicollinearity (Correlation Matrix Determinant in Table 1). All correlation matrix determinants were well above the minimum threshold of 1e-05 Field (2009), with the lowest at 0.040. We also examined the relevant statistics for

Table 2 Factors for Cohort 1

Survey question	Apprehension	Feelings of inadequacy	Zoning out
Stressed to take maths subject	0.41	**0.60**	
Afraid to ask questions in maths class		**0.50**	
Just don't understand maths		**0.82**	
Worry it will become difficult	**0.65**		
Zone out in maths class			**0.80**
Don't know how to study for maths tests	**0.54**	0.41	
Clear in class, not when I go home	**0.70**		
Afraid can't keep up with the subject	**0.76**		

Significant loadings are in bold

Bartlett's test of sphericity which yielded a significant P-value for each data set, indicating each correlation matrix was significantly different to the Identity matrix (in which correlations between items would be strictly 0). The results of Bartlett's test of sphericity indicate the existence of relationships between survey items in all cohorts Field (2009). The only preliminary concern were the Kaiser-Meyer-Olkin measure of sampling adequacy statistics for all cohorts, all of which were 'miserable' at 0.5, which we note is exactly the minimum accepted value Kaiser (1974). As will be later demonstrated, although the KMO statistics for each cohort are minimally acceptable, there will be sufficient communality in the final factor solutions to warrant the factor analysis of the data.

Exploratory factor analyses were conducted in R with Principal Component estimation and Varimax rotation for all cohorts. Scree plots for each cohort (not shown here for space considerations) were used as a guide for the number of factors to extract. Generally the scree plots indicated extracting between two and four factors. We ultimately selected three factors in each case. When arriving at this decision, we took into account the suggestion of the scree plots in conjunction with statistics as advised in Hair et al. (1998), including: (1) a minimum of 60 % cumulative percentage of variance explained by the proposed factor solution; (2) the interpretability of the factors (including cross-loadings); (3) an absolute upper limit on the number of factors extracted as one-third of the total number of items; and (4) the results of post hoc tests regarding the suitability of the proposed rotation method, as well as the size of the contribution of each factor item to the solution (communality). The results of the post hoc tests will be noted in due course in what follows.

The three factor solutions for the individual cohorts are presented in Tables 2, 3 and 4. In each case, the solutions explain 60 %, 67 % and 70 % of the variance in the data, respectively. Factor loadings less than 0.4 were considered non-significant due to the sample sizes involved and labels for each factor were assigned to reflect the items with the highest loading Hair et al. (1998).

In each case there is a general consistency across the factor solutions with slight variation due to items moving between factors. The three themes arising from the factor solutions suggest *Apprehension, Feelings of Inadequacy* and *Zoning Out*. The emergence of the *Apprehension* factor ties in with the results reported in Fig. 7

Table 3　Factors for Cohort 2

Survey question	Apprehension	Feelings of inadequacy	Zoning out
Stressed to take maths subject	**0.64**	0.41	
Afraid to ask questions in maths class		**0.90**	
Just don't understand maths		**0.60**	0.48
Worry it will become difficult	**0.76**		
Zone out in maths class			**0.94**
Don't know how to study for maths tests	**0.64**		
Clear in class, not when I go home	**0.66**		
Afraid can't keep up with the subject	**0.73**		

Significant loadings are in bold

Table 4　Factors for Cohort 3

Survey question	Apprehension	Feelings of inadequacy	Zoning out
Stressed to take maths subject	**0.62**	0.56	
Afraid to ask questions in maths class	**0.63**		
Just don't understand maths	0.41	**0.63**	
Worry it will become difficult	**0.84**		
Zone out in maths class			**0.94**
Don't know how to study for maths tests		**0.88**	
Clear in class, not when I go home		**0.58**	
Afraid can't keep up with the subject	**0.73**	0.42	

Significant loadings are in bold

as well as the results of Lyons and Beilock (2012) with regard to the anticipation of a mathematical task. On the other hand, the factor corresponding to *Feelings of Inadequacy* seems to encapsulate despondence and might explain why we see students effectively resigning themselves to struggling through the course content. Comments such as *I don't even know where to start in this subject* indicates just how helpless some students can feel. The third distinct factor, *Zoning Out*, encapsulates the commonly heard perception that quantitative subjects are 'boring'. Overall, these factors suggest an underlying complexity to the notion of 'maths anxiety' as experienced by first year university students in quantitative courses.

The average communalities of the eight items for the three factor solutions were 0.60, 0.67 and 0.70, identical to the respective cumulative proportion of variance explained by each model. Field (2009) indicates that for large sample sizes exceeding an average communality of 0.6 is desirable, while Costello and Osborne (2005) suggest that a communality range between 0.4 and 0.7 is common for the social sciences. Finally, to justify the use of Varimax (orthogonal) rotation, for each cohort we determined the component score covariance matrices of the three factors, with each yielding a 3×3 Identity matrix, supporting the use of an orthogonal rotation method Field (2009).

4   End-of-Semester Analysis

At the end of semester, students are also asked to respond to an anonymous and voluntary *Course Experience Questionnaire*. This questionnaire includes a set of 5-point Likert-scaled items together with a question about best aspects of the course, as well as an opportunity to provide further comments. Results for questions most directly relating to the objectives of the study presented here, broken down by question and student cohort, are shown in Fig. 8. In all three cohorts, the overwhelming majority of respondents either strongly agreed or agreed with the given statements. Therefore, in terms of gaining an understanding of the subject matter and being provided with a supportive learning environment, their experience in the course appears to have been very positive. Generally, the distributions of responses to each question were similar across all cohorts, with the highest proportion of students choosing 'Strongly Agree' in relation to staff showing genuine interest in teaching.

As a way to validate the effectiveness of *The Golden Arches of Teaching and Learning*, responses to open-ended questions included in the *Course Experience Questionnaire* were analysed in the same way as the responses to the *Maths Anxiety* survey presented earlier. The results of this analysis are summarised in Fig. 9.

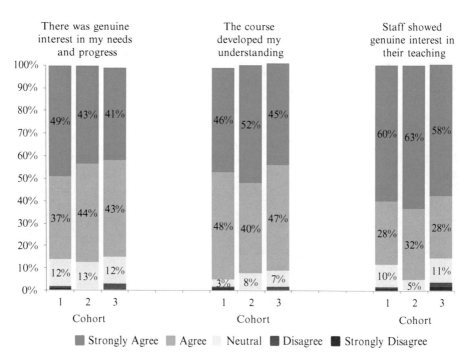

Fig. 8 Student responses to selected questions from the end-of-semester course evaluation questionnaire

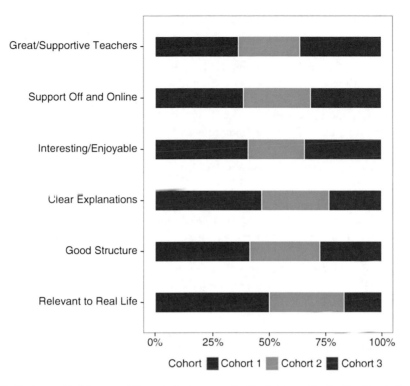

Fig. 9 Best aspects of the course from student comments. The total sample size for all coded units is 232

The responses per cohort reveal a number of notable differences. Cohort 2 students appear to represent approximately one third of each category, with the exception of *Interesting/Enjoyable*, which has been slightly less valued. In contrast, Cohorts 1 and 3 have placed varying degrees of importance on the nominated aspects of their course experience. Cohort 3 seems to have valued teacher quality, support resources and interesting content over good explanations or relevance to real life. Further, compared to results summarised in Fig. 4, Cohort 3 students maintain that having a good and supportive teacher is important. However, contrary to their initial expectations in the prior perception analysis, they appear to have valued an enjoyable experience in the course more than explanations and examples. Cohort 1 on the other hand appears to have appreciated good explanations, relevance to real life and good course structure more than the other two groups. All three cohorts appear to have placed equal importance on good teaching quality and availability of support resources.

It is worth noting that the quality of teaching and the level of support provided accounted for 53 % of all coded units. In contrast, relevance to real life, which upon entry into the course appeared to be the key to motivating students and as such was one of the considerations in course material redesign, was much less frequently cited (8 % of all coded units). Nevertheless, we believe that exposing students to real-life applications indirectly contributed to making the course more interesting and enjoyable. The course being interesting and/or enjoyable was in fact frequently mentioned in response to 'Any other comments?' and accounted for 14 % of all coded units. Therefore, an approachable, friendly and supportive teaching team wins the day, and fun, enjoyment and better understanding of the applications of quantitative methods to personal and professional life are simply an added bonus:

> Thank you for providing a terrific learning environment. I have never been a lover of math but I genuinely had fun in this course and felt like my opinions and questions were being heard and answered! (Student comment)

Finally, to investigate whether there was any change in the overall pass rate following the introduction of the *Golden Arches* approach, we conducted a hypothesis test on pass rates pre- and post-implementation, which were 71 % ($n = 1059$) and 77 % ($n = 1070$), respectively. The difference between these two proportions was statistically significant (P-value $= 0.00$), intimating a marked improvement in student results and thus, in terms of *The Golden Arches of Teaching and Learning*, both teacher and student success.

5 Conclusion

Service courses provide many challenges for both students and their teachers. To help students succeed, a lecturer must first acknowledge and then address the maths anxiety. *The Golden Arches of Teaching and Learning* embodies a teaching approach that can lead to delivery of an enjoyable and successful learning experience. Key to these strategies is a solid course design coupled with a genuine interest in the students' progress.

6 Note

Developed from a paper presented at Eighth Australian Conference on Teaching Statistics, July 2012, Adelaide, Australia.
   This chapter is refereed.

References

Costello, A. B., & Osborne, J. W. (2005). Best practices in exploratory factor analysis: Four recommendations for getting the most from your analysis. *Practical Assessment, Research and Evaluation, 10*(7), 1–9.

Field, A. (2009). *Discovering statistics using SPSS*. London: Sage.

Gordon, S. (2004). Understanding students' experiences of statistics in a service course. *Statistics Education Research Journal, 3*(1), 40–59.

Hair, J. F., Black, B., Babin, B., Anderson, R. E., & Tatham, R. L. (1998). *Multivariate data analysis* (5th ed.). Upper Saddle River, NJ: Prentice Hall.

Kaiser, H. F. (1974). An index of factorial simplicity. *Psychometrika, 39*, 31–36.

Lyons, I. M., & Beilock, S. L. (2012). When math hurts: Math anxiety predicts pain network activation in anticipation of doing math. *PLoS ONE, 7*(10), e48076. doi:10.1371/journal. pone.0048076.

Moore, D. (2005). Quality and relevance in the first statistics course. *International Statistical Review, 73*(2), 205–206.

Nankervis, K. (2008). Service teaching: Student experiences, issues and future directions at RMIT, Ed. *The RMIT Learning and Teaching Journal, 3*(1).

Nicholls, D. F. (2001). Future directions for the teaching and learning of statistics at the tertiary level. *International Statistical Review, 69*(1), 11–15.

Nunnally, J., & Bernstein, I. (1994). *Psychometric theory* (3rd ed.). New York: McGraw Hill.

Preis, C., & Biggs, B. T. (2001). Can instructors help learners overcome math anxiety? *ATEA Journal, 28*(4), 6–10.

Scutter, S., Luzeckyj, A., Burke da Silva, K., Palmer, E., & Brinkworth, R. (2011). What do commencing undergraduate students expect from first year university? *The International Journal of the First Year in Higher Education, 2*(1), 8–20.

Thompson, R., LeBard, R., & Quinnell, R. (2012). *In IISME forum on preparedness for first year mathematics: Issues and strategies for dealing with diverse cohorts.* Sydney, NSW, Australia: University of Sydney.

Part C
Undergraduate Learning

How Do Students Learn Statistical Packages?
A Qualitative Study

James Baglin and Cliff Da Costa

Abstract Technological skills are increasingly necessary for modern statistical literacy. The ability to operate a statistical package is perhaps the best example. Surprisingly, very little is known about the development of technological skills in statistics education and its impact on statistics courses through the acquisition of these skills. This chapter reports the qualitative findings of a mixed methods study comparing error-management training to guided training for learning to operate a statistical package in an introductory statistics course. Qualitative data was obtained from 15 semi-structured interviews exploring a range of topics, which included students' attitudes, confidence, emotions, difficulties, need for assistance, problem-solving and suggested improvements. Audio-recordings of interviews conducted face-to-face and over the telephone were transcribed verbatim and analysed using thematic analysis. The primary aim of the thematic analysis was to explore the overall student experience and, secondly, to compare the experience of students under the different training approaches, including the interaction with their statistical learning. The outcomes of the thematic analysis are discussed in terms of future research directions.

Keywords Technological skills • Statistics software • Statistical packages • Technology training • Error-management training • Guided training

J. Baglin (✉) • C. Da Costa
School of Mathematical and Geospatial Sciences, RMIT University,
Bundoora Campus East, Plenty Road, Bundoora, VIC 3083, Australia
e-mail: james.baglin@rmit.edu.au; cliff.dacosta@rmit.edu.au

H. MacGillivray et al. (eds.), *Topics from Australian Conferences on Teaching Statistics:* 169
OZCOTS 2008-2012, Springer Proceedings in Mathematics & Statistics 81,
DOI 10.1007/978-1-4939-0603-1_10, © Springer Science+Business Media New York 2014

1   Introduction

The use of technology in statistics education has rapidly evolved over the last couple
of decades, spurred on by the increased availability of cheap and powerful computing technology. In the 1980s and 1990s, statistics instructors were beginning to
realise the possibilities that computing technology might offer for improving the
teaching of statistics (e.g. Barr 1986; Holmes 1986; McKenzie et al. 1986).
Simulation (Stirling 1990), spreadsheets (Soper and Lee 1985) and statistical packages
(Lee and Seber 1990) were beginning to be extensively explored by the statistics
education community. Reflecting the positive outcomes of this work, statistics education reform in the USA during the early 1990s highlighted the use of technology
as a key recommendation for the improvement of statistics education (Cobb 1992).
Around the same time in the UK, many projects were funded to facilitate the use of
computers for the teaching of statistics in university courses (Bowman 1990). For
example, the Statistics Education through Problem Solving (STEPS) Consortium
developed computer-based learning material through a major grant awarded by the
Teaching and Learning Technology Programme (Bibby and Davies 1995; Bowman
and Gilmour 1998).

The use of technology in statistics education continued to flourish throughout the
late 1990s and into the twenty-first century in developed countries. Surveys from
the USA regarding reform efforts showed that technology was being implemented
in the majority of statistics courses surveyed (Garfield et al. 2002). Later surveys
showed that this trend would continue (Hassad 2012). Examples of the wide range
of technologies used today include statistical packages (e.g. *SPSS, STATA, Minitab,
SAS* and *R*), educational software, spreadsheets, applets, graphics calculators, multimedia material, Internet data repositories (Chance et al. 2007) and online virtual
environments (Bulmer and Haladyn 2011). Today, statistics instructors continue to
advance the use of technology in statistics education, even to the point where technology may very well shape the nature of statistical inference that is taught to future
students (e.g. Wild et al. 2011).

The core of technology's use in statistics education has been based on the belief
that it can be used to support student learning outcomes (Ben-Zvi 2000). When
used appropriately, this outcome can be achieved. However, there is more to technology use in statistics education. Technology skills may very well interact with
students' learning outcomes. Consider the widespread use of statistical packages.
Statistical packages include any software used for the specific purpose of carrying
out statistical analysis (Chance et al. 2007). Statistical packages are an excellent
tool for automating statistical calculations and producing graphical displays. In
addition, they are useful to instructors for demonstrating statistical concepts and
building practical research-based skills in students. The growing importance of
statistical package skills, or technology skills in general, relates to the changing
nature of statistical practice. Currently, technology is arguably inseparable from
modern statistical practice (Gould 2010), and as such, modern statistics courses
should be equipping students with the necessary skills to use this technology

(Nolan and Temple Lang 2010). As Gould explains, this will inevitably lead to some degree of "teaching the technology". This view contrasts with the conventional belief of some statistics educators to teach "the content, and not the tool" (Chance et al. 2007, p. 4). Regardless of an instructor's view on this issue, there is no denying the role of technology in statistics education and the fact that students of modern statistics courses will need to develop some level of technological skill alongside statistical knowledge. As such, a better understanding of how this skill development interacts with statistics courses becomes important.

Most studies in statistics education that discuss the use of technology focus on how technology can be used to enhance students' conceptual understanding. Studies that focus on understanding the development of statistics technological skill, e.g. statistical packages, are few. The studies that do exist focus on the impact of different training approaches on training transfer. Training transfer is evidenced by a student's ability to execute learnt skills outside of training situations (Hesketh 1997). There are two major types. Transferring the same skills covered in training is termed *analogical transfer* and adapting training skills to negotiate novel situations not covered during training is termed *adaptive transfer* (Keith et al. 2010). Adaptive transfer is considered the more important type of transfer as it promotes sustainable learning and the ability to approach new tasks that were not covered due to the brevity afforded by most training programs.

A meta-analysis of 24 studies by Keith and Frese (2008) evaluating the effect of different training approaches for the development of general software skills (e.g. word processors, spreadsheets and presentation software) compared guided training (GT) to error-management training (EMT) approaches. Guided training assumes that students are passive participants who progress through training by practising step-by-step, comprehensive instructions. The goal is to avoid operational errors which waste time. GT is based on the behavioural programme learning method developed by Skinner (1968). On the other hand, EMT is a type of active-exploratory training that uses a combination of minimal instruction to encourage exploratory behaviour and promote the positive functions of errors as valuable learning opportunities (Frese et al. 1991). EMT uses emotional control strategies for dealing with the negative emotions associated with errors and the use of minimal instruction. EMT achieves this by framing errors in a positive light through positive reinforcement by trainers and heuristics presented throughout training material, e.g. *Errors are a natural part of learning. They point out what you can still learn!* (Dormann and Frese 1994, p. 368). EMT assumes that students are active participants during training (Bell and Kozlowski 2008) and, therefore broadly reside within the constructivist theory of learning.

The results of the meta-analysis by Keith and Frese (2008) found that EMT was marginally better than GT for analogical transfer and moderately better for adaptive transfer. EMT has been posited to work through the development of self-regulatory skills, i.e. emotional control and metacognition. Self-regulatory skills have been defined as the ability to direct one's engagement in a task by controlling cognition, mood, behaviour and focus (Karoly 1993, p. 25). EMT's use of minimal instruction and emotional control strategies develops students' self-regulation throughout training more so than the passive environment of GT (Keith et al. 2010). When students are required to adaptively transfer their skills to novel situations, self-regulatory skills become important in predicting performance. EMT students are better off because they have been practising these skills throughout training.

A study by Dormann and Frese (1994) compared GT to EMT on training transfer performance for operating the statistical package *SPSS*. They argued that by minimising operational instructions, operational errors would be increased, which would lead to exploratory behaviour and to a deeper learning experience when compared to GT that used comprehensive instructions. Thirty psychology students were randomly allocated to a 2 h EMT or GT session covering data entry and correlation in *SPSS*. Immediately following this session, the students were given a task to evaluate training transfer performance. The results of the study found that the EMT condition significantly outperformed the GT condition on moderate and difficult (adaptive) transfer tasks. However, the study was limited by the use of a small sample, a single one-off training session outside of a real introductory statistics course, and the use of an immediate follow-up that ignored temporal stability of transfer. The positive outcome for EMT suggested by this study was not followed up with a subsequent study until nearly two decades later. This chapter reports on the qualitative component of a study investigating EMT compared with GT in learning a statistical package throughout a real introductory statistics course.

2   Background for the Qualitative Study

The qualitative data reported in this chapter formed part of a larger mixed methods study that evaluated EMT and GT for learning to operate a statistical package in an introductory statistics course. The quantitative results of the first phase are reported in Baglin and Da Costa (2012a). The outcomes of this phase are summarised to provide a context to the qualitative data and establish the exploratory nature of this follow-up analysis. In the quantitative phase, Baglin and Da Costa randomly allocated 100 psychology students enrolled in an introductory statistics course to either GT or EMT 1-h fortnightly *SPSS* computer laboratory training sessions. Topics covered included the basics of *SPSS*, frequencies, bar charts, cross-tabulations, Chi-square tests, correlation and regression. Computer laboratory sessions were supervised by tutors trained to implement either of the training approaches. Training material and exercises were delivered through a web-based assessment system similar to Blackboard. Each session would require students to work through exercises that introduced, practised and assessed their ability to use *SPSS*. This web-based material was manipulated to be delivered from either an EMT or GT perspective. The EMT approach encompassed minimal instruction, the promotion of errors, and the use of error-management heuristics. Tutors were trained to steer students back on track if they ventured too far off. They were also asked to avoid giving them direct instructions. The GT approach received comprehensive step-by-step instructions and screen shots that explicitly guided students through learning *SPSS*. Tutors in GT were instructed to assist students as much as necessary.

Self-assessment exercises in the final weeks of the first semester were used to measure training transfer. These exercises required students to answer randomly selected questions about the statistical analysis of a dataset using *SPSS*. It was assumed that training transfer would be evident in a student's ability to successfully analyse data in *SPSS* and answer the self-assessment exercises correctly. The tasks were set as self-assessment because no summative grade was given. Students gained a participation grade for attempting the exercises during their scheduled session.

In the final week of the first semester, students were also given a post-training questionnaire to measure the validity of training approaches' manipulations and other training outcomes including students' perceived training difficulty, satisfaction, self-efficacy and anxiety. Manipulation checks incorporated questions requiring students to self-rate the degree to which they engaged in behaviours consistent with their allocated training approach. Self-ratings were given on a Likert-scale ranging from (1) "strongly disagree" to (7) "strongly agree". For example, if training manipulations were valid, GT students would be expected to rate the use of step-by-step instructions and the use of assistance from tutors higher than EMT, whereas EMT would be expected to rate their level of exploratory behaviour and learning from errors higher than GT. The final question in this questionnaire invited students to participate in the in-depth qualitative interviews during the exam period before the second semester. In the first 2 weeks of second semester, 79 of the original participants were followed up to complete the same self-assessment exercises.

The quantitative results of the study found no statistically significant difference between the EMT and GT approaches on measures of analogical and adaptive training transfer both at post-training in the first semester and follow-up in the second semester. The quantitative results also found a lack of evidence from student self-reports supporting the validity of training approach manipulation. Only one manipulation check, the use of step-by-step instructions, was rated significantly higher on average in the GT approach when compared to EMT. No difference in self-reported ratings of exploratory behaviour and reliance on assistance from instructors was found. The quantitative study also found no significant difference between training approaches on other self-rated outcomes, which included perceived training difficulty, overall training satisfaction, *SPSS* self-efficacy, and *SPSS* anxiety. However, Baglin and Da Costa (2012a) identified a number of limitations that prevented drawing definite conclusions.

As Baglin and Da Costa report, issues with internal validity included unblinded participants, IT issues, limited computer laboratory resources, student non-compliance, and poor student engagement with the "self"-assessment exercises. These issues highlighted the challenges of embedding randomised experiments in real education settings. Baglin and Da Costa also reported possible problems with the manipulation of training approaches suggested by the lack of significant differences between average self-ratings of manipulation checks. The authors believed that time pressure (1 h sessions) may have negatively impacted the validity of the EMT approach as rushed students had little time to explore and experiment.

The validity of the self-assessment tasks as measures of training transfer were also called into question. As the tasks required students to make statistical decisions about the types of analysis to conduct in *SPSS* and because the self-assessment task scores were highly correlated with students' test and exam performance, the exercises may not have been measuring a student's *SPSS* training transfer.

The quantitative findings were inconclusive and suggested that different training approaches did not impact students' statistical package skills development. This raised the question as to what factor did. Inclusion of the qualitative phase to this study allowed further exploration of the quantitative results as well as the opportunity to explore the student experience of technology training from a more general perspective. This would help raise questions for future research. This chapter reports the results of the qualitative phase of the Baglin and Da Costa (2012a) trial. The primary aims were as follows:

- To document an in-depth exploration of the overall student experience of statistical package training
- To explore the students' perceptions of different training approaches used in computer laboratory sessions

An earlier version of this qualitative analysis was presented at the Eighth Australian Conference on Teaching Statistics in Adelaide, Australia (Baglin and Da Costa 2012b).

3 Method

The full study employed an explanatory sequential mixed methods approach (Creswell and Plano Clark 2011) which involved initially gathering quantitative data and then following up with qualitative data to explain the quantitative results. This method section will summarise the training sessions, training conditions and qualitative data collection process.

Computer laboratory training: Training was delivered in an introductory statistics course for psychology students that ran concurrently across two campuses. Course content covered exploratory data analysis, hypothesis testing, tests of association, correlation and regression. Contact hours included 2 h of lectures and 6 fortnightly 1-h *SPSS* training laboratories. Students were allocated to odd and even weekly computer laboratory sessions due to limited laboratory space. This fortnightly rotation was used to manipulate the GT and EMT approaches, i.e. GT was for odd weeks and EMT for even weeks. This alternating order was reversed for each campus to control for possible time effects. For example, computer laboratory attendance drops off during weeks when tests are scheduled. A time effect may have been created by having EMT or GT in only odd or even weeks across both campuses. Therefore, it was better to counter balance this effect across campuses. The *SPSS* training topics covered the basics, frequencies, cross tabulation,

Chi-square tests, correlation, regression and an *SPSS* self-assessment. Completion of each session contributed to a participation grade. The students had to satisfactorily pass (75 %) each session to attain their mark, but were permitted multiple attempts (formative assessment).

During computer laboratories, training material was delivered online in place of handouts using a proprietary online assessment tool known as *WebLearn*. Each training session introduced a research scenario and a set of *SPSS* exercises that would be used to analyse the scenario's data. Each exercise included *SPSS* operational instructions and an online auto-marking question whose purpose was to confirm that the exercise had been completed correctly (e.g. enter the mean for the treatment group). Each exercise was done several times as practice and reviewed in later laboratory sessions. A self-assessment lab was provided towards the end of the semester to challenge students to operate *SPSS* without instruction by answering statistical questions based on a research scenario and an accompanying data file.

The training delivered by *WebLearn* aligned with the training approach that each student was randomly allocated to at the beginning of the semester. The GT condition received comprehensive step-by-step instructions and screenshots that guided students through each training topic. Students were instructed to follow the steps as closely as possible and to avoid making errors. If students got into difficulty, they were instructed to seek assistance from the tutors that supervised each computer laboratory training session. Prior to the commencement of the semester, all tutors were trained for 1 h by the lead researcher to enable them to deliver either training approach. The EMT training condition received only minimal instruction, typically being the location of the *SPSS* procedure or command that would enable them to complete the task. For more difficult exercises, hints were also provided to help students work through difficult procedures. Students in the EMT condition were required to play around and explore each *SPSS* procedure to figure out how it operated. Therefore, there were no step-by-step instructions or screenshots. Error-management heuristics were embedded in the training material to help students frame errors in a positive light and learn from their mistakes, e.g. *Don't discount your errors. Acknowledge and learn from them.* Tutors were instructed not to guide students, but to help steer them back on track if they got into difficulty. Tutors were there to motivate, not to direct.

Interviews and data analysis: 15 interviewees out of 100 participants from the first semester (GT $N=9$ [Campus A $=5$, Campus B $=4$] and EMT $N=6$ [Campus A $=4$, Campus B $=1$]) volunteered to participate in semi-structured interviews. Interviews were conducted face-to-face and over the telephone during the first semester exam period. Interview questions covered a range of topics including attitudes towards training, confidence in operating *SPSS*, emotions experienced during training, training difficulties, assistance required, problem-solving and suggested improvements. All interviews were audio-recorded and transcribed verbatim. Qualitative data was analysed using a six-step inductive thematic analysis method described by Braun and Clarke (2006). The six steps included: (1) data familiarisation, (2) initial code generation, (3) theme searching, (4) theme revision, (5) theme definition and naming, and (6) reporting. Once the overall analysis had been completed, coded extracts for each main theme were compared across the different training approaches to consider possible moderating effects. Any differences in the theme trends between the approaches were noted.

4   Results and Discussion

Eight major themes emerged from summarising the qualitative data via the thematic analysis. A thematic map of these themes is provided in Fig. 1. Each theme will now be defined and discussed in the following sections. It's important to note that these themes were gleaned from a small sample of volunteers. While they cannot be considered representative of all students who went through the training, they do provide a starting point for the exploration of technological skills in statistics education and the establishment of future research questions. Participant quotes are labelled using identification codes (e.g. EMT—14 refers to interviewee 14 from the EMT approach).

1. *It has utility*: This theme referred to the students' perceptions of the utility of training. Almost all volunteer interviewees, regardless of approach, agreed that learning to use *SPSS* was important for their future academic careers and it would make doing statistical analysis easier:

   > Because it [*SPSS*] makes it easier in the future if we have lab reports and stuff like that without having to manually input the data and make up our own graphs; the system will do it more accurately than I guess we would. [GT—2]

   This was good news as instructors typically spend a lot of time justifying statistical package utility, not to mention the need to learn statistical concepts itself. The participants appeared to have recognised the importance of learning *SPSS*. However, a closer inspection of this theme revealed an interesting trend.

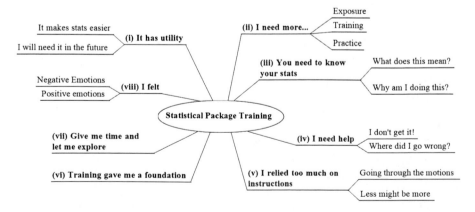

Fig. 1 Map of thematic analysis

Students continually referred to the "future" applicability of this skill. They did not appear to see its current relevance. One student questioned whether these skills could wait until later in their degree.

[E]ventually we'll start to need to know how to use all this researchy [sic] stuff. And when we do […] experiments we need this, so I think it'll be good for the future, but I think it would better if we had this later on instead of now, I think. [EMT—15]

This was an important insight into the motivation of students and the likely level of engagement that they will exhibit during training. Perhaps more effort is needed to enhance the immediate perceived relevance of statistical package skills. For example, if these skills were necessary to complete data analysis projects early in the course. There is no denying that utility is an important determinant of attitude and motivation (e.g. Venkatesh et al. 2003). Future research is needed to understand how students' appraisal of the utility of statistics technology impacts skill development in statistics education and identify methods that can be used to improve perceived utility should utility prove to be an important factor.

2. *I need more* … : Almost all participants expressed the need for further exposure, training and practice using *SPSS*:

I think if we had more labs, that would be very helpful, because that would give us more exposure to the actual product. Because, outside of the labs, we don't really use *SPSS* or we can't really see it at home, but if we had more exposure to it, I think we would learn it a lot better [GT—5].

This theme was also consistent across training approaches, with neither approach being more or less likely to express this need. Students suggested that more training sessions would have been beneficial, and some participants proposed embedding *SPSS* demonstrations into lectures, laboratory sessions or tutorials to increase exposure and familiarity of the package.

I suppose maybe with the tutor showing us how to do it first, rather than just using the instructions and getting their help if needed.[GT—9]

Doing so might also help students to see a stronger link between statistical concepts covered in lectures and the exercises covered during training.

When discussing the need for further practice, a few participants raised the inaccessibility of *SPSS* from home as being a major limitation, or as one student explained:

I don't have access to it [SPSS] anywhere else so that's kind of the only practice I got with it and I don't think that's enough. [EMT—10]

The price of a personal licence for many industry-based packages is a barrier to students. Students in this study only had access to *SPSS* on campus computers through a site-based licence. Lack of external access may have decreased the effectiveness of training as students had little opportunity to consolidate their skills between laboratory sessions. Personal access would provide valuable practice opportunities outside of regular training.

The eventual goal of statistical package training or any technology training in statistics education should be to provide students with the necessary skills and dispositions required to master the technology and make it a part of the students' regular repertoire of technology skills. A statistically literate person should be as conversant with the ability to operate a statistical package as a computer literate person is with using a word processor. Access can present a major barrier to this eventual goal. Perhaps instructors have grossly underestimated the importance of access and its impact on students' skill development. It's interesting to consider if free packages, such as R, promote skill development outside regular course activities. However, while R is considered accessible in terms of its free licence, the accessibility of its user interface is not on par with a package like *SPSS*. Understanding the impact of all types of accessibility, including personal access and ease of use, on the development of technological skills in statistics education should be another objective for future research.

3. *You need to know your stats*: This theme consistently appeared when discussing students' experiences during training. Participants from both EMT and GT acknowledged a strong dependency between understanding statistical concepts and understanding the *SPSS* training:

> I think maybe because my confidence for maths and statistics anyway is pretty low, I was just like 'I don't understand this, so how am I going to understand the program? [EMT—10]

Another participant explained how a strong understanding of the content made training easier:

> I think I have a [sic] better confidence than my friends, but I think that's mainly because I have a grasp on the actual theory behind it, rather than just the steps. [GT—3]

Participants also talked about their difficulty linking the training with their lecture content. They sometimes failed to understand not only what they were doing, but more importantly putting it all together to understand why:

> I especially realised when I completed the quizzes, the self-review quizzes, how much I didn't actually understand it. I sort of just basically learned how to follow the steps but I didn't have a good foundation of understanding as to why I was doing it. So when I was given the task of doing it without the steps I realised how much I didn't get it. [EMT—12]

This finding was not surprising having been observed by Baglin and Da Costa (2012a), but it does have a very important implication. It implies a strong dependency between knowledge of statistics and the ability to operate a statistical package. As Chance et al. (2007) explains, introducing technology too early can overwhelm students who are still developing their understanding of statistical concepts. Students can become lost in the technology and lose sight of the bigger statistical picture:

> I sort of really didn't get the lectures and then I'd go to the lab and I felt like it was completely different and I didn't really get that. [EMT—12]

The views expressed by the interviewees evidently suggest that statistical understanding must precede the development of statistical package skills. However, delaying the introduction of technology is probably not a feasible option given the importance of fostering these skills. Statistics is increasingly been taught with the aid of technology to reduce computational burden and to enhance a focus on a student's conceptual understanding (American Statistical Association 2005). When does a student have a suitable level of statistical knowledge to not feel overwhelmed by statistical technology? Can instructors wait that long given the time constraints of most courses? Are feelings of being overwhelmed normal and something that just needs to be worked through? How long does it take to do so? Will we degrade students' attitudes towards statistics before this happens?

This dependency between statistical knowledge and the skills required to use statistics technology requires further investigation. The first question that needs to be addressed is whether statistical knowledge and statistics technology skills can be meaningfully separated at all? Where do students' statistical knowledge stop and their technological knowledge begin? How do knowledge and skill interact? Can both be used to complement each other or is their development linear, i.e. knowledge first, skills later? Does this interaction change over time as students' gain more experience? Moore's (1997) conjecture that effective learning emerges from the right balance and alignment of content, pedagogy and technology reiterates the importance of finding answers to these questions.

4. *I need help*: The majority of participants reported seeking assistance during the lab training sessions mostly from the supervisor and sometimes from their peers. The degree of reliance on assistance varied between students:

> I asked for help straight away, which is probably not a good thing because I could have worked it out for myself but I'd just sort of look at it and think "ok that doesn't marry up" and then I'd look again quickly and freak out a little bit and then put up my hand and the teacher would come over [GT—2].
>
> I just asked, I didn't even bother trying to figure it out myself because the one experience I did have of trying to fix it myself I made it worse. So I learnt and put my hand up and [the tutor] would be like "I'll be with you in a minute" [GT—1].
>
> I asked the tutor sometimes, when I was really stuck. If I was just kind of stuck I'd still try to do it myself. But only if it was a really difficult situation, I'd ask the tutor [EMT—15].

The first and second quote reflects a disposition in trainees that should not be reinforced. These students were clearly not engaged with training and perhaps the easy access to assistance from the supervisor enabled or exacerbated this poor engagement. The third quote is more aligned with a desirable work-ready disposition. This student persisted in the face of difficulties, but knew when it was time to seek help. The most common reason stated for seeking assistance was to help identify where participants had gone wrong and when participants didn't understand the exercise or didn't understand the output:

> And a lot of the time, when you have to look at tables and stuff, I wouldn't know which number I was meant to be looking at. So I guess it was more to do with the theory. [GT—3]

An apparent difference between the volunteer EMT and GT interviewees emerged in this theme. The GT respondents were more likely to seek immediate assistance for the problems they faced. Respondents from the EMT condition were more inclined to identify themselves as "problem solvers" who would only seek help after first giving it a try.

I asked a couple of questions, of tutors just when I had no idea what I was doing and had tried about a million times [EMT—10].

No, look I'm a problem solver, so I really wanted to try and do it myself, so I only asked for help from the tutor as a last resort [EMT—11].

Not really. The first few I sat by myself and did it all by myself. I asked the tutor one or two questions but mainly just worked it out myself [EMT 13].

This difference between approaches in reports of seeking assistance suggested the interviewees may have been adopting behaviours largely consistent with their allocated training approaches. However, this difference suggests a bias in respondents as it was not consistent with manipulation checks reported in the quantitative phase (Baglin and Da Costa 2012a). No statistically significant difference was found between mean self-reported ratings of agreement to a questionnaire item that asked students whether they sought immediate assistance when faced with a problem.

Understanding the difficulties and assistance required by students when learning to use statistics technology provides insight into how these skills are developed. Some students feel the need for close supervision, while others are happy to carry on in a largely self-directed manner. Eventually, all students must develop the ability to transfer skills outside of the training environment without external assistance. Further research is needed to identify the most effective methods for developing this ability in students.

5. *I relied too much on instructions*: When asked about how they managed the self-assessment tasks at the end of the semester, many interviewees from GT talked about how difficult some of the tasks were after their instructions were taken away:

Yes. When we didn't have the exact instructions I was a bit lost, so that just said to me I relied too much on the instructions before. [GT—2]

I've been going by the instructions, and when there was no instruction I found it really difficult, like I realised I hadn't really remembered how to do it on my own, and sometimes I could figure it out and obviously, that was fine, but then there were times when I just had no idea what I was doing. [GT—8]

I don't think I actually learned how to use it. I think I learned how to follow steps but if I were to sit down in front of the package now I could probably do one thing that we did in the first lab and then continued it throughout, comparing the means or something, but I could not do anything else because basically, what I found, all you were doing was looking at the steps and then just following it one step at a time, not as a whole. [GT—3]

A few students explained that they had developed an overreliance on the instructions which resulted in them just going through the motions during training:

I was just learning how to follow the steps and just try to get a sufficient amount of right answers to pass each time. [EMT—12]

With the training there was a little bit of step by step and, personally, I didn't THINK a lot about it. [GT—6]

As a result, one participant from the GT condition suggested using less instructions as a way for improving training [GT—2]. Another GT student proposed to use instructions initially and then stop them later in the semester.

I think the best way was what we did this semester—give instructions, follow certain instructions to do certain things and, after a while, just stop the instructions and see how the students go [GT—4].

However, it's not hard to imagine how unpopular this would be. Building on the student's suggestion, a better approach might be slowly easing the instructions off as the semester progresses.

Statistical package training, given time, should allow students to transfer their skills outside of training without the need for comprehensive instructions used in training. Too much instruction may prove to be a bad thing. Finding the right balance of instruction might be the goal, or as one interviewee explained when asked about the difficulty of training:

Not really, they sort of, I think they [the computer laboratories] were actually quite good. They were a good level, they weren't sort of like giving you exactly step by step, there was enough room to actually have a play around yourself I think. Yeah, I think they were actually at quite a good level, sort of that middle point where it wasn't too hard but it wasn't just take the steps [EMT—10].

Regardless, the balance will never be right for all students. One interviewee from EMT felt that even the removal of minimal instruction made the self-assessment tasks more challenging.

[Training] was positive when the instructions were there, like they were telling me what to do but, without it I don't think I can cope unless I have more training and get used to it more [EMT—14].

While it was clear that the students needed more practice in this study, too much instruction may be inducing dependency and disengagement during technology training. These qualitative results suggest that interviewees varied in their sense of reliance on instructions. Finding the right balance of instruction and guidance that maximises student engagement and training transfer remains a challenge. Perhaps methods that allow the student to self-adjust the level of instruction throughout training might prove more effective than a traditional "one size fits all" approach.

6. *Training gave me a foundation*: Participants discussing their level of preparation for the self-assessment task and use of *SPSS* outside of training had mixed perceptions about their ability to transfer their skills. However, a general perception that training had provided them with a basic foundation emerged. When asked if they felt they were ready to use *SPSS* beyond training, one student commented:

A little bit, at least I'm a little bit more familiar, but I wouldn't say that I would be confident in going into an assignment where I would be expected to use *SPSS* for a lab report. I think I'd struggle a little bit. [EMT—12]

This perception was mostly explained by the relative shortness of the training delivered in this course. At the very least, the training did manage to familiarise the students with the basic operations of the package and provided a foundation for future development.

When comparing the responses between approaches, interviewees from the GT approach were more likely to initially present as confident users, but then be quick to point out that they were confident only in the basics.

After the training I feel that I'm very confident in it, in the subjects we actually did I think I'm pretty good, but if I was asked to do something off that, maybe I would have a little bit of trouble—I'd have to find my way around but the basics I think I've got down pat [GT—5].

Respondents from the EMT condition were not as certain:

Well, I'm more confident than I was at the beginning, but I'm not very confident. [EMT—16]

Once again, this difference observed in this theme conflicted with self-reported post-training questionnaire items measuring statistical package self-efficacy in the quantitative phase. No statistically significant difference in mean ratings was found between approaches (Baglin and Da Costa 2012a). It is possible that interviewees from EMT were underestimating their ability, perhaps because active-exploration had allowed them to see a much bigger picture of *SPSS* when compared to GT. Interviewees from the GT approach may have been more confident about the smaller and more controlled "snapshots" of *SPSS* they had experienced, detached from the bigger, and possibly overwhelming, picture observed by those in EMT. Interviewees and participants from the different training approaches in both the quantitative and qualitative phases may have been measuring their confidence on a different scale.

7. *Give me time and let me explore*: Interviewees were asked how they went about solving problems that arose during training and a hypothetical question about how they would approach a novel statistical analysis not covered in training. Many participants reflected on an innate propensity to explore *SPSS* to solve their future problems:

I guess I just tried to do it a different way, just kind of cover every possible option of doing something. [EMT—12]

In terms of hypothetically figuring out how to do an analysis not covered during training, many participants were quietly confident they could figure it out for themselves if given enough time to explore:

Given a reasonable amount of time, yes. If I had time to sort of play with it and make mistakes, cause that's how I've taught myself with everything else on a computer is I've had time to sit there and put things in and try different things, yeah I think I could. It would take me time, but I would get there [GT—1].

Regardless of the underlying nature of the training approaches, many interviewees from GT reported using exploratory behaviours to solve problems that they faced:

I started playing around with certain things—example, if you gave me a certain question and I had no idea and I just started playing around and I actually got it, that actually helped me to learn how to get that [GT—4].

    Yeah, so you'd just try and apply a bit of logic; try and make an educated guess of what it would be and just go from there [GT—6].

It's just me—to learn on computers I just click every button to see what it does; that's how I learn, whereas when I'm doing a lab I'm not sure if I should do that because I may stuff up the test, so to me, if I'm doing it by myself ... I don't know—I get distracted [GT—7]

Students from both approaches reported using exploratory behaviour. Even in the presence of comprehensive instructions, many students appeared to be at ease with playing around with technology and exploring the technology on their own terms. In the previous "I need help" theme many students from GT reported a tendency to seek immediate assistance, while in this theme many GT interviews also reported using exploratory behaviour. It would be valuable to know what factors explain why some students choose to seek help while others appear happy to figure things out for themselves.

Given the open-ended nature of this question, it was surprising to find most students would first explore to see if they could figure out how to conduct a new analysis procedure. In retrospect, it was possible that this trend reflected student's inexperience with statistics and an attempt to find any solution that seems correct (Chance et al. 2007). However, the impression from the data was that the students were expressing a general approach to the use of technology.

While some instructors might be concerned about the thought of their students stumbling around a little trying to find the correct method, perhaps there is a way to take advantage of this innate propensity to explore technology. It's not entirely clear at his stage just how this could be achieved; however, future research along these lines might prove fruitful.

8. *I felt* ... : Participants reported experiencing a wide range of positive and negative emotions during training. The similarity in experiences between the approaches was strong. Regardless of conditions, training was a very emotionally rich environment. Negative emotions were mostly related to anxiety or a fear of failure:

Emotions? A bit of nervousness. A bit of an attitude of "what happens when I fail?" Fear that I won't actually understand the instructions and I'll have to constantly put up my hand for help. Just fear of really not understanding the questions, basically, and the instructions. [GT—4]

As training progressed, anxiety shifted towards feelings of frustration, stress and annoyance. In contrast, many other participants expressed positive attitudes and emotions towards training:

I kind of did enjoy it actually. It was fun trying to solve the damn things, even though it was difficult, but still, I liked it I guess. [EMT—15]

Some participants explained that their emotions helped them engage. When asked if their frustration was distracting, one participant explained:

No I wouldn't say distracting, I think it just gives me motivation to knuckle down and do it again. [GT—3]

Another interviewee answered when reflecting on the effect of their anxiety:

Yes, probably beneficial because it was more motivating and it sort of encouraged me to take my time and read it slowly and work out what I'm doing without just rushing ahead which caused the anxiety to begin with [GT—2].

Other participants explained how they used emotional control skills to deal with negative emotions as they arose during training:

It was just ... "I hate it [*SPSS*]" and then I'd go "this is ridiculous, I don't want to do this" and then sort of I'd have to talk myself into "well, you have to do it. Slow down, let's go back, let's have a look at why you've picked the wrong one" or, you know. And sometimes it was just that I had misread a step or skipped over a step, so I was thinking I was doing everything but I'd missed something in the instructions [EMT—11].

There didn't appear to be any perceivable association between EMT and the qualitative data extracts related to emotional control. This finding was supportive of the results obtained from the quantitative phase. Baglin and Da Costa (2012a) found no statistically significant difference in mean ratings of self-reported emotional control during training between EMT and GT approaches.

While the primary focus of training is to promote transfer, training should also be a time when students can develop their self-regulatory skills, i.e. meta-cognition and self-regulation. Training methods that promote the development of such skills alongside transfer may prove to be highly beneficial to students during their studies and later in their careers. Too much of anything can also be a bad thing. Understanding factors that lead to levels of emotions that interfere with skill development should also be explored. For example, some participants reflected on the negative impact of time pressure. This brings the discussion back to the issue of personal access and having the ability to practice outside of scheduled training sessions.

5 Limitations

A number of limitations to this in-depth qualitative analysis must be raised. The data presented was obtained from a larger mixed method study with a primary quantitative phase. Out of the 100 students that participated in the quantitative experiment, 15 (9 GT and 6 EMT) volunteered to participate in the interviews. The extent to which these volunteers represented the views and experiences of the remaining students cannot be reliably judged. It is possible that the views presented by the interviews may be biased and not reflective of the entire sample. This would provide a plausible explanation for why some of the quantitative findings from students' self-reported ratings differed from the findings of the qualitative analysis. Perhaps, students who volunteered for the qualitative study were those students who benefited the most from their respective training approach and were therefore more likely to report their experiences.

This study also disregards the baseline knowledge, attitudes and confidence of students that may have influenced their perceptions and experience of training. For example, how confident were the students with technology before the study commenced? By knowing something about these factors prior to training, a better understanding of the differences between interviewees from the GT and EMT approaches may have helped explain some of the differences in perceptions reported.

In addition, this study only considered the experiences and perceptions of learning to use statistical technology in psychology students. Statistics technology is taught to a wide variety of other disciplines and the extent to which the perceptions reported in this study are shared with students in other disciplines is unknown.

6   Recommendations and Conclusions

The results of this qualitative phase of a mixed methods study have been insightful. Being the first qualitative study to look specifically at the development of technology skills in statistics education, many interesting findings have emerged. The first point relates to the merit of employing mixed-method research in statistics education. Had only one method of research been employed an opportunity to gain further valuable insight into the quantitative study would have been missed. Sometimes the evidence obtained using both qualitative and quantitative methods did not converge, but in most cases a large degree of agreement was detected. This was mostly evident in the overall high degree of similarity in students' experiences of statistical package training between approaches. This overall shared experience provided further support for the quantitative phases' major finding of no difference between training approaches on the major outcomes.

The overall general themes that emerged from the in-depth qualitative analysis provided thought-provoking insight into how statistical package training is perceived by students. These overall themes are summarised as follows. (1) Students understand the future utility of statistical package training but an effort should be made to make this utility felt sooner. (2) Instructors should not underestimate the time required for students to develop a sense of proficiency with a statistical package. Providing access to the statistical package and increasing training opportunities is important. (3) A single course is unlikely to develop a sense of proficiency, but instead will lay a foundation to be built upon. (4) Students feel they need to better understand statistical concepts before getting the most out of statistical package training. (5) When training instructions are taken away, students realised they have unintentionally developed a reliance on them. This was reported to impact students' training transfer. (6) Students discussed a natural urge to explore and problem solve when confronting novel situations with statistics packages without instructions. (7) Training is an emotionally rich environment. Students who reported to be in control of these emotions seemed to benefit the most from training.

The role of technology in statistics education has advanced beyond helping students to understand statistics to being a necessary skill required to engage in modern statistical practice. The need for technological skills in statistics education will only continue to grow and statistics education must begin the process of understanding how these skills are developed and how these skills impact on statistics courses. This study demonstrates that, even at an introductory level, the situation is quite complex. Statistics education needs to continue its research efforts in this area so that students can equip themselves with the necessary skills required to engage in modern statistical practice.

7 Note

Developed from a paper presented at Eighth Australian Conference on Teaching Statistics, July 2012, Adelaide, Australia.

This chapter is refereed.

References

American Statistical Association. (2005). *Guidelines for assessment and instruction in statistics education*. Washington, DC: Author. Retrieved from http://www.amstat.org/education/gaise/GAISECollege.htm.

Baglin, J., & Da Costa, C. (2012a). An experimental study evaluating error management training for learning to operate a statistical package in an introductory statistics course: Is less guidance more? *International Journal of Innovation in Science and Mathematics Education, 20*, 48–67. Retrieved from http://ojs-prod.library.usyd.edu.au/index.php/CAL/article/view/ 5809/6529

Baglin, J., & Da Costa, C. (2012b, July). Students' thoughts and perceptions of training to use statistical packages in introductory statistics courses: A qualitative study. In MacGillivray, H. & Phillips, B. (Eds.), *Proceedings of the Eighth Australian Conference on Teaching Statistics*. Adelaide, SA.

Barr, D. R. (1986). Use of computers in teaching statistics. In *Proceedings of the Second International Conference on Teaching Statistics* (p. 157). Victoria, BC, Canada. Retrieved from http://iase-web.org/documents/papers/icots2/Barr.pdf

Bell, B. S., & Kozlowski, S. W. J. (2008). Active learning: Effects of core training design elements on self-regulatory processes, learning, and adaptability. *Journal of Applied Psychology, 93*, 296–316.

Ben-Zvi, D. (2000). Toward understanding the role of technological tools in statistical learning. *Mathematical Thinking and Learning, 2*, 127–155.

Bibby, K., & Davies, N. (1995). STEPS for learning statistics. *Teaching Statistics, 17*, 107–110.

Bowman, A. (1990). Two projects from the British "Computers in Teaching Initiative". In *Proceedings of the Third International Conference on Teaching Statistics*. Dunedin, New Zealand. Retrieved from http://www.stat.auckland.ac.nz/~iase/publications/18/BOOK2/B3-2.pdf

Bowman, A., & Gilmour, W. H. (1998). Computer-based learning in statistics: A problem solving approach. *The Statistician, 47*, 349–364.

Braun, V., & Clarke, V. (2006). Using thematic analysis in psychology. *Qualitative Research in Psychology, 3*, 77–101.

Bulmer, M., & Haladyn, J. K. (2011). Life on an island: A simulated population to support student projects in statistics. In *Technology Innovations in Statistics Education, 5*. Retrieved from http://escholarship.org/uc/item/2q0740hv

Chance, B. L., Ben-Zvi, D., Garfield, J. B., & Medina, E. (2007). The role of technology in improving student learning of statistics. In *Technology Innovations in Statistics Education, 1*. Retrieved from http://escholarship.org/uc/item/8sd2t4rr

Cobb, G. W. (1992). Teaching statistics. In L. A. Steen (Ed.), *Heeding the call for change: Suggestions for curricular action* (pp. 3–43). Washington, DC: Mathematical Association of America.

Creswell, J. W., & Plano Clark, V. L. (2011). *Designing and conducting mixed methods research* (2nd ed.). Thousand Oaks, CA: Sage Publications, Inc.

Dormann, T., & Frese, M. (1994). Error training: Replication and the function of exploratory behavior. *International Journal of Human-Computer Interaction, 6*, 365–372.

Frese, M., Brodbeck, F., Heinbokel, T., Mooser, C., Schleiffenbaum, E., & Thiemann, P. (1991). Errors in training computer skills: On the positive function of errors. *Human-Computer Interaction, 6*, 77–93.

Garfield, J. B., Hogg, B., Schau, C., & Whittinghill, D. (2002). First courses in statistical science: The status of educational reform efforts. *Journal of Statistics Education, 10*(2). Retrieved from http://www.amstat.org/publications/jse/v10n2/garfield.html

Gould, R. (2010). Statistics and the modern student. *International Statistical Review, 78*, 297–315.

Hassad, R. A. (2012, July). Faculty attitude toward technology-assisted instruction for introductory statistics in the context of educational reform. In *Proceedings of the International Association of Statistics Education 2012 Roundtable Conference: Technology in Statistics Education: Virtualities and Realities*. Cebu City, Philippines. Retrieved from http://icots.net/roundtable/docs/Thursday/IASE2012_Hassad.pdf

Hesketh, B. (1997). Dilemmas in training for transfer and retention. *Applied Psychology, 46*, 317–339.

Holmes, P. (1986). The impact of computers on school statistics teaching. In *Proceedings of the Second International Conference on Teaching Statistics* (pp. 194–196). Victoria, BC, Canada. Retrieved from http://iase-web.org/documents/papers/icots2/Holmes-3.pdf

Karoly, P. (1993). Mechanisms of self-regulation: A systems view. *Annual Review of Psychology, 44*, 23–52.

Keith, N., & Frese, M. (2008). Effectiveness of error management training: A meta-analysis. *Journal of Applied Psychology, 93*, 59–69.

Keith, N., Richter, T., & Naumann, J. (2010). Active/exploratory training promotes transfer even in learners with low motivation and cognitive ability. *Applied Psychology: An International Review, 59*, 97–123.

Lee, A. J., & Seber, G. A. F. (1990). The role of package driven statistics courses. In *Proceedings of the Third International Conference on Teaching Statistics*. (pp. 217–221). Dunedin, New Zealand. Retrieved from http://iase-web.org/documents/papers/icots3/BOOK2/B3-13.pdf

McKenzie, J. D., Rybolt, W. H., & Kopcso, D. P. (1986). A study of the interaction between the use of statistical software and the data analysis process. In *Proceedings of the Second International Conference on Teaching Statistics* (pp. 190–193). Victoria, BC, Canada. Retrieved from http://iase-web.org/documents/papers/icots2/McKenzie.Kopcso.pdf

Moore, D. S. (1997). New pedagogy and new content: The case of statistics. *International Statistical Review, 65*, 123–137.

Nolan, D., & Temple Lang, D. (2010). Computing in the statistics curricula. *The American Statistician, 64*, 97–107. Retrieved from http://www.tandfonline.com/doi/abs/10.1198/tast.2010.09132.

Skinner, B. F. (1968). *The technology of teaching*. New York: Appleton-Century-Crofts.

Soper, J. B., & Lee, M. P. (1985). Spreadsheets in teaching statistics. *The Statistician, 34*, 317–321.

Stirling, W. D. (1990). Teaching elementary inference with computer-based simulations. Proceedings of the Third International Conference on the Teaching of Statistics, 1990 (pp. 171–177). Dunedin, New Zealand. Retrieved from http://iase-web.org/documents/papers/icots3/BOOK2/B3-6.pdf

Venkatesh, V., Morris, M. G., Davis, G. B., & Davis, F. D. (2003). User acceptance of information technology: Toward a unified view. *MIS Quarterly, 27*, 425–478.

Wild, C., Pfannkuch, M., Regan, M., & Horton, N. J. (2011). Towards more accessible conceptions of statistical inference. *Journal of the Royal Statistical Society, 174*, 247–295.

A Comparison of First Year Statistics Units' Content and Contexts in a Multinational Study, with a Case Study for the Validation of ASSIST in Australia

Ayse Aysin Bilgin, Caterina Primi, Francesca Chiesi, Maria Virginia Lopez, Maria del Carmen Fabrizio, Veronica Frances Quinn, Tamas Gantner, and Petra L. Graham

Abstract The study of statistics has become widespread throughout many degrees around the world in many universities, as the emphasis on evidence-based decision making has gained momentum in the business world. Students' approaches to their learning bear significant weight over the skills and understanding that students acquire during their studies. Three distinct learning approaches have been identified by researchers over the last three decades: deep, surface (British Journal of Educational Psychology 46:115–127, 1976) and strategic (Educational Research Journal 5:18–28, 1990). The discrepancy between desired learning outcomes and the aptitude and skills that students of statistics acquire (e.g. International Statistical Review 63:25–34, 1995) is well documented but the underlying reasons for choosing different learning approaches in statistics has only been investigated in limited studies and only from the perspective of a student's demographics. It is therefore important to understand how unit and student characteristics might encourage students to utilise certain approaches, especially students who do not major in statistics.

A.A. Bilgin (✉) • T. Gantner • P.L. Graham
Department of Statistics, Faculty of Science, Macquarie University,
Building E4A, Room 515, North Ryde, NSW 2109, Australia
e-mail: ayse.bilgin@mq.edu.au

C. Primi • F. Chiesi
Department of NEUROFARBA—Section of Psychology, University of Florence,
via S.Salvi 12—Padiglione 26, Florence 50135, Italy

M.V. Lopez • M.d.C. Fabrizio
Department of Quantitative Methods and Information Systems, Faculty of Agriculture,
University of Buenos Aires, Av. San Martin 4453, Buenos Aires C1417DSE, Argentina

V.F. Quinn
University of Sydney, Sydney, NSW, Australia

Macquarie University, North Ryde, NSW 2109, Australia

H. MacGillivray et al. (eds.), *Topics from Australian Conferences on Teaching Statistics: OZCOTS 2008-2012*, Springer Proceedings in Mathematics & Statistics 81, DOI 10.1007/978-1-4939-0603-1_11, © Springer Science+Business Media New York 2014

The aims of the current chapter are therefore to provide a brief review of learning approaches, a detailed description of the multinational study and validation of the Approaches and Study Skills Inventory for Students (ASSIST) as a measure of the learning approaches utilised by a cohort of Australian students of statistics.

Keywords Learning approaches • Statistics education • Multinational • ASSIST

1 Introduction

The study of statistics has become an integral feature of tertiary education across multiple disciplines, and in many countries. Despite this, there is often a discrepancy between the learning outcomes desired by educators and the aptitude and skills that students of statistics acquire (Garfield 1995). The economic and cultural globalisation of higher education necessitates that students be proficient in their understanding and application of these statistical skills as the pressure for individual institutions to meet international standards increases (Marginson and Van der Wende 2007). In addition it has been documented that many developed countries will face shortages of graduates from mathematical and statistical sciences (Australian Academy of Science 2006). The primary reason for including statistics into curricula is to enable students to make judgments about data, or about data interpretation, using multiple tools (Gal and Garfield 1997; Cobanovic 2002). It is therefore imperative for research to determine the nature of barriers faced by students, especially students who do not major in statistics, and potential attenuating or accentuating variables within these relationships. Barriers faced by students may include their approach to learning, which may in turn be correlated with their success in a statistics course.

There have been a vast number of research projects carried out to understand the underlying reasons for different ways of learning (termed approaches). Although earlier research was mainly qualitative, later both qualitative and quantitative research followed. Marton and Saljo (1976a, b) first asked participants to memorise passages, finding that some students tended to focus on the general meaning of the passage, and others on specific words. Inferring from this evidence of a greater discrepancy in students' approach to learning, they asked these students open-ended questions, such as "What do you mean by 'learning'?" They found evidence for two distinct approaches that students possessed towards learning which were clearly associated with differences in the levels of understanding achieved: deep and surface. The concept of learning approaches now entails both a student's motives for learning and their resulting strategies for achieving this learning, and generally identifies a third approach termed strategic (achievement) (Biggs 1990). Students adopting deep approaches do so in order to understand and internalise concepts for later use, and therefore often interact more critically with subject content and endeavour to relate these concepts back to their prior knowledge and experience (Ramsden 1992). In contrast, those adopting a surface approach do not personally engage with the learning process, and focus on memorisation of concepts without attempting to relate them first (Marton and Saljo 1976a). The focus on evaluation and assessment within learning institutions has been seen to produce the third type of learner: individuals adopting strategic approaches in an endeavour to maximise their marks and comply with academic requirements rather than to holistically understand course materials (Biggs 1987; Entwistle 1991; Tait and Entwistle 1996).

The aim of this chapter is to describe the multinational project designed to explore characteristics of students' learning approaches across three countries and then to validate the ASSIST survey tool as being a good measure of students' learning approaches for the Australian data only. Providing a context we describe the similarities and differences between three first year service statistics units (courses) on which data are currently being collected. We restrict our examination to the contents being covered and the learning environments. The learning environments of the units being studied will be used for analysis in the second stage of this research as predictors of the students' learning approaches. We also describe the survey tools used in the project. To validate ASSIST in the Australian data, we apply factor analysis to determine whether the factors identified load appropriately according to the ASSIST model. Validation of ASSIST in the Argentinian and Italian data and other stages of this project are beyond the scope of this chapter.

The ethical aspects of this research were approved by Macquarie University Human Ethics Committee (Reference Number 5201100809) in 2011.

2　The Multinational Learning Approaches Project

The approaches to learning have been researched in Australian Universities, and the existence of these same three approaches is now well established (Scouller 1998; Bilgin 2010). The cultural background of students has been the focus of considerable enquiry in Australia (Ballard and Clanchy 1984; Donald and Jackling 2007; Kember 2000; Kember and Gow 1991); possibly because of a reasonably large proportion of (predominantly Asian) international enrolments.

In Italy, the Bologna process identified the need for "lifelong learning" practices, similar to the concept of the deep approach to learning, and caused an intense reform of Italian tertiary education (Jakobi and Rusconi 2009). Considering its relevance, the approaches to learning theory has yet to be well established in any Italian University to the authors' knowledge.

Salim (2006) established that the three approaches to learning also exist within Argentinian students, and found that students adopting achievement approaches had significantly higher grades than those adopting deep approaches. It was suggested that the large percentage of students adopting achievement and surface approaches was due to the focus of tertiary education in the sample being for job opportunities rather than a desire to learn more, as well as the structure of the course in rewarding this approach (Salim 2006). However, this sample was of biochemistry students, and was deliberately selected to over-represent both academically successful students and those who had failed repeatedly. A similar pattern may not exist for students studying statistics rather than a content-based subject, as success is contingent upon the thorough understanding of both practice and theory.

This multinational study is the first of its kind (to the authors' knowledge) that aims to explore the utilisation of learning approaches in statistical education and compare these findings in a multinational setting, not just looking at the student characteristics but also course characteristics.

The core research team (the first five authors of this chapter) for this study was formed in 2010 in Ljubljana, Slovenia during the Eighth International Conference on Teaching Statistics (ICOTS), "Data and context in statistics education: Towards an evidence-based society", after one of the team member's presentation on learning approaches. The team developed a research framework until 2011, and then started collecting data from students in Australia, Italy and Argentina. Since then two researchers from Turkey joined our research group and have surveyed around 500 students in six different universities in Turkey. In addition, one researcher from Vietnam who has surveyed nearly 700 students in his university has also joined the project. For the future we aim to add researchers from North America and Africa to our research team so that we can cover all continents except Antarctica. A main aim is to shed light on the factors underlying students' choice of different learning approaches in statistics units, so that we can better inform educators of statistics to be aware of these factors when they are designing their curriculum. Of course we would also like to provide guidelines to creating better statistical learning environments for all so that the societies we live in become more statistically literate.

Our research project has three stages. Stage one will cover understanding the learning context and unit contents as well as student characteristics by data collection from the students and from the course coordinators. We will document similarities and differences between unit characteristics: the content—what has been taught and how the units are delivered. We will also need to test the validity of ASSIST in each country to ensure that we are using a reliable tool. This is important before we can use ASSIST to identify learning approaches of students in these three countries. After validating ASSIST we will start developing models to identify characteristics of students and units for different learning approaches (stage two). Finally stage three will involve providing resources and recommendations for statistics educators on statistical units' designs, which enable deep approaches to learning. We envisage this study is a long-term study and expect to move to second stage within 12 months.

2.1 The Approaches and Study Skills Inventory for Students

The Approaches and Study Skills Inventory for Students (ASSIST) (Tait et al. 1998) was developed to assess students' approaches to learning across the three types using a five-point Likert scale for 52 statements relevant to learning. ASSIST aims to offer a mechanism through which educators and researchers across countries and disciplines can gain an understanding of the approaches utilised by students, and potentially the influence of contextual and personal variables on these approaches.

ASSIST has three parts. Part A includes six statements to describe "what is learning?" in students' eyes. Part B consists of 52 statements which are used to identify the learning approaches of students. Finally, Part C of the survey helps researchers to identify student preferences for different types of courses and teaching, that support understanding (related to deep approach) or that transmit information (related to a surface approach). This is done by using eight statements together with a question asking students how well they think they have been doing their assessed work so far. This survey tool is publicly available (Centre for Research on Learning and Instruction 1997), although it can only be used where the educational language is English. Therefore the core authors translated the survey into Italian and Spanish for the purpose of this study. The validation of ASSIST has been reported for UK data by Entwistle et al. (2000) and for Egyptian data by Gadelrab (2011). In Ireland, ASSIST has been used to assess students' learning approaches for accounting and science students (Byrne et al. 2002, 2010). As such it is of interest to validate ASSIST on the data from each country in this project.

2.2 The Demographic Survey

An additional survey was also developed to gather information regarding the demographics of students (e.g. gender, age, language spoken at home, their parents' educational background), the students' official university identification number (Student ID) so that it was possible to access their official final grade for the unit, their educational background (e.g. where they completed high school, what kind of high school they attended), their current circumstances (e.g. where they live, whether they work) and their future educational plans (e.g. whether they intend to enrol a higher degree). Finally after providing a brief description of the three learning approaches (deep, surface and strategic), the survey asked students to identify their learning approach for the statistics unit they were studying and write a few sentences regarding why they used this specific approach in this statistics unit. The full demographic survey used in Australia is provided in the Appendix. In Italy and Argentina minor modifications were made to the demographic survey to address differences in high school and tertiary education systems.

To be able to have an acceptable assessment of students' learning approaches in statistics, we surveyed our students towards the end of their study period so that they had been exposed to almost all of the concepts to be covered in the semester. By this time they would have been assessed in some aspects of their learning and they would have been given feedback on their assessment tasks.

2.3 The Learning Environment and Unit Characteristics in Three Countries

The characteristics of three first year statistics units offered in Italy to Psychology students, in Argentina to Agricultural Engineering and Environmental Sciences students and in Australia to mainly Business students will be compared in this section.

The characteristics are based on 2011 offerings of these units since the data collection from the students started in 2011.

In Australia, the focus of the research was on an introductory statistics unit within Macquarie University. Although not compulsory for all students, many degrees in the University have this unit as a prerequisite for further study, including Bachelor of Applied Finance, Bachelor of Business Administration, Bachelor of Economics, Bachelor of Marketing and Media, Bachelor of Biodiversity and Conservation, Bachelor of Marine Science and Bachelor of Medical Sciences (Macquarie University 2012a). Since there is no assumed knowledge, the course begins with an introduction to variable types, study designs and the relationship between a sample and population. Graphical and descriptive statistics are covered in detail, followed by probability and sampling distributions. Hypothesis testing and confidence intervals are a main focus of this course which includes methods for known and unknown variances. The second half of the course explores the concepts and applications of correlation and regression, which is followed by categorical data analysis. The final week is dedicated to the review of all the contents covered and a reminder of how the topics relate to each other.

The unit teaching team for 2011 consisted of three academic staff members, all with more than 10 years of experience, from the Department of Statistics. The teaching team was headed by a Senior Lecturer. There were more than 900 student enrolments which necessitated four lecture streams, each 2 h a week for 13 weeks. One of them was taken by the lecturer in charge, two of them by one academic staff member and the final one by the other academic staff member. The largest lecture class size within this course consisted of 334 seats tiered theatre.

As well as attending a 2 h lecture each week, students were required to attend a 1 h tutorial and a 1 h practical (both with up to 50 students per group) each week. Tutorials involved guided problem solving using pen and paper and manual calculations. Practicals helped students to learn how to solve these problems using a statistical software package. Both tutorials and practical classes started in the second week and continued until the last week. They were mainly run by higher degree students (i.e. Ph.D., Honours). In second semester 2011, there were nine tutors running 23 tutorial classes and seven practical demonstrators running 23 practical classes. Consultation times were offered by all academic staff in the department and covered most hours between 9 am and 5 pm each working day.

Assessments for the unit included online quizzes (15 %), three group-based assignments each worth 5 %, a class test run under exam conditions organised during tutorials just before the midsemester break (15 %), and a final examination worth 55 % (Table 1). Unless students were able to prove through a special consideration application (Macquarie University 2012b) that serious and unavoidable disruption to their studies resulted from an event or set of circumstances (i.e. illness) that prevented them from being able to sit the final exam on the allocated date and time, students were not given an opportunity to attempt to pass the examination again.

The course of interest in Italy was an introductory statistics course for psychology students of the University of Florence. Students are introduced to the concept of measurement and introductory statistics using examples and data from

Table 1 Assessment characteristics of units in Australia, Italy and Argentina

Country	Australia	Italy	Argentina
Online quizzes[a]	9 (15 %)	–	–
Assignments or quizzes[a]	3 (15 %)	1 (10 %)	2–12 (10 %)
Written exam[a]	2 (15 and 55 %)	1 (70 %)	2 (40–50 %)
Oral exam[a]	–	1 (20 %)	–
Attempts allowed for exams	1 with justification	5 no justification	4 no justification

[a]The numbers represent how many and the percentage means the weighting of the assessment tasks towards the final grade

Table 2 Face-to-face hours per semester in Australia, Italy and Argentina

Country	Australia	Italy	Argentina
Number of students	970	400	450
The lecture class sizes (seats available)	334, 320, 250, 216	250, 250	115, 115, 115, 110
Teaching team (lecturer + tutor + demonstrator)	4+9+7	4	4+8
Total face-to-face lecture hours per semester	26	40	32
Maximum number of students per tutorial	50	120	36
Total face-to-face tutorial hours per semester	12	20	42
Maximum number of students per practical	80	–	36
Total face-to-face practical hours per semester	12	–	6
Total face-to-face hours per semester	50	60	80

psychological literature and research. The course is designed to provide students with sufficient theoretical and practical knowledge of descriptive statistics and probability theory to then study hypothesis testing and confidence intervals, as well as descriptive statistics and other inferential statistical analyses. This course is compulsory for first year students. There were 400 students in the course in 2011 (Table 2). As with the Australian course, the teaching required a team effort and was headed by a senior lecturer with more than 10 years of experience. The teaching team consisted of only two academics who ran all of the lectures and tutorials. The course runs for 10 weeks, and consists of one 4 h lecture and one 2 h tutorial (with students working in groups) each week. The largest lecture class size within this course consisted of 250 seats (i.e. maximum number of students could be 250 but it is possible that there were less number of students in any given week in any given lecture time). Consultation hours were also offered to students for one-on-one help with exercises. Classes were based around the discussion of theoretical issues, followed by practical examples and exercises undertaken with pen and paper, rather than using computer packages in tutorials. The assessment for this course consisted of a group report (10 %), an ungraded assignment for providing students with formative feedback, and written (70 %) and oral (20 %) final examinations (Table 1). The tasks in these examinations consisted of solving problems (numerical answers) and open-ended questions in which students had to apply and explain concepts acquired during the course. In contrast to the Australian sample, students were allowed to sit the examination up to five times in the year.

The sample chosen in Argentina was from a general statistics unit which is compulsory for students studying towards Agricultural Engineering or Environmental Sciences Degrees. As for the Australian and Italian first year units the teaching team in 2011 consisted of several academic staff members and the team was headed by a Senior Lecturer with more than 10 years of experience. There were 450 students in the course in 2011. The course ran for 16 weeks, and consisted of one 2 h lecture per week, one 3 h tutorial (with up to 36 students per group) a week and two practicals a term. The largest lecture class size within this course consisted of 115 students. The first half of the course covered topics related to descriptive statistics, probability and random variables (particularly binomial and normal variables). The second half of the course introduced sampling distributions, hypothesis testing and confidence intervals for the mean and mean differences. In the final weeks of the semester, simple linear regression and categorical data analysis were taught.

The performance of the students was assessed through continuous evaluation (with assignments that were submitted in every class) and two midterm tests (four or five problem-solving exercises). If the students gained 70 % or above in the midterm tests, then they passed the course, they did not need to sit a separate final exam. If their performance during the semester fell below 40 %, then they failed the unit and they were not allowed to sit the final exam. Students with intermediate performances—achievement between 40 and 70 %—were required to sit a final, integrated examination, which consisted of multiple choice questions (Table 1). These students were given four attempts to sit the final examination.

The total face-to-face contact hours during the semester for the three learning environments are provided in Table 2. While Australian students had 50 h total face-to-face contact with academics, half of these were with junior academics (i.e. current higher degree students), and all contact hours in Italy (60 h per semester) and in Argentina (80 h per semester) with academic staff members. Future work will explore whether this aspect of the units had any relationship with students' learning approaches and success in statistics.

3   Validation of ASSIST in the Australian Data

3.1   Data Collection in Australia

In Australia in the second semester of 2011, students were surveyed in week 13 in their practical classes. Unfortunately, the practical attendance by that time was very low. Instead of 50 enrolled students per practical class, on average there were no more than 10 students in each class. Dr. Bilgin went to each practical class, introduced the study to students and distributed the survey forms. She left the class and surveys were collected by practical demonstrators. Although the total number of enrolments for the unit was close to 1,000, due to the drop in the practical class attendance, only 68 students returned a completed survey. This is not surprising in Australian higher education institutions. Students are usually asked to complete surveys all through the semester such as Learner Evaluation of Unit, Learner Evaluation of Teacher and educational research undertaken by academics or higher degree students. For a recent study, Dr. Bilgin only managed to get 4.9 % response rate from 10,000 randomly selected local students for an online survey after two reminders.

To be able to achieve a reasonable number of student responses for the two surveys, we repeated the survey in first and second semesters in 2012. At the time of writing, the data for the second semester 2012 was in paper form and not available for analysis. In total, we had 68 responses in 2011 and 67 responses in 2012 available for analysis. These 135 responses were used in factor analysis to verify the factorial structure of ASSIST in the Australian cohort.

3.2 Participants in Australia

The sample was evenly split between male (48.9 %) and female (51.1 %) participants. The average age of students was 21 (SD = 5.4) years. Ninety-three per cent of the students were aged 26 years or younger. One in five students identified themselves as international students (19.8 %). The international students were mainly from China (35 %) and other Asian countries (54 %). Although only 20 % of the students were international students, 52 % of sampled students indicated that they spoke a language other than English at home and only 59 % of the students stated that English is their first language. One-third of the students completed secondary education through a private or independent high school, and a further 18.5 % in Catholic high schools, while nearly half of the students (49 %) graduated from a government high school (including selective high schools—only 5 %). Forty per cent of the students attended coaching for more than 30 h in a year prior to starting university. For nearly 40 % of students (39 %) neither parent had a university degree, while one-third had both parents with a university degree (29 %), 13 % only the father and 13 % only the mother had a university degree. Only a few students did not know whether either of their parents had a university degree (6 %). The majority of the students lived with their parents (63.6 %), others lived in shared accommodation (14.4 %), with partner/husband/wife (8.3 %), alone (7.6 %), residential college (1.5 %) or in other accommodation (4.5 %). Sixty-one per cent of the students had a job during their studies where they worked from 3 h per week to 60 h per week. On average they worked 17.2 (SD = 12.6) hours each week.

Two-thirds of students indicated that they liked studying in general however, only 52 % of students stated that they liked studying mathematics in their high school years. The percentage of students who considered statistics to be useful for their future work (55 %) was slightly lower than the percentage of students who intended to enrol in a higher degree after completing their current degree (64 %).

Their self-identified learning approaches were for deep, surface and strategic approaches, 33 %, 44 % and 23 % respectively. The higher percentage of surface approach and the lower percentage of deep approach might be due to students' immaturity at university studies. Seventy-six per cent of the students provided an explanation for why they had chosen a certain learning approach to study their current statistics units.

Deep and surface approaches consist of four subscales each with four statements; therefore if a student chooses the highest possible value for each statement, the highest possible score for deep and surface approaches are 80. The strategic approach consists of five subscales each with four statements; therefore if a student chooses the highest possible value for each statement, the highest possible score for the strategic approach is 100. The mean deep, surface and strategic approach scores for the 135 students were 55.2 (SD = 10.4), 52.6 (SD = 11.7) and 68.1 (SD = 13.4), respectively. To be able to graphically display all three learning approaches so that they are comparable, scores were standardised prior to creating the boxplot in Fig. 1.

Figure 1a clearly shows that the distributions of the three learning approaches scores were very similar for this sample. Although there did not seem to be any relationship between surface and deep approaches and between surface and strategic approaches, it is visible in Fig. 1b that deep and strategic approaches had a positive linear relationship and they were significantly correlated with each other (Pearson correlation coefficient = 0.715, $p < 0.0001$), while the correlation between deep and surface (Pearson correlation coefficient = 0.03, $p = 0.73$) and surface and strategic (Pearson correlation coefficient = −0.01, $p = 0.89$) approaches were insignificant.

3.3   In Students Words

We have used word clouds to identify student stated reasons for choosing certain learning approaches in their statistical learning (Figs. 2, 3 and 4). We found relevant quotes for most observed words from students' responses to provide students' perspectives. There were 42, 37 and 24 student responses to open-ended questions on why they have chosen surface, deep and strategic approaches to their learning, respectively.

While the students' explanations for choosing a surface approach to their learning in statistics ranged from (Fig. 2):

> I have other important subjects to focus on at the moment, and only really need to pass stats.
>   Only need to pass this unit for science-only need some key information for science tests.

to,

> I find it easier to cope with statistics by focusing on what I have learned each week and sometimes I find it difficult relating different week's information to each other.
>   I like to be told exactly what to learn and where to find it. I don't think creative thinking in stats (for me anyway) will give me good marks

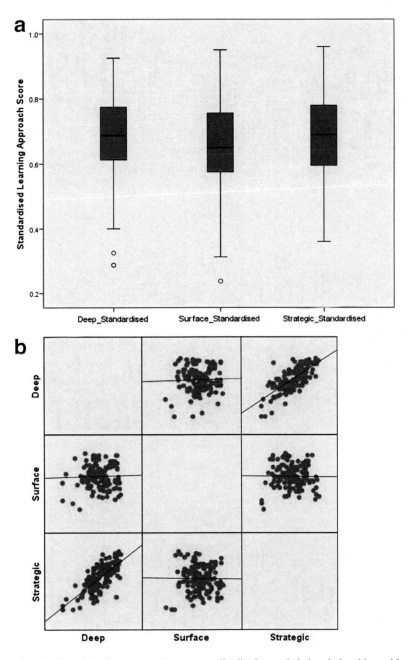

Fig. 1 Standardised learning approaches scores distribution and their relationships with each other for the Australian data set

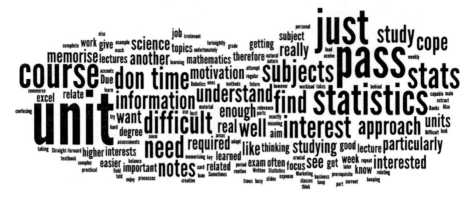

Fig. 2 Word cloud for surface learning approaches

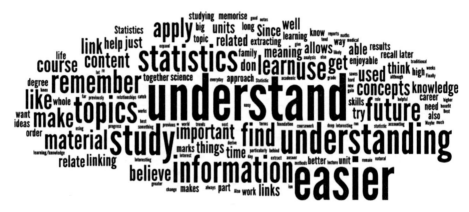

Fig. 3 Word cloud for deep learning approaches

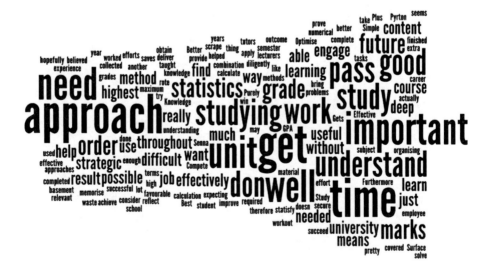

Fig. 4 Word cloud for strategic learning approaches

The reasons for them choosing a deep approach to their learning in statistics were more in line with the definition of the deep approach (Fig. 3), i.e. to understand and internalise the concepts:

> Understanding and linking concepts makes it easier for me to recall things, and to derive anything else. I'm more likely to remember something if I think it is meaningful.
>
> It's easy to remember things if you understand them in your own context link them and draw relationships between each other. It saves time and tends to stay in the memory longer.
>
> That is how I like to approach everything, the joy of learning.
>
> Because it's important I can apply my statistical analysis skills to other units now and in the future. Statistic has a big role in science (doing a medical science degree).
>
> If you extract meaning from what you are learning then it is much easier to understand.

Finally, the strategic approach to learning in statistics appears to have been made for obvious (strategic) decisions:

> I use this approach because without organising time diligently then I find it difficult to complete tasks and study. Furthermore, studying effectively means I don't waste time.
>
> Optimise the time then the work is finished and completed. Study method means all content is covered.
>
> Simple. Effective. Gets the job done. I don't have time to engage in the work. Plus, there is way too much content to engage with
>
> This was the approach was taught and used throughout my high school years. I believed it has worked for me which is why I continue to use it in university.
>
> The grade is really important to me as I need to get a really good grade as I'm a sponsored student. And the approach can help me to succeed in this unit.

In summary students stated that their main reason for choosing a surface approach was because their only aim was to pass the unit and the unit was potentially difficult for them to pass. Others wanted to understand the content of the unit because that made it easier for them to remember and relate what they have learnt to their future studies therefore they are choosing deep approach to their learning. The remaining students pointed out that they do not have enough time to use any other approach than a strategic approach in order to get high grades which was very important to them.

3.4 Exploratory Factor Analysis of the ASSIST Model

The purpose of carrying out an exploratory factor analysis on the learning approaches data was to reveal the underlying relationship between the three learning approaches (i.e. factors) and the component subscales of ASSIST without any prespecifications and restrictions on cross-loadings. The method used here involved a principal components method (PCA) for performing the exploratory factor analysis. We wish to validate ASSIST for the sample of Australian statistics students.

Component subscales loading high on the learning approach they represent with minimal cross-loadings are an indication of the good discriminant validity of the factors, and thereby reinforce that the ASSIST subscales efficiently measure the individual learning approaches of the Australian first year statistics students.

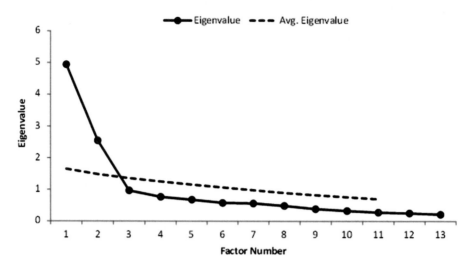

Fig. 5 Scree plot for the Australian data set

We began by examining the correlations between 13 subscales and found that most of the correlations were greater than 0.3 except for the correlations between surface approach subscales, and either deep or strategic approach subscales. The statements relating to the surface approach had weak correlations with the other two learning approaches. The value of the determinant was 0.002, which is small, but still acceptable since it is not zero and hence the correlation matrix can be inversed. The Kaiser-Meyer-Olkin (KMO) measure of sampling adequacy for this data set was 0.835, which is much great than required minimum (0.5) to be able to apply factor analysis. Furthermore, Bartlett's test of sphericity returned a highly significant p-value (<0.001), meaning that the off-diagonal entries in the correlation matrix were significantly greater than zero. Based on these findings, it can be concluded that it is worthwhile to carry out an exploratory factor analysis.

To decide how many factors to extract we used three measures:

1. A rule of thumb is to extract the factors where the eigenvalues are approximately greater than 1. The first four eigenvalues for this data set were 4.946, 2.550, 0.970 and 0.770 which suggests the extraction of two or three factors.
2. As a visual guide, we examined the scree plot (Fig. 5, solid line) to decide how many factors to extract. The elbow appears to be at the third factor, suggesting retention of two or maybe three factors.
3. We also applied the so-called "parallel procedure" first introduced by Horn (1965). In short, this procedure uses Monte Carlo simulation to generate eigenvalues a large number of times on the same sample size as the data set, averages them, and compares the actual eigenvalues to these standards. The cut-off is set where the average eigenvalue exceeds the actual eigenvalue, i.e. the variance is greater than the average variance in random samples. In our case, with $n = 135$ and $p = 13$, the first four average eigenvalues are as follows: 1.654, 1.487, 1.360 and 1.255. Our third actual eigenvalue was 0.970 and the third average eigenvalue

Table 3 Rotated component matrix

| | Components | | |
	1 (Deep)	2 (Strategic)	3 (Surface)
ST: Organised studying	0.431	0.744	−0.021
ST: Time management	0.233	0.865	−0.006
ST: Alertness to assessment demands	0.532	0.358	0.111
ST: Achieving	0.412	0.774	−0.041
ST: Monitoring effectiveness	0.607	0.447	0.012
SU: Lack of purpose	0.113	−0.320	0.720
SU: Unrelated memorising	−0.060	0.063	0.879
SU: Syllabus-boundness	0.023	0.002	0.787
SU: Fear of failure	0.036	0.117	0.774
D: Seeking meaning	0.772	0.256	−0.013
D: Relating ideas	0.785	0.123	−0.004
D: Use of evidence	0.799	0.166	0.085
D: Interest in ideas	0.607	0.420	−0.025

of 1.360 from the simulation is slightly exceeding this. As the cross-over point (Fig. 5, dashed line versus solid line) is closer to the third eigenvalue, the method of the parallel procedure argues for the extraction of three factors.

While, strictly speaking, the first two guidelines suggest the extraction of two factors only, we know that these guidelines are very rough and the third eigenvalue of 0.970 was very close to the cut-off value of 1. Lastly, the results of the parallel procedure also support the extraction of three factors rather than two.

The PCA approach explained nearly 65 % of the variation with three factors extracted and it is worthwhile to note that the first two factors only accounted for 47 % of the variance. As such, our decision to extract three factors appears further justified. The factor analysis distinctly separated the component subscales of the three learning approaches with most of them having a positive loading of above 0.7, although some subscales ("alertness to assessment demands" and "monitoring effectiveness") appeared to have some overlap between the strategic and deep approaches by loading more strongly on the deep approach (Table 3). It is interesting to note that these two subscales were also found to have inappropriate loadings in a research project that aimed to validate ASSIST on a sample of Norwegian undergraduate students (Diseth 2001). Diseth (2001, p. 382) argues that the subscale "alertness to assessment demands", which includes items focusing on the utilisation of feedback, is less applicable to the Norwegian sample as Norwegian students get little feedback during the semester and it more relates to studies beyond the first year. With regard to the subscale "monitoring effectiveness", Diseth (2001) notes that it is a related subscale, which might not be applicable in all contexts as is the case with the Norwegian sample, and it might be more relevant to graduate studies as noted by Entwistle et al. (2000 as cited in Diseth 2001). These arguments are likely to hold in relation to the factor analysis carried out on the Australian first year statistics students. Furthermore, the fact that unrotated loadings (not presented) of these two subscales were greater than 0.7 on the strategic approach supports the theoretical argument that they are primarily related to the strategic approach.

In summary, the exploratory factor analysis appears to validate the ASSIST model by loading appropriately onto the correct component for all but a couple of subscales.

4   Conclusion

The study of statistics has become widespread throughout many degrees in Australian universities, such that many statistical courses are now prerequisites for further study. The use of surface approaches within statistical study could be considered detrimental to both the student and the field, as only deep approaches award students with sufficient knowledge to progress and be capable of future statistical enquiry. However, to date there have been few empirical studies exploring the relationship between learning approaches and course outcomes in statistics courses. It would be of interest to determine if these relationships differ within students who are progressing towards a degree other than statistics, such as in psychology, biology, agricultural engineering or accounting, where statistical knowledge is essential but often overlooked. Comparisons of course characteristics and student demographics across countries might allow educators to gain a better understanding of the influence of contextual (i.e. course related) and demographic (i.e. personal) variables on students' learning approaches.

Research concerning individual characteristics of students and their relationship to learning approaches has limited applicability for academic reform, and therefore there is greater need for clarification of the role played by dynamic factors. As such, extensive tutoring prior to university entry (i.e. high school years), hours of work undertaken that is unrelated to what has been studied, language of education (i.e. native or second language) and features of the courses are of particular interest. It has been found that previous experiences and predispositions to learning shape a student's approach to learning in tertiary settings, with students who have come to rely on surface approaches preferring university materials that allow surface learning (Entwistle 1991). It would be of interest to determine how preparation for high school completion examinations, a set of examinations largely focused around content memorisation, prior to university admission would be related to the learning approaches subsequently utilised. The relationship between a student's hours of paid work and their learning approaches has also been of interest to researchers. Zhang (2000) found that work experience was positively associated with deep and achieving approaches, and negatively with surface approach. However this result may only hold for work relevant to the student's study, and could otherwise detract from the student's ability to utilise deep approaches to learning. Further work is therefore warranted to measure the hours a student works, the necessity of this work, and its relation to their learning at university, to shed light on this matter.

The comparison of the three units offered in Australia, Italy and Argentina, and the educational systems of the countries in which they are taught, offers researchers a perspective through which contextual and contents of the first year statistics units are taught. Along with demographic variables, these context variables might be able to give us an indication why students of statistics are using certain learning approaches in their studies.

Several differences documented here might help us to identify relationships between these variables and learning approaches. Practical classes, statistical packages and online resources are used in teaching the Australian sample, and it is of interest as to how these additions to statistical teaching may alter the approaches used by students. In light of the big push to Massive Online Open Courses (MOOCs), the findings might be of interest to the wider educational community.

The impact of class sizes could also prove to be important as the Argentinian classes consisted of far fewer students than were in both the Italian and Australian samples. The assessment styles of the three groups were similar in terms of their mix of intermittent assignments, group work and examinations, but with more emphasis given to the final examination in the Italian and during semester assessments in the Argentinian samples. In addition, more opportunities were given to students to pass the final examination without justifying why they should be given a second chance to sit it in Italy and Argentina. Using one-to-one oral final examination is unheard of in the undergraduate Australian higher education environment; due to the large numbers of enrolled students, it seems an unlikely option for future Australian undergraduate education.

By using the 135 responses from the Australian students, we were able to verify that ASSIST was a valid measure of students' learning approaches (deep, surface and strategic). We are applying similar analysis to Italian and Argentinian data sets to verify the effectiveness of the translated versions of ASSIST. The results will be presented in future publications.

We acknowledge that the Australian sample might be biased due to the timing of data collection and because we were unable to reach a reasonable proportion of students. Nevertheless, even with this small sample we were able to show the validity of ASSIST in this sample which may be useful for future Australian studies. In the future we might be able to increase response rates by surveying students in the middle of their studies and by utilising online tools.

A comprehensive study of the learning approaches utilised by students in (at least) these three countries, Australia, Italy and Argentina, as well as identifying the relationships between the learning approaches and student background variables and features of the learning environment, will potentially provide avenues for academic reform in statistics education by highlighting the needs for curriculum changes.

5   Note

Developed from a paper presented at Eighth Australian Conference on Teaching Statistics, July 2012, Adelaide, Australia.

This chapter is refereed.

Acknowledgments   The authors wish to thank all of the students who took part in this study by completing the survey and colleagues who allowed us to survey their students.

Appendix: Demographic Survey Used in Australia

MACQUARIE UNIVERSITY | FACULTY OF SCIENCE

Department of Statistics
Learning Approaches in Statistics Demographic Survey

Welcome to the **Multinational study on learning approaches variability of university students in Statistics courses research.** As a lecturer in the department of statistics, I would like to better understand our students' approaches to learning. This survey is used to collect demographic information as well as information on your approach to learning. Please complete the questions below, either by writing your answer or ticking the appropriate option. Any confidential information pertaining to individuals will not be released to anybody outside the research team. Only the aggregated results will be presented and/or published.

Thank you very much for taking the time to read this information

---

Student ID: _____

Q1. What is your birth date? _____/_____/_____

Q2. What gender are you?   ○ Female   ○ Male

Q3. Are you an international student?   ○ Yes   ○ No
  1. If yes, which country do you come from? _____
  2. How many years have you lived in Australia? _____

Q4. Do you speak another language at home? ○ Yes   ○ No
  If yes, which language do you speak? _____

Q5. Is English your first language?   ○Yes   ○No
  If no, what is your first language? _____

Q6. In which degree are you enrolled? _____

Q7. Which year did you finish high school? _____

Q8. What type of high school did you attend before university?
  ○ Government   ○ Private/Independent   ○ Catholic system
  ○ Government Selective   ○ Other, please specify_____

Q9. Where did you complete your schooling?
  ○ Sydney region   ○ Overseas - South East Asia
  ○ Other Australian capital/major city   ○ Overseas – Other (please specify below)
  ○ Rural Australian school   _____

Q10. Did you attend coaching for more than 30 hours in a year before you came to university (i.e. preparation for HSC)?   O Yes   O No

Q11. What was your university admission mark (ATAR in Australia) (if applicable)?

_____

Q12. Do either of your parents have a university degree?
O Yes, both   O Yes, only my mother O Yes, only my father O No O Don't know

Q13. What are your current living arrangements?
- O With parents   O Alone
- O With husband/wife/partner   O Residential college
- O Shared accommodation   O Other_____

Q14. Are you a full-time or part time student? O Full time   O Part time

Q15. Is this your first, second, third, fourth or more year at the university?
O (1)   O (2)   O (3)   O (4)   O More_____

Q16. How many units of study are you taking this semester?
O (1)   O (2)   O (3)   O (4)   O (5)

Q17. Will you have a job during semester?   O Yes   O No
        If yes, please answer the following questions.
- a. Do you have to work to support yourself?   O Yes   O No
- b. Will this job be on campus?   O Yes   O No
- c. Will you be using what you are learning or learnt at university for your work?
        O Yes   O No
- d. How many hours per week will you work (on average)?_____
- e. Do you think that work will negatively impact on your learning?
        O Yes   O No

Q18. Do you like studying?   O Yes   O No

Q19. Did you like studying mathematics in high school?   O Yes   O No

Q20. Do you consider statistics useful for your future work? O Yes   O No

Q21. Do you consider enrolling in a higher degree after completing your Bachelor degree?

        O Yes   O No

Q22. Which grade do you expect to achieve for this unit?

O HD or D   O Credit   O Pass   O Fail

Q23. Why have you chosen to study this statistic unit?

Q24. Is there anything else that you would like to add?

_____

_____

_____

_____

Q25. Please read the information provided below before you answer parts a and b.

Possible Approaches to Learning	Deep Approach	Surface Approach	Strategic Approach
What might be related to each	The intention is to extract meaning from what is learnt and relate ideas to prior learning. The course topics are linked to each other as well as linking to prior learning.	The intention is to cope with the task at hand. The course topics seems unrelated to each other. Memorise certain bits to answer questions in assessment tasks.	The intention is to achieve the highest possible grades by using organised study methods and good time-management. Monitor your study effectiveness to optimise the time to study.
a) **Which learning approach would you place yourself in Statistics Unit that you are studying?** **(Tick One box on the right)**			
b) **Why do you use this approach in this Statistics unit?** **(Please write a few sentences)**			

References

Australian Academy of Science. (2006). _Mathematics and statistics: Critical skills for Australia's future._ University of Melbourne. Retrieved Nov 28, 2012, from http://www.review.ms.unimelb.edu.au/Report.html

Ballard, B., & Clanchy, J. (1984). _Study abroad: A manual for Asian students._ Kuala Lumpur, Malaysia: Longman.

Biggs, J. (1987). _Student approaches to learning and studying._ Hawthorne, VIC, Australia: Australian Council for Educational Research.

Biggs, J. (1990). Effects of language medium of instruction on approaches to learning. *Educational Research Journal, 5*, 18–28.

Bilgin, A. (2010). Non-statisticians learning statistics. In C. Reading (Ed.), *Proceedings of the eighth international conference on teaching statistics: Data and context in statistics education: Towards an evidence-based society*, Ljubljana, Slovenia.

Byrne, M., Finlayson, O., Flood, B., Lyons, O., & Willis, P. (2010). A comparison of the learning approaches of accounting and science students at an Irish university. *Journal of Further and Higher Education, 34*(3), 369–383.

Byrne, M., Flood, B., Lyons, O., & Willis, P. (2002). The relationship between learning approaches and learning outcomes: A study of Irish accounting students. *Accounting Education: An International Journal, 11*(1), 27–42.

Centre for Research on Learning and Instruction. (1997). *Approaches and study skills inventory for students (ASSIST)*. Retrieved Nov 20, 2012, from http://www.etl.tla.ed.ac.uk/questionnaires/ASSIST.pdf

Cobanovic, K. (2002). Role of statistics in the education of agricultural science students. In B. Phillips (Ed.), *Proceedings of the sixth international conference on teaching statistics*, Cape Town, South Africa.

Diseth, A. (2001). Validation of a Norwegian version of the Approaches and Study Skills Inventory for Students (ASSIST): Application of structural equation modelling. *Scandinavian Journal of Educational Research, 45*, 381–394.

Donald, J., & Jackling, B. (2007). Approaches to learning accounting: A cross-cultural study. *Asian Review of Accounting, 15*(2), 100–121.

Entwistle, N. (1991). Approaches to learning and perceptions of the learning environment. *Higher Education, 22*, 201–204.

Entwistle, N., Tait, H., & McCune, V. (2000). Patterns of response to an approaches to studying inventory across contrasting groups and contexts. *European Journal of Psychology of Education, 15*(1), 33–48.

Gadelrab, H. F. (2011). Factorial structure and predictive validity of Approaches and Study Skills Inventory for Students (ASSIST) in Egypt: A confirmatory factor analysis approach. *Electronic Journal of Research in Educational Psychology, 9*(3), 1197–1218.

Gal, I., & Garfield, J. B. (1997). *The assessment challenge in statistics education*. Amsterdam: IOS Press & International Statistical Institute.

Garfield, J. (1995). How students learn statistics. *International Statistical Review, 63*, 25–34.

Horn, J. L. (1965). A rationale and a test for the number of factors in factor analysis. *Psychometrika, 30*, 179–185.

Jakobi, A., & Rusconi, A. (2009). Lifelong learning in the Bologna process: European developments in higher education. *A Journal of Comparative and International Education, 39*, 51–65.

Kember, D. (2000). Misconceptions about the learning approaches, motivation and study practices of Asian students. *Higher Education, 40*, 99–121.

Kember, D., & Gow, L. (1991). A challenge to the anecdotal stereotype of the Asian student. *Studies in Higher Education, 16*(2), 117–128.

Macquarie University. (2012a). *2013 Handbook: Undergraduate degrees and diplomas*. Retrieved Nov 28, 2012, from http://www.handbook.mq.edu.au/2013/DegreesDiplomas/UGDegrees

Macquarie University. (2012b). *Special consideration policy*. Retrieved Nov 28, 2012, from http://www.mq.edu.au/policy/docs/special_consideration/policy.html

Marginson, S., & Van der Wende, M. (2007). Globalisation and higher education. *OECD Education Working Papers, 8*, 1–86.

Marton, F., & Saljo, R. (1976a). On qualitative differences in learning, I—Outcomes and process. *British Journal of Educational Psychology, 46*, 4–11.

Marton, F., & Saljo, R. (1976b). On qualitative differences in learning, II: Outcome as a function of the learner's conception of the task. *British Journal of Educational Psychology, 46*, 115–127.

Ramsden, P. (1992). *Learning to teach in higher education*. London: Routledge.

Salim, S. (2006). Motivations, learning, approaches, and strategies in biochemistry students at a public university in Argentina. *Revista Electronica de Investigacion Educativa, 8*, 1–17.

Scouller, K. (1998). The influence of assessment method on students' learning approaches: Multiple choice question examination versus assignment essay. *Higher Education, 35*, 453–472.

Tait, H., & Entwistle, N. J. (1996). Identifying students at risk through ineffective study strategies. *Higher Education, 31*, 97–116.

Tait, H., Entwistle, N. J., & McCune, V. (1998). ASSIST: a reconceptualisation of the Approaches to Studying Inventory. In C. Rust (Ed.), *Improving students as learners*. Oxford: Oxford Brookes University, The Oxford Centre for Staff and Learning Development.

Zhang, L. (2000). University students' learning approaches in three cultures: An investigation of Biggs's 3P model. *The Journal of Psychology, 134*, 37–55.

Understanding the Quantitative Skill Base on Introductory Statistics: A Case Study from Business Statistics

Joanne Elizabeth Fuller

Abstract Basic mathematical skills are critical to a student's ability to successfully undertake an introductory statistics course. Yet in business education this vitally important area of mathematics and statistics education is under-researched. The question therefore arises as to what level of mathematical skill a typical business studies student will possess as they enter the tertiary environment, and whether there are any common deficiencies that we can identify with a view to tackling the problem. This paper will focus on a study designed to measure the level of mathematical ability of first year business students. The results provide timely insight into a growing problem faced by many tertiary educators in this field.

Keywords Statistics education • Business studies • Quantitative skills

1 Introduction

Traditionally the teaching of business statistics has received less attention from educators and has been generally perceived as "hard to teach". Yet these skills need to be taught if students are to be successful in their degree studies and in professional life. Students entering into business studies at a tertiary level will typically undertake a statistics course as part of the core curriculum. The importance of basic mathematical skills for success in such courses is clear. Students are, however, fearful in this area; anxiety that in many instances comes from a lack of confidence as to their prior level of mathematics skills. A dislike of mathematics is another factor. Though a common problem at all levels of education, for business studies students

J.E. Fuller (✉)
School of Economics and Finance, Queensland University of Technology,
Gardens Point Campus, Brisbane, QLD 4001, Australia
e-mail: j.fuller@qut.edu.au

H. MacGillivray et al. (eds.), *Topics from Australian Conferences on Teaching Statistics:* 211
OZCOTS 2008-2012, Springer Proceedings in Mathematics & Statistics 81,
DOI 10.1007/978-1-4939-0603-1_12, © Springer Science+Business Media New York 2014

mathematics is also frequently perceived as being completely redundant. Regardless of the underlying cause, the lack of basic mathematical skills represents a significant problem.

The challenges arising as a consequence of the lack of prior basic mathematical skill are immense and it is therefore vital that we have a clear understanding of the nature and scope of the problem (Wilson 1992; Wilson and MacGillivray 2007). The relationship between mathematics and statistical reasoning has been considered in the literature, for example see Garfield (2003), Moore (1997), Vere-Jones (1995) and Watson and Callingham (2003); though most research has focused on anxiety. More recent work, such as that of Wilson and MacGillivray (2005, 2006), has provided essential insight into the importance of numeracy to statistical reasoning. Yet an opportunity exists to examine these issues further, specifically in the context of business education. The questions of what level of mathematical skill a typical business studies student will possess as they enter the tertiary environment and whether there are any common deficiencies that we can identify are critical to being able to assess the extent of the problem and the development of future strategies. While the longer term objective is to raise the level of mathematics awareness and ensure that students do indeed begin their business studies with the requisite level of mathematical knowledge, the pursuit of this goal begins by assessing the current tertiary landscape.

The focus in this paper are first year business studies students. The results of a study designed to measure the level of mathematical ability of these students as they enter a compulsory introductory statistics course will be investigated. The details of the survey will be presented, followed by the results collected over more than 3 years. It will be shown that there are several areas of concern where students do not possess the requisite knowledge required for successful completion of the course. The consequences of these deficiencies will be considered with regard to the impact upon a student in being able to perform various statistical tasks. Possible strategies to conquer such deficiencies will also be discussed, including a workbook strategy, as well as ideas for further refinement and focus.

2   The Quiz

At the Queensland University of Technology (QUT) business studies students complete eight core compulsory subjects. The eight core subjects are designed to ensure that all students achieve a minimum standard of skills considered necessary for a successful career within the business environment, regardless of the major selected (i.e. majors include economics, finance, accounting, marketing, advertising, public relations, international business and management). Data Analysis is one of the core subjects and it provides a standard first year university statistics course, including descriptive statistics and inferential statistics as major streams within the 13 weeks of classes. With an enrollment of approximately 1,000 students each semester, the course attracts students from a wide range of backgrounds (for example, school leavers, as well as mature age students) and the level of mathematical ability is correspondingly broad. It is also important to note that students are able to gain entry to the course without having studied senior mathematics, as is now common-place for many tertiary level business courses.

Involvement with the course and therefore such a large and diverse group of students provided an opportunity to consider the level of mathematical ability of first year business studies students as they enter a compulsory introductory statistics subject. In 2007 a diagnostic mathematics quiz was developed as part of the Data Analysis curriculum. The quiz, developed in conjunction with the QUT School of Mathematics, was designed to assess core mathematics skills applicable to the course, such as basic algebraic manipulation which would be required to solve problems involving formulas. It was also launched with a view to supplementing an existing workbook (intended to revise such basic mathematics skills) by directing students to relevant workbook sections for any incorrect answers. The aim was therefore to assist students to focus their efforts so as to develop the quantitative skills required for success and thus build student confidence in the early weeks of the semester. The quiz also provided an opportunity to collect the much needed data to examine the problem in greater depth so as to reflect on possible further avenues of support.

The quiz was made available to students in an online format via the Blackboard site so as to increase flexibility and therefore encourage high levels of participation. The quiz featured in Orientation Week messages as part of a strategy designed to make students aware of the potential for concern in this area, as well as encouraging all students to self access their level of mathematical aptitude and therefore be pro-active with their studies. This initial promotion was then followed by discussion in the first week of classes, both lectures and tutorials, so as to maximise awareness in this optional activity. The quiz consists of eight short questions in a multiple choice format to ensure ease of participation. Questions 1 and 2 involve conversions between decimals and percentages. Questions 3, 4 and 5 involve arithmetic calculations. Questions 6, 7 and 8 involve algebra. The questions and answer options are the same for all students who undertake the quiz. The quiz questions and answer options are shown in Table 1.

3   Student Performance

The quiz was trialled in the second semester of 2007 and has been implemented every semester since, thus providing a large amount of data with which to analyse. It should be noted that for each quiz question, students were given four answer options, as well as a "don't know" option. The results have been collated to observe the longer term trends and thereby examine whether initial findings were cohort based or indicative of a more systemic problem. The database of results currently

Table 1 The quiz questions and answer options

1. Written as a decimal, 6.5 % is equal to
   (a) 0.0065
   (b) 0.065
   (c) 0.65
   (d) 6.5
   (e) Don't know
2. Written as a percentage, 0.0178 is equal to
   (a) 0.0178 %
   (b) 0.178 %
   (c) 1.78 %
   (d) 17.8 %
   (e) Don't know
3. $2 + 3 \times 0.5$ is equal to
   (a) 2.5
   (b) 3.5
   (c) 5.5
   (d) 8
   (e) Don't know
4. $(5 - 7) / 0.5$ is equal to
   (a) – 4
   (b) 4
   (c) 1
   (d) – 1
   (e) Don't know
5. $100(1 + 0.06)^3$ is equal to
   (a) 100.0216
   (b) 112.36
   (c) 119.1016
   (d) 318
   (e) Don't know
6. The solution for b to the equation $2a - b = c$ is given by
   (a) 2a – c
   (b) c – 2a
   (c) $-(2a + c)$
   (d) $-(2a - c)$
   (e) Don't know
7. The solution for i to the equation $F = P(1 + in)$ is given by
   (a) $[(F/P) - 1]/n$
   (b) $(F-P)/n$
   (c) $(F/nP) - 1$
   (d) F/nP
   (e) Don't know
8. The solution for x to the equation $a = (x - y) / z$ is given by
   (a) az + y
   (b) az – y
   (c) y – az
   (d) (a + y)z
   (e) Don't know

Table 2 Number of students who completed the quiz each semester during the period from Semester 1, 2008 to Semester 1, 2011

Semester, Year	Number of students
1, 2008	245
2, 2008	220
1, 2009	166
2, 2009	278
1, 2010	97
2, 2010	148
1, 2011	189

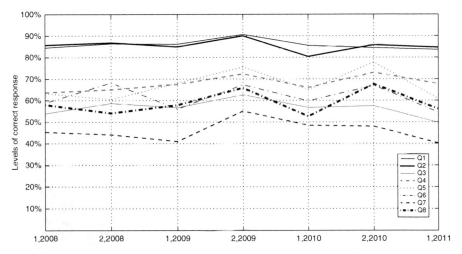

Fig. 1 Percentage of students who answered correctly for each question in each semester during the period from Semester 1, 2008 to Semester 1, 2011

covers from Semester 1, 2009 to Semester 1, 2011. Table 2 lists the number of students who completed the quiz each semester during the period being considered. Figure 1 provides the percentage of students who answered correctly for each question, in each semester. Figure 2 provides the percentage of students who answered "don't know" for each question, in each semester. Figure 3 summarises the average percentage of students who answered "don't know" for each question across the full period.

The levels of "don't know" response are particularly useful in assessing the reason for an incorrect answer. This answer option allows us to consider whether students thought they knew what to do and got the answer wrong, or whether they simply did not know how to tackle the problem.

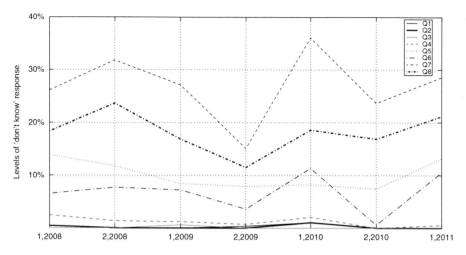

Fig. 2 Percentage of students who answered "don't know" for each question in each semester during the period from Semester 1, 2008 to Semester 1, 2011

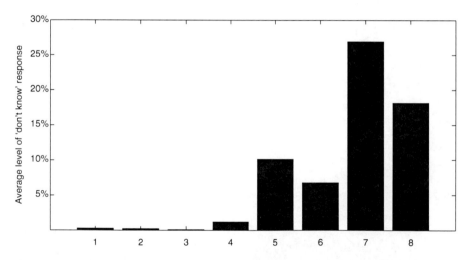

Fig. 3 Average percentage of students who answered "don't know" for each question during the period from Semester 1, 2008 to Semester 1, 2011

The results presented demonstrate that there are several areas in which large numbers of students lack the requisite mathematical skill, deficiencies which will undoubtedly cause distress for students while studying. The results, as well as the implications, will be considered in more detail below.

3.1   Conversion

Questions 1 and 2 on the quiz test a student's ability to convert between decimal and percentage format. The results in this area are the most positive overall. As shown in Fig. 1, competence to perform these types of conversions ranges between 80 and 90 %. Despite being the strongest of all areas, it is still important to note that this tells us there will generally be 1 out of every 10 students who are not able to understand and implement these basic level conversion problems. In a statistics course this can have significant consequences. For example, consider the study of the normal distribution and, in particular, the understanding of problems relating to the standard normal distribution tables. Students frequently need to be able to switch between probabilities written in decimals and percentages and thus an inability to master this skill could have a substantial impact on the subsequent understanding and analysis of results.

Interestingly, for both questions 1 and 2 we see very low levels of acknowledgment by the students as to potential concern or lack of understanding, as evidenced by the small percentage of "don't know" response shown in Figs. 2 and 3. During most semesters no student selected this option and occasionally one student selected this option. These results tell us that students are more confident with their knowledge to solve these types of problems, and also that students who were not able to get the correct answer were generally unaware of their inability in this area.

3.2   Arithmetic

Questions 3, 4 and 5 on the quiz test a student's ability to perform relatively straight-forward arithmetic calculations. In particular question 3 was designed to test a student's knowledge of the order of operations; question 4 was designed to see whether using brackets had any impact; and question 5 was designed to examine the effect of a power within the calculation.

If we begin with a closer inspection of questions 3 and 4, we can observe some informative results. Questions 3 and 4 can be primarily differentiated on the basis of whether brackets have been used to guide the processing of the calculation. Question 3 involved only addition and multiplication, without the use of brackets. The question was designed to test whether a student could apply the order of operation rules. Subtraction and division were included in question 4; however, in this question brackets were used to indicate that the subtraction component should be performed first. The highest percentage of correct response for question 3 during any semester was 63 %, while for question 4 it was 73 %. These results show that students have a lower level of competence for these types of arithmetic calculations than for the more simple conversion problems given in questions 1 and 2. Interestingly, the higher success rate for question 4 was evident during every semester examined. The difference ranged from 6 to 18 % (as shown in Fig. 1), suggesting that the use of brackets reduced the potential for error and therefore that understanding of the order of operations was not as strong as we would hope. A common mistake for question 3 was to calculate the addition first and therefore work left to right without regard to the rules of the order of operations. In question 4 the use of brackets removed the need for students to demonstrate such knowledge. The consequences are of great concern. Even if students are able to understand the higher level statistical concepts and applications being taught, when they are working on problems or assessment tasks approximately 1 out of 3 will not be able to determine the correct final calculation values. On the basis of the level of "don't know" response (as shown in Figs. 2 and 3) students perceived question 4 as being more difficult than question 3, despite the lower number of correct answers in question 3. This result suggests that students are also not aware of the importance of order of operations and of their potential deficiency in this area.

Question 5 of the quiz extends from the more basic arithmetic calculations of questions 3 and 4 to introduce a power component. Although in an introductory statistics course the more common power calculation to which students would be exposed is that of squaring, it seemed prudent to consider a different power calculation to establish whether students possessed a more general understanding of the concept. The results are quite surprising and it should be noted that although the levels of correct response were similar to that of question 4 (as shown in Fig. 1), the level of "don't know" response spiked to an average of 10% which represents the third highest on average across the eight quiz questions (as shown in Figs. 2 and 3). This result can be compared to what was observed in question 3 (the of order of operations problem). Similar to the results for question 3, students seemed either to be able to successfully calculate the power or, if not, admit they didn't know. Thus the percentage of incorrect answers was significantly lower.

3.3   Algebra

Questions 6, 7 and 8 on the quiz test a student's ability to perform relatively simple algebraic manipulation. In particular, questions 6 and 8 were written using more standard notation (i.e. a, b, c and x, y, z), as opposed to question 7 which uses less common notation, despite being the equation for the future value of an amount invested under simple interest.

Both questions 6 and 8 involved algebraic expressions using the more typical representations of unknowns. In each question students were required to rearrange the expression to find the answer. The rearrangement in question 6 required an addition followed by a subtraction; the rearrangement in question 8 required a multiplication followed by an addition. Similar levels of success were observed for each of these questions, with question 6 obtaining an average rate of 62% correct answers and question 8 obtaining an average rate of 59% correct answers (as shown in Fig. 1). Interestingly, question 8 had the slightly lower figure which may be attributed to the use of brackets. It appears that while brackets assist students when performing arithmetic, as discussed above, the inclusion of brackets adds complexity when algebraic manipulation is required. It is also particularly interesting to note the levels of "don't know" response across these two questions (as shown in Figs. 2 and 3). For question 6 the level of "don't know" response was on average 7%, where as for question 8, the level was 18%. Again, it appears that in the less complicated question 6 there were many students unable to get the correct answer that did not believe they did not understand the process, as represented by the lower "don't know" response level. This result can be contrast directly with question 8 in which almost half of the students who were unable to get the correct answer admitted that they didn't know.

Finally, if we consider the results of question 7, an even more bleak state is evident. Figure 1 shows that question 7 has the lowest success rates across all eight questions. This is despite being a more practical application of algebra in which students were asked to manipulate the equation for the future value of an amount invested under simple interest so as to solve for interest rate. On average only 46% of students were able to achieve a correct answer, thus indicating that just over half of the class each semester is unable to perform such a task. The lowest individual semester recorded a mere 40% of students able to give the correct answer. The highest individual semester saw 55% of students able to give the correct answer. These results clearly demonstrate that this was the most challenging aspect of the quiz for students. The level of "don't know" response, as shown in Figs. 2 and 3, also reached a new high for this question with an average of 27% of students admitting they didn't know how to answer. As the steps involved in rearranging the expression in this question are of a comparable level of difficulty to that of questions 6 and 8, these results suggest that the use of less common mathematical notation (i.e. no x) has been a factor in the perceived difficulty. Further, such a finding must therefore also raise the question of whether students have fully appreciated the nature of algebra in earlier studies or whether this is indicative of rote-style learning in which there has been a failure to fully understand the core algebra concepts.

These results and the corresponding implications are cause for concern. It is therefore absolutely vital that as educators we are aware of and consequently factor such levels of prior knowledge into the design of course curriculum. To ignore such deficiencies would serve no purpose other than to prevent students from achieving at the tertiary level. While it may be simpler to assume prior knowledge and leave the responsibility of which to the students themselves, a more appropriate approach is to understand the problems so as to work toward productive and effective solutions.

4   Strategies for Improvement

In a statistics course for psychologists, Gnaldi (2003) showed that the statistical understanding of students at the end of the course depended on students' basic numeracy rather than the number or level of previous mathematics courses the student had undertaken. The importance of understanding the level of basic mathematics skills is therefore critical to the subsequent development and customisation of focused strategies to combat observed deficiencies. The growth of accessibility to faster Internet connections has given rise to an opportunity to broaden existing teaching methodologies and therefore enhance learning outcomes for students. In today's learning environment there are wonderful opportunities to incorporate technology and therefore offer blended learning and flexible delivery, thus creating an environment where we can seek to tackle fundamental mathematics deficiencies in conjunction with the tertiary course so as not to detract or reduce the scope for higher level skills and competencies.

Prior to the offering of the quiz, a workbook was developed to provide Data Analysis students with an extra resource to revise their core mathematics skills, with a view to ensuring that all students reach the desired level of prior knowledge. The optional workbook consisted of several chapters, including arithmetic, algebra, summation, arithmetic progressions (APs), geometric progressions (GPs), and linear functions, available online to ensure ease of access. The workbook was designed to introduce each topic as gently as possible with the use of examples and then a comprehensive series of questions, to be completed in a student's own time. Although at more than 100 pages in length, the feedback from students included that it was too broad and therefore too time consuming. As well many students indicated that, while they initially had good intentions of working through the resources, they did not complete more than the first one or two chapters because of its length and a perceived lack of relevance.

The quiz provided an opportunity to refocus efforts towards the mathematics workbook by linking specific sections to questions within the quiz. Further, the online mode of the quiz was used to provide customised feedback in which upon getting questions incorrect the relevant sections of the workbook would be recommended. This strategy for improvement has been successful in providing students a more focused approach to the immense range of course resources, as evidenced by student feedback such as "I was particularly impressed by the large number of extra curriculum work available. For a student who normally struggles with mathematics, it was comforting to have material to help me improve" (Student feedback, QUT Learning Experience (LEX) survey, 2009).

The initial quiz results also yielded an understanding of the areas in which students experience the highest level of difficultly, such as arithmetic and algebra, as discussed above. This knowledge was able to be used to source additional resources including online videos to assist with the development of core mathematics concepts and include the relevant links within a *Math Help* section of the Data Analysis Blackboard site. Further, the longer term consistency of such trends provides the motivation to continue to seek additional resources to assist students. For example, a future project will involve the development of more business-specific instructional videos so as to still cover the core knowledge but to do so in a manner that is customised for the student and their learning objectives. This, as such, also represents an exciting opportunity for further research in this important area of teaching scholarship.

5 Conclusion

At the tertiary level a dislike of mathematics is a common problem due to the prolonged failure to master core skills, as well as potential avoidance of such education. This is especially evident for business studies students, whom in many instances perceive mathematics as being completely redundant to their studies. This paper focused on the results of a long-term study to measure the level of mathematical ability of first year business students as they enter a compulsory statistics subject by means of a mathematics quiz.

The quiz though primarily designed and implemented to support student learning, offered a valuable opportunity to measure performance, providing valuable information about the entry level of mathematical ability. The results highlighted several areas of concern where students do not possess the requisite knowledge required for successful completion of the subject. The results, though not necessarily surprising to business educators, should raise questions about the adequacy of pre-tertiary mathematics education. The consequences of such deficiencies were discussed and in particular the surprisingly low levels of arithmetic and algebra competence were examined in more detail. Strategies to improve core mathematical ability were discussed, including a workbook strategy, as well as ideas for further refinement and focus. Student feedback so far shows these approaches to be successful in raising levels of engagement and even enjoyment of this traditionally unpopular type of course.

6 Note

Developed from a paper presented at Eighth Australian Conference on Teaching Statistics, July 2012, Adelaide, Australia.

This chapter is refereed.

References

Garfield, J. (2003). Assessing statistical reasoning. *Statistics Education Research Journal*, 2, 22–38.

Gnaldi, M. (2003). Students' Numeracy and Their Achievement of Learning Outcomes in a Statistics Course for Psychologists. Unpublished M.Sc., University of Glasgow, Faculty of Statistics.

Moore, D. (1997). New pedagogy and new content: The case of statistics (with discussion). *International Statistical Review*, 65, 123–137.

Vere-Jones, D. (1995). The coming of age of statistical education. *International Statistical Review*, 63, 3–23.

Watson, J., & Callingham, R. (2003). Statistical literacy: A complex hierarchical construct. *Statistics Education Research Journal*, 2, 3–46.

Wilson, M. (1992). Measuring levels of mathematical understanding. In T. A. Romberg (Ed.), *Mathematics assessment and evaluation*. Albany: State University of New York Press.

Wilson, T., & MacGillivray, H. (2005). Numeracy counts in the statistical reasoning equation. In *55th Session of the International Statistical Institute*, Sydney, Australia.

Wilson, T. M., & MacGillivray, H. L. (2006). Numeracy and statistical reasoning on entering university. In A. Rossman, & B. Chance (Eds.), *The Proceedings IASE/ISI 7th International Conference on Teaching Statistics*, Brazil. Voorburg: ISI.

Wilson, T. M., & MacGillivray, H. L. (2007). Counting on the basics: Mathematical skills amongst tertiary entrants. *International Journal of Mathematical Education in Science and Technology, 38*(1), 19–41.

Part D
Workplace Learning and Professional Development

Square PEGs in Round Holes: Academics Teaching Statistics in Industry

Peter Martin

Abstract The teaching of statistical techniques to people in industry, as part of quality control or process improvement programmes, can be a rewarding but somewhat daunting process for academics. To be effective in this arena the academic needs to be demand-driven, and client-focused, as well as being capable of building client relationships, ensuring customer responsiveness, and supporting flexible delivery. As well as using training methods that are likely to be effective in promoting long-lasting learning, allowance needs to be made for the evaluation of both the training programme and its outcomes in terms of the objectives of the organisation. The new knowledge and the skills imparted by the training programme must relate to real workplace needs. Successful industrial training involving statistical techniques depends primarily upon targeting the right people at the right time and providing appropriate content. This chapter explores some of the requirements and conditions that contribute to the successful teaching of statistics by academics to adult students in industry settings. Whilst the spin-offs for undergraduate teaching are numerous, the biggest gains include the cooperation between the specific industry and the academic institution concerned.

Keywords Industrial training • Training programmes • Training material • Participants • Adult learning • Understanding • Analysis • Skills • Knowledge • Context • Needs • Statistics educators • Evaluation • Assessment • Projects

P. Martin (✉)
University of Ballarat, Ballarat, VIC, Australia
e-mail: p.martin@ballarat.edu.au

H. MacGillivray et al. (eds.), *Topics from Australian Conferences on Teaching Statistics:* 225
OZCOTS 2008-2012, Springer Proceedings in Mathematics & Statistics 81,
DOI 10.1007/978-1-4939-0603-1_13, © Springer Science+Business Media New York 2014

1   Introduction

Industrial training generally differs significantly from school or university teaching. The primary purpose of an educational institution such as a school or university is to educate, which according to Azevedo (1990) implies the bringing out or developing of qualities or capacities latent in the individual, or regarded as essential to his or her position in life. Whilst an industrial training programme is also about educating, the usual focus is to provide training in immediately useful skills. This is clearly evidenced in quality control or process improvement training programmes such as those offered by providers of Six Sigma training. Such programmes typically involve training in skills that are used immediately in projects designed to optimise processes by reducing scrap rates, downtime, and so on. Industrial training almost invariably suggests a distinct end or aim which guides the facilitators and instructors. Industrial training programmes are typically paid for by industries willing to train their employees for specific purposes, whereas university students often pay for their own courses which offer a broad base of skills for a career but often as yet unspecified job.

Successful industrial training involving statistical techniques depends primarily upon targeting the right people at the right time and providing appropriate content. Getting the right "who" trained in the right "what" is crucial (Morgan and Deutschmann 2003). These authors claim that it is not enough to ask senior managers what they think their staff should know. In fact senior managers may not be aware of the range of possibilities and certainly not aware of the range of barriers, and/or misconceptions to be overcome. Attention needs to be paid to matching worker competencies with the organisational needs. As well as using training methods that are likely to be effective, promoting long-lasting learning, allowance needs to be made for the evaluation of both the training programme and its outcomes in terms of the objectives of the organisation. The new knowledge and the skills imparted by the training programme must relate to real workplace needs.

Whereas training emphasises the effective use of certain techniques and materials in order to answer immediate needs, it usually fails to meet the requirements of long-range professional development, which is one of the goals of an educational institution (Azevedo 1990). It is in this context then that training may be considered different to education. This therefore represents a potential barrier (hurdle) for academics as they wrestle with switching between teaching university students and conducting appropriate industrial training programmes. In academic settings there is a freedom to provide students with skills and techniques that they hopefully utilise either in their future studies or in their careers. With industrial training there is no such freedom as the emphasis or primary focus is typically on the company and optimising processes and profits. So it is this focus mismatch that the academic trainer has to resolve.

From an educational perspective, what is it that makes an industrial training programme successful? According to Ramsden (1992) and Biggs and Moore (1993), successful training programmes invariably promote a deep approach to learning.

Typically this involves relating existing knowledge to a project in hand and drawing on knowledge from as many sources as possible via project teams. These authors also argue that deep learning will be enhanced by placing theoretical statistical ideas into the realm of shop-floor experience. When statistical ideas are transformed into real-world examples, meaningful context is established for the user, thereby enabling better understanding and appreciation of a situation. The importance of meaningful context in interpreting graphical information, for example, has been argued by various researchers (Makar and Confrey 2002; Wild and Pfannkuch 1998; Roth and Bowen 2001).

Another key to success involves recognition of learning style. Anxiety levels of those involved in such training programmes need careful consideration, for as Pretty et al. (1995) stated, "adults learn best when their own motivation is supported, their active participation is encouraged, their experience is valued and the content is relevant to their daily work". Werner and Bower (1995) emphasised the need to focus upon participant learning rather than teaching style.

As adults we tend to learn best when support is provided for our own personal motivation; when our experiences are valued; when we are encouraged to participate; and when the training material delivered is perceived as being relevant to our daily work (Pretty et al. 1995). Self-directed learning opportunities and interactive learning environments will shift the focus from the style of the trainer or teacher to the trainee's learning. Therefore, it is important to match participant competencies with the needs of the organisation.

A National Forum on Building Networks in Statistics Education was held in Brisbane in 2009. A major aim of this forum was to identify common issues, interests, challenges, and resources in the learning and teaching of statistics. A breakaway discussion group comprising statistics educators and employers was asked to consider the potential for cooperative interaction between the two groups.

The statistics educators found the discussion exercise challenging and focused primarily upon the difficulties associated with establishing initial contact with employers and with integration into existing tertiary programmes. The main concerns raised by the statistics educators were as follows:

- Initial Contact: Establishing initial contact with employer groups was seen as a major concern. Various possibilities were suggested including the role of existing organisations such as The International Association for Statistical Education (IASE) and the Statistical Society of Australia Inc (SSAI). Other suggestions were to set up Advisory Committees made up of employers and academic staff with regular meetings to establish networks. Out of such groups potential contact between academic staff and employer groups could be established and potentially used to enhance existing academic programmes.
- University Recognition: There was considerable concern that too often much of this external interaction work is on top of normal university workload and may be seen as detracting from research and teaching. It was strongly felt that formal recognition of such interactions with employer groups is needed, perhaps such as reduced teaching loads, or payments in kind.

- Integration with existing programmes: Any interaction needs to be integrated with the university teaching and/or research programmes. Issues of potential assessment possibilities were discussed along with relevant visits to and from various employer groups. Projects need to fit into academic programmes and be a "unit of learning" as such, as well as being manageable, pitched at appropriate levels, and not absorb too much time.
- Staff suitability: The right sort of people need to be available; not everybody's cup of tea as various interpersonal skills are required as well as relevant statistical expertise.

The employer group, on the other hand, knew precisely what they wanted from such interactions with academic staff. Their main concerns included the following:

- Relevance: Any collaboration/interaction needs to be relevant to the employers' work programme, including objectives and mission statements. The nature of the relationship should be aligned with company objectives and be within the grasp of those involved.
- Intellectual Property: Ownership needs to be established at the beginning of any collaboration, and protocols for actions such as data collection, storage, access, and publishing need to be agreed upon by all stake holders.
- Connection: The connection between industry and researchers will be enhanced by showcasing from both sides and also by ensuring continuity of researchers throughout the project life.
- Competence: Employer groups need to be assured of the competence of the researchers to undertake the task at hand—things like outcome/output orientation, availability to do the work, track record of project completion, and reputation would be important factors here. If students are involved some consideration would need to be given to their academic capabilities as well as their interpersonal skills.
- Value Add—Win-Win situation: Employer groups offer real data problems to motivate researchers (and in turn their students). At the same time there is value added to employer research, particularly for methodological skills transfer from the researcher to the employer group, where cutting-edge analytical techniques are used.

We can see here the mismatch between the sentiments of academics and employers. There was a discomfort associated with the academic concerns that reflected a campus focus, whereas those expressed by employers were more general and wide ranging. While these comments related to interactions between the two groups, of which training might only be a small part, the differences represent real obstacles for academics to effectively engage in teaching in this arena.

Whilst specific educational research on learning statistics as part of industry training is very limited, there are many studies, papers, presentations, and reports dealing with aspects of potential interest and/or application to statistics education in industrial contexts. These studies and reports typically consider the more general issues associated with training, learning, relevance, and many provide insights into the requirements for success in both the teaching and learning of statistics as part of vocational education training.

The most prevalent types of learning opportunities provided for Australian workers consist of informal learning experiences, followed by non-formal short courses and in-house training (Allen Consulting Group 2006). This is supported by other studies (Richardson 2004; Mawer and Jackson 2005), which conclude that after informal on-the-job training, in-house training, or its equivalent, was the most common form of training for Australian workers and was highly valued because it was immediately relevant and could be put into practice.

The Australian Industry Group (Allen Consulting Group 2006) reported an increasing realisation by industry that their future access to the skills they felt neces sary for future growth could only be achieved by increasing their own training efforts in-house to ensure the supply of skills. This concern regarding the meeting of industry needs is also reflected in a paper by McMahon et al. (2009) who report industry's concern that its standards are devalued through delivery and assessment not conducted in accordance with their requirements. McMahon et al. (2009) state that these issues are often related to the ways in which training packages have been implemented, with insufficient attention to the institutional and human resource requirements needed to support effective use.

Employers are primarily interested in essential technical skills and knowledge required for jobs and compliance with legislative requirements (Misko 2006; Smith and Oczkowski 2009). Companies do not much care if workers did or did not have any formal qualifications, unless they were required to for compliance reasons (Mawer and Jackson 2005). These attitudes were also shared by employees. Most Australian workers want training that is informal, immediate, and delivered on the job by peers or supervisors, rather than anything which reminds them of the school environment (Marr and Hagston 2007). Industry representatives preferred a combination of on- and off-floor training that had immediate workplace application and also incorporated opportunities for practice and reflection. Cost of training was not seen as a deterrent provided the training was of perceived value, clearly focused upon business-specific needs, delivered with a personal approach (such as by a known facilitator), and delivered flexibly (Dawe and Nguyen 2007).

Training and learning strategies that are needs-based, just-in-time, and very interactive are highly valued approaches to facilitating learning in enterprise-based environments (Harris et al. 2005). The selection of the "right" people from universities or TAFE institutes for collaborative training linkages with business organisations was seen as important. In particular, the need for these practitioners to quickly familiarise themselves with the environment, culture, and networks relevant to the particular enterprise was emphasised. Practitioners must also be able to determine the specific requirements of an enterprise and be able to identify skill deficits and options for "top-up" training should the need arise. In addition, they need to be flexible and able to adapt training approaches to the workflow of the enterprise, work collaboratively, and sensitively customise training methods and materials.

Mitchell (2008) identified several critical issues associated with addressing the needs of both industry clients and individual learners including,

- Customising and personalising training services
- Developing a deeper understanding of individuals' learning styles and preferences
- Effectively providing services and support for different learner groups (equity, e-learning, for example)
- Understanding the different ways learning can occur in workplaces
- Developing partnerships between external teachers and enterprise-based managers and trainers

Mitchell (2008) suggested industrial trainers need to be demand-driven, client-focused, responsive to industry, and be able to build client relationships, ensure customer responsiveness, and support flexible delivery. Such practitioners will have a deeper knowledge of education and industry, and be able to customise training and devise business solutions required by enterprises and individuals.

The importance of context and of adult learning principles was mentioned by Gibb (2004) in relation to vocational education training. These have also been reported in other studies (Rogers 1986; Ramsden 1992; Biggs and Moore 1993; Martin 1997, 2005, 2006, 2008). Ramsden (1992) and Biggs and Moore (1993) argued that successful training programmes invariably promote a deep approach to learning by relating existing knowledge to a project in hand or drawing on knowledge from as many sources as possible via project teams. Martin (1997, 2008) concluded that the embodiment of context, adult learning, and deep learning, into the structure of a training programme, were key factors contributing to the success of that training programme. When statistical theory is placed into the realm of workplace experience, deep learning will be enhanced. Transforming the theory into practice establishes a meaningful context for the user, thereby enabling better understanding and appreciation of a situation.

Smith et al. (2007) reminded us that mature-aged people are the fastest growing segment of the market for vocational education and training. Further, with respect to teaching, learning, and assessment issues, they noted that mature-aged workers often prefer training to be workplace-based and practical rather than classroom-based and theoretical, and to be task-related and not necessarily leading to a qualification. Mature-age workers expect a more personal relationship from the trainer as well as respect for their experience.

Those of us who want to become involved in teaching statistics in industry programmes can learn from the research already described, particularly if we want what we do to be perceived as of value and have lasting effect. The clear message coming from this research is that as statistics educators in this arena we need to be strongly client focused rather than textbook focused or University bound. It is not good enough for us to have a deep knowledge of statistics education; we also need to have knowledge of the industry concerned, and be able to customise training content and devise appropriate business solutions required by enterprises and individuals.

2   Teaching Statistics in Industrial Settings

Over the past 15 years or so the author has been fortunate enough to have acted as a consultant to more than a dozen companies/organisations, primarily with the view to providing statistical training programmes for their employees in an attempt to improve productivity and increase efficiencies. These companies have included food manufacturing, industrial manufacturing (vehicles, parts, and related components), utility industries, service industries, and computing industries. In each case the training provided consisted primarily of statistical content related to collecting and analysing data, analysis of measurement systems, capability studies, statistical process control, hypothesis testing, and aspects of experimental design. The training was often linked with Tertiary and Further Education (TAFE) competency certificate awards, as well as a nationally accredited university postgraduate award. Some programmes were obviously more successful than others, and after programme evaluations, and post-mortem discussions with industry personnel, various factors emerged that were felt to have an important impact upon the success or otherwise of each training programme. The following summary provides an outline of those factors thought to be important for academics interacting with employers in this context:

- Credibility: Sometimes for an academic this is not easy because so often the question is "What do you know about what we do here that can benefit us? After all, you're just an academic!" To establish credibility the following points may be useful:

  - History—i.e. track record; provide evidence of involvement in previous projects, particularly ones that have been successfully completed.
  - Homework—do some research on the company or organisation concerned; visit their website and read about their goals and objectives and what they do.
  - Hear what they say—i.e. let them talk, listen carefully, and take notes.

- Establish requirements: Determine precisely what the employer really wants. This may be difficult at times because employers may not know this themselves—particularly when it comes to issues such as learning, training material, and selection of participants. It is also important for both parties to ascertain each other's respective capabilities. For in-house training employers need to be able to provide adequate facilities and time release for the trainees. On the other hand, the capability of the academic to do the training is just as important. For both parties it is important to know their limitations and when it is in the best interests of all concerned to postpone, or even walk away from the programme.
- Plan effectively: Considerable attention needs to be given to timetabling and scheduling. Often this needs to be flexible, as issues will always arise. Clarify how payment is to be made, whether to the university, or personally, as this impacts workloads and scheduling. Prepare for possible publications or other research opportunities by introducing this option early. Such activity requires IP agreement from the beginning. Academics collecting data from participants about their responses, etc., to the training will need to obtain ethics approval.

• Value add: This is always an important part of the process of making the client feel they are in safe and reliable hands. What happens here is what goes down the grapevine so to speak. Compromises made and shortcuts taken will be simply reflected in future clients. Four key characteristics here are availability, flexibility, transparency, and generosity. To be available simply means always responding to emails and phone calls. Flexibility applies to changes in scheduling, client requirements, development of training materials, and so on. Transparency refers to openness and honesty in communication, and generosity implies being prepared to go the extra mile—doing the extra analysis, or spending extra time coaching someone.

The need to be demand-driven, client-focused, and responsive to clients may well be daunting for many academics, making them feel awkward, out of their comfort zone, feeling like a square peg in a round hole so to speak. The points described above represent potential difficulties for academics trying to teach in industrial settings, especially when the primary focus of industry is about the company and to maximise profit. In this sense an academic history of publications and research may take second place to previous practical experience with other industries. This then creates the dilemma for the academic of how to get started. With regard to the points mentioned above, for this author it was initially a case of doing some background homework on the company, and mostly listening to what was wanted, rather than describing what could be done.

While these training programmes have been a challenge to all concerned they have emphasised the need to be familiar with the features underlying adult learning as well as the workplace context within which the training will apply. In addition, if commitment on the part of the participants is to be achieved the training material needs to be relevant to their specific work practices. The following case study is presented as a successful example of an industrial training programme that embodied many of the features described in this section.

3 Case Study: Process Improvement in a Dairy Industry

The routine use of statistics within the company was recognised as an essential core capability to be systematically developed throughout the company. Therefore, the primary objective of the programme was to provide various levels of statistical training for specific groups of employees within the company so as to build up a capability within the organisation to make better use of statistical tools on a day-to-day basis. A secondary objective was to train the employees to utilise best practice methods in technically based projects and for performing relevant research.

As part of one of the company's key strategic aims, training in statistics was to underpin not only the success of research and development projects across all key strategic aims, but was also expected to lift the company's performance as a whole. Successful utilisation appropriate statistical tools and analytical methods throughout the company was seen to be important in helping to secure a future in an increasingly dynamic and competitive national and global industry.

The author was not only to provide one-to-one mentoring throughout the programme but would also be available on an as needed basis to help support staff to apply their learnings to their business functions well beyond the completion of the training programme. At the time of writing the author is involved in a project with this company 7 years after the training programme.

The programme was tailored to the needs and roles of individuals within the company. In particular it was to provide the background, statistical skills, and tools needed to better use data to improve business performance whether in operations, research and development, marketing, finance, sales, accounting, laboratory, or anywhere else. This was achieved by taking into consideration various aspects relating to organisational needs:

- The range of entry levels into the programme in terms of personal capabilities and interest.
- The range of capability levels in terms of level of awareness and applications/use of statistics.
- The need for the training material to be practically based involving parallel application on projects of relevance to an individual's job function in order to enable immediate application.
- The need to have ongoing support by and access to the trainer for key people to effectively utilise the knowledge and skills developed.

To this end the final programme consisted of 15 modules (see Table 1) delivered to three groups of participants in half-day sessions of 3–4 h. The groups were made up as follows:

- Group A consisted of 8 participants mainly drawn from R&D and laboratories.
- Group B consisted of 11 participants, mainly comprising process supervisors and people from the production line.
- Group C consisted of 9 participants from middle management positions

Table 1 List of training modules and the groups to receive the training

Module	Groups
Intro & Overview	A + B + C
Process Characterisation	A + B + C
Exploratory Data Analysis	A + B
SPC—Basics	A + B
Gauge R&R	A + B
Multi-Vari Studies	A + B
Process Capability	A + B
Correlation & Regression	A
Hypothesis Testing & CIs	A
ANOVA	A
Experimental Design I	A
Experimental Design II	A
SPC—Advanced	A

The allocation of staff to these groups was based largely on the perceived extent and depth of statistics to be used by that person in performing their business function. For those in Group A who daily used statistics for analytical purposes, attendance was required at all training sessions. For those in Group B who regularly used statistics for maintaining an awareness of the current state of the production process, attendance was required at 7 of the training sessions. For those in Group C who only occasionally used statistics for purposes such as reporting, support, or question decisions based on data analysis, attendance at the first two sessions only was required.

Each session was delivered by the same presenter (the author) and comprised a lecture component, as well as a practical component and group work where necessary. Participants were expected to complete set tasks after each session involving computer worksheets requiring analysis of specific data sets, collection and analysis of workplace data, and/or group presentations of analyses carried out during the training sessions. All participants were provided with a comprehensive set of training notes and web-based reference readings, as well as self-paced computer worksheets involving analysis of work-based data sets using Minitab.

All participants were required to apply their new knowledge and skills, gained throughout the course, to projects drawn from their everyday work in order to secure their understanding and appreciation of statistics in their workplace. These workplace projects were seen as the keystone of the programme. The PPDAC cycle (problem, plan, data, analysis, conclusions) developed by Wild and Pfannkuch (1999) to model statistical thinking was clearly evident. At each step of the cycle participants were required to present a report to the group responsible for overseeing their particular project. This resulted in an ever-increasing appreciation of the importance of interpretation and communication of findings for each participant, as noted by Forster et al. (2005) on the benefits of getting students to write about statistics. This enhanced appreciation was reflected in some of the participants' comments, as indicated. These comments also showed an appreciation of the learning of good statistical practice, discussed by Svennson (2007), particularly in the context of the projects in which they were involved.

- I found it very useful, and feel confident to use the tools to get what we want, particularly GR&R, EDA, and Normal Dist. The material was useful and a valuable resource, but needs project running alongside.
- Value for $; gets to the nuts & bolts; enables you to see differences between issues that do & don't count; useful to apply in project.
- Changing work behaviour—now see need for planning what data I need to collect; see real need for EDA, GR&R, Capability, & ANOVA.

4 Assessment

Upon successful completion of the training programme and individual projects, Group A and B members were to be assessed, and if successful, awarded two certificates: a certificate of participation and a certificate of competence.

This assessment took place 9 months after the completion of the training to allow for the finalisation of projects. The nature and design of the assessment were determined by the trainer in cooperation with the company programme organisers.

The aim was for the assessment to be seen as fair and relevant to the purposes of the training programme. It was deemed important that the assessment measure both procedural and conceptual understanding. While it was important for the participants to be able to correctly perform a task, it was considered equally important that they have some knowledge of what was being done and why. Lipson (2007) and Garfield et al. (2003) have discussed the importance of such aspects in assessing statistical knowledge, admittedly in the context of tertiary students; the same certainly applies for workplace-based training involving statistics. In fact, the very applied nature of the workplace context probably makes it easier for this type of assessment to be applied.

A certificate of participation was awarded to participants who satisfied the following requirements for their Group:

- Attendance at 75 % of the training sessions
- Satisfactory completion of 80 % of the practical worksheets involving computer analysis of specific data sets
- Active involvement in at least one of the projects designed to accompany the training programme
- Participation in the preparation and presentation of project reports at various stages throughout the life of the project

Assessment of the training material for each topic required participants to submit practical computer worksheets involving analysis of specific data sets using Minitab and/or Excel and to collect and analyse some data obtained from their particular work environment. This often involved individual and/or group presentations of analyses carried out during the training sessions, or more formal preparations of written reports based upon results of experimental studies carried out at the workplace. Assessment of the worksheets was graded as either satisfactory or not satisfactory, and was included as participation as it was an indicator of involvement and commitment in addition to knowledge gained.

Assessment for the certificate of competency was only undertaken by request, and took the form of a 45 min one-on-one "interview" with the trainer, in effect an oral examination. Whilst this was recognised as a potentially daunting process for the participants, it was mitigated to some extent in that both trainer and participants had known each other over a period of 2–3 years and that each participant was provided with an outline of the types of questions and topics to be covered. Each "interview" was structured to account for the differing levels of experience of the participants by selecting appropriate topics related directly to their work-related responsibilities. The procedure that was followed was the same in each case and is outlined in Table 2. During the "interview" participants were required to:

- Speak to a topic of their choosing as well as one selected by the interviewer
- Answer various questions about any of the topics covered
- Interpret computer output in the context of the workplace
- Use Minitab to perform various analytical tasks given some workplace data

Table 2 Sample questions for competency certification (done one on one with the interviewer)

• The following list of topics includes those covered in the training programme. Select one and tell me what you know about it.

Levels A & B topics	Level A topics
Project plan	*t*-Tests
Process map	Paired & independent *t*-tests
Exploratory data analysis (EDA)	Correlation & regression
Measurement systems analysis (Gauge R&R)	Chi-square test
Capability analysis	ANOVA & experimental design

- EDA (A & B)
- Gauge R&R (A & B)
- Capability study (A & B)
- Histogram (A & B)
- Pareto chart (A & B)
- Time series plot (A & B)

- • What is the purpose of …
- • Describe how you would do …
- • Outline a work situation where it might be appropriate to do …

- • Give me an example that shows the difference between categorical data and continuous data
- • When might it be appropriate to do a time series analysis

- Paired *t*-test (A)
- ANOVA (A)
- Chi-square test (A)
- Correlation/regression analysis (A)

- • When would you do …
- • What are you looking for when you do …

- • Here is some Minitab output from an analysis of data collected from on-site
  - What statistical tool is this an example of?
  - What does it tell you?
  - When might you use a tool such as this?
- • Here is some data. Use MINITAB to obtain
  - Descriptive stats
  - Boxplots
  - Histogram
  - Normality test
  - Scatterplot

An outline of a sample of the types of questions asked and the topics covered in the interview is provided in Table 2. Each participant was provided with a copy of this same outline and was expected to provide a satisfactory response to at least one question from each topic covered during the training programme. No access to any references was permitted during the "interview".

The questions asked of a participant were selected according to the designated level of that participant and the results of each "interview" were summarised on a recording sheet. Table 3 shows an example of the recording sheet used with results for 2 of the participants from each of Groups A and B. A star rating of one to three was used to indicate the extent of each participant's knowledge and understanding of the topics discussed (see Table 3). There was also provision on the recording sheet for comments by the trainer regarding individual performances.

From the company's perspective, evaluation of the overall programme was always going to be based primarily upon the successful completion of the projects

Table 3 Recording summary of assessment details for two participants from groups A & B

		Gp A		Gp B	
General topic	Comments	CH	MS	AD	BC
	Self-choice	ANV	GRR	Cap	GRR
Familiar topic	Trainer choice	Nest	Res	EDA	EDA
Exploratory data analysis	Where is categorical data used?				**
	Where is continuous data used?				
	What does the stdev measure?		***	**	***
	What does the mean measure?				*
	How do we know if the data is normal?		***	***	
	Why do we want to know if the data is normal?	***	***		
	What's a histogram? Boxplot? Bar Graph?		***	**	**
Hypothesis tests & CIs	When would you use a one sample t-test?	**			
	When would you use a two-sample t-test?		*		
	What does a 95 % CI tell you?	**	**		
	What test would I use if …?				
	What is a p-value?				

*Needed help; **good working knowledge; ***strong understanding

Minitab tasks	***Unassisted; **needed some help				
	Descriptive Stats (Milkfill by Machine)			***	***
	Boxplots of Moisture by Pallet	***	***	***	***
	Pareto (Stoppages)			***	**
	1-Way ANOVA (MilkFill by Machine)	***+	**		

Extra comments by trainer

CH: Strong u/standing & appreciation of tools; realises power of applying tools; awkward expression at times

MS: V. good u/standing & appreciation of tools; quiet confidence in use; good role model/mentor potential

AD: Good u/standing & appreciation of tools; needs more application e.g. to increase awareness of application

BC: V. Good u/standing & appreciation of tools; excellent attitude; keen to apply where applicable; excellent mentor

allocated at the commencement of the training, and in the savings or efficiencies gained by the improvements made therein. While such successes are an important factor in the continuing successful application of the initial training received, they also play a vital role in maintaining a competitive edge in national and global markets. Any researcher will be buoyed by initial success and will most likely be better equipped to continue with the experimental process. Evidence of this continuation for these participants can be seen in the wide ranging new projects that have been completed in the 7 years post-training.

For example there have been several measurement systems analyses involving colour recognition and taste testing procedures. There have also been projects aimed at developing more efficient predictive models related to several important product components, as well as statistically designed survey procedures to obtain information from both suppliers and customers. At the time of writing there is a project under way to make adjustments to an existing process that will enhance efficiencies and free employees from considerably overloaded workloads.

From the participants' perspective, we get some idea of their perception of the success or otherwise of the programme from their comments made on evaluation surveys completed at various stages during the programme, and also by their continuing involvement and use of the statistical tools in projects post-training. As shown in the following comments, not all their comments were positive, and neither have all continued their involvement post-training.

- Useful to apply in projects, but having no computer made it difficult
- Didn't know objectives well enough to find a suitable project; can see the value, but felt a bit left behind; need a slower pace
- Great value personally; appropriate content, excellent reference when reviewing post-training. The A/B break-up was very good & good timetabling of 1/2 day sessions

Such comments highlight the importance of the initial planning and structuring of such a programme, as well as the relevance of the projects, and the need for provision of adequate facilities for use by the participants. Their critical comments could generally be classified as "constructive", or "destructive" as shown by the following two comments:

- Project needs to done during sessions
- Don't use tools in my role; not worth the $

All of the "destructive" comments came from the Group B participants, i.e. process supervisors and production line people, and tended to avoid taking personal responsibility for difficulties encountered during the programme (there was a tendency to blame everybody and everything else). The constructive comments came equally from both groups, tended to realise the value of the programme and offered suggestions for enhancing the programme.

5   Case Study Conclusion

There is no doubt that this particular training programme was successful in meeting the original objectives stated by the company concerned. The key characteristics of this programme were as follows:

- Before the training programme commenced the company began with clear aims and objectives in place.
- Key company personnel were identified to lead the programme.
- Strong positive cooperation between trainer and key company personnel.
- Careful selection of programme participants.

- Training material specifically tailored to suit both company needs and those of the participants.
- Projects were carefully selected to be strategically aligned with company objectives, and were to be a key feature of the programme.
- A realisation by the company of the need for some form of assessment to establish the extent to which their participants had mastered the substance of the training and to assess their capability to make better use of statistical tools on a day-to-day basis in their respective workplaces.
- Acknowledgement of success in the form of endorsed certificates awarded to participants upon successful completion of the programme.

The assessment of the statistical knowledge gained from this programme had to be taken outside the traditional format that only tests procedural understanding. Tasks that assessed both procedural and conceptual knowledge were required almost by default for the programme to "fit the workplace" in a meaningful and cost-effective way. While it was important for the participants to be able to correctly perform a task, it was considered equally important that they have some knowledge of what was being done and why. It was strongly felt by all concerned that the certificates had been well earned by the participants, the presence of both company and university logos on the certificates lending further credibility to their worth.

However, the key focal points from the company's perspective were always going to be the successful completion of carefully selected projects allocated at the commencement of the training, and in the savings or efficiencies gained by the improvements made therein.

6 The Importance of Evaluation

There seems to be an increasing interest in assessing the knowledge and skills of those trainees participating in industrial training programmes (McMahon et al. 2009). However, the evaluation of a training programme needs a broader focus other than testing of procedural and conceptual knowledge. While the teaching of statistical techniques to people in industry can be a valuable experience, the evaluation of such training programmes and their outcomes must take into account the objectives of the organisation, as well as the personal needs of all the participants. The new knowledge and the skills imparted by the training programme must relate to real workplace needs.

6.1 Programme Evaluation Tool

The programme evaluation tool described in this section has been modified from a lean manufacturing tool devised by US Company Strategos, Inc (2005) to investigate, evaluate, and measure nine key areas of manufacturing. The modification involved identifying nine key features thought to be required for achieving success in industry-based training programmes involving statistical

Table 4 Questions for pre-planning and likert-style scoring

1.0	Pre-planning	Response	✓	
1.1	To what extent has senior management been involved in the initial pre-planning of the training programme	None		*0*
		Very little		*1*
		Some involvement	✓	**2**
		Keen interest		*3*
		Strong active involvement		*4*
1.2	Was a project manager appointed and if so to whom does the project manager report	No appointment		*0*
		Peers & colleagues		*1*
		Middle management	✓	**2**
		Senior management		*3*
		CEO		*4*
1.3	Upon what basis was the need for training identified	It wasn't—just happened		*0*
		Told to do it		*1*
		As result of measured need		2
		Result of previous experience	✓	**3**
		Part of improvement programme		*4*
		Pre-planning score		**7**

content. The resulting programme evaluation tool groups were called PEGs and arose out of the concerns expressed in the literature, the discussions arising from the 2009 National Forum in Brisbane, and practical experiences from a wide range of consultancies. The PEGs include the following:

- Pre-planning and preparation
- Project selection
- The training material
- The venues and facilities
- The trainees
- The trainer
- The training
- The level of internal support
- The degree of accountability

They are in no particular order and form the basis of a tool that may be used for evaluating the success or otherwise of such training programmes. Each PEG has a series of related questions or statements to be answered and scored using Likert-type scales. Table 4 shows an example of responses to questions associated with the first PEG as listed. The responses were based upon a trial of the evaluation tool during a recent industrial training programme.

The scored responses are aggregated to provide a score for each PEG. In this instance, the values for each response are 2, 2, and 3, respectively. This provides an aggregated score of 7 for the PEG entitled Pre-Planning. Note that the responses for each subsection are vertically ranked from least desirable (score=0) to most desirable (score=4, the maximum).

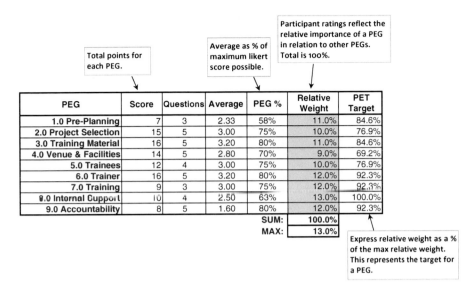

Fig. 1 Summary table of calculations for the evaluation of a training programme

Dividing the aggregate score by the number of questions for the PEG provides an average for that group. The PEG average is then expressed as a percentage of the maximum score (in this case 4) to provide the final score for that particular PEG.

The participants in the training programme are required to determine the relative importance of each PEG. This is done by consensus, and the result is expressed as a percentage where the total for all nine PEGs adds to 100 %. This then represents a weighted measure of the relative importance of each PEG in relation to other PEGs, perceived to be required for the successful implementation of such a training programme. Figure 1 shows the table of calculations associated with one particular training programme. Note that the target score for each PEG is the Relative Weight percentage divided by the maximum of the relative weights, which in this case is 13 %. In this way the targets represent perceived levels of importance relative to the PEG section considered to be the most important.

The final result is presented as an XL graphic display (Radar Chart) showing where expectations have been exceeded or otherwise. The blue-shaded region represents participants' assessments of the current training programme while the red trace represents their collective opinion of the relative importance of each PEG (Fig. 2).

The gaps between the target and the actual are assumed to represent perceived differences or shortcomings of the current training programme being evaluated. The magnitudes of the differences between the two traces are where potential improvements in the training programme might need to be made. In this case it seems that the gaps associated with internal support as well as pre-planning were largest, which suggests that it might be worthwhile critically appraising these areas to see how they might be done better.

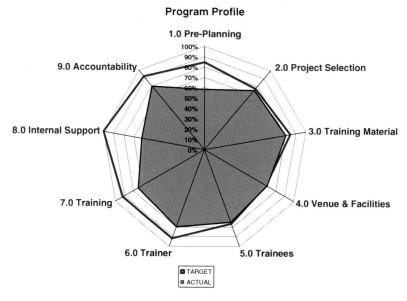

Fig. 2 Radar chart comparing actual rating of training programme with desired evaluation rating

This remains a work in progress, and much still needs to be done to establish the validity and reliability of the items that comprise such analysis. As an evaluation procedure it attempts to consider aspects of industrial training programmes considered to be important from both a workplace and a pedagogical perspective by both participants and educators.

7 Conclusion

So how then do we go about fitting square pegs into round holes? It is not easy for academics to teach statistics in industrial training programmes, probably because we have not been required to be so customer focused. As evidenced in the case study discussed previously, there is a requirement for the trainer to be demand-driven, client-focused, and responsive to their clients. There is also a strong need to be capable of building positive client relationships, ensuring customer responsiveness and supporting flexible delivery. This will require a deeper knowledge of both education and industry, and more skills in both customising training and devising business solutions. Rarely have these attributes been seen on academic job descriptions. In this author's experience the most effective way of overcoming these barriers is to spend considerable time and effort in the initial planning stages and to be very focused upon establishing a positive cooperative relationship.

Many academics involved in teaching probably tend to teach in a manner that reflects their own level of comfort and perhaps their own preferred learning style. However, those with knowledge of adult learning principles, and theories about student-centred learning and constructivism, are more likely to be more comfortable in focusing upon the participant's learning. Such educators tend to use self-directed learning opportunities and interactive learning environments rather than the traditional lecture format.

For an academic to teach statistics in industry it is not enough to be familiar with the features underlying adult learning. It is also necessary to be familiar with the workplace context within which the knowledge/training will apply. This is particularly important to win the confidence of the participants. As such, all learning material needs to be relevant to participants' work practices if commitment is to be achieved, and as educators we need to think about the role of evaluation of participation, knowledge gained, and application. For industry-based training programmes to be effective any new knowledge and skills must relate to real workplace needs.

From the trainer's perspective the overall success of a programme may be evidenced by participant pride upon completion of the training, reports of cost savings, and efficiencies resulting from the successful completion of projects, and in continued involvement to this day in post-training project work, in both an advisory and consultative capacity. The author (at the time of writing) is involved in a project to improve the efficiency, reliability, and validity of a particular process work practice. The result will be the automation of a process step based upon extending some statistical tools previously used in the training programme as described. As a result of the changes proposed the process step will be characterised by statistical evidence and therefore be more scientifically sound.

There is a considerable body of literature dealing with teaching, learning, and assessment issues related to industrial training and adult (mature-age) learning. Whilst there is little educational research, if any, on the learning of statistics as part of industrial training, there is a commonality between the generic studies described in this article and the current body of knowledge related to statistics education in general, a commonality that may be applied to the learning of statistics as part of industrial training.

A training programme that is based upon recognised adult learning characteristics will maximise the opportunity for achieving long-lasting learning and success for both participants and the organisation as a whole. With careful pre-planning and commitment from the senior levels of an organisation, the industrial training experience can be a rewarding experience for all concerned, not the least being the undergraduate students back at home base.

In the Dairy Industry case study the assessment programme was demanding and time-consuming for all, and probably not appropriate for large groups of university students. There were some aspects, however, that might be able to be adopted for small, interest-based university classes. Such instances might include, for example, small, one-off classes for research students, for students from other faculties, occupational health and safety programmes, postgraduate programmes.

The potential benefit that such industrial teaching experiences might have with respect to University teaching should not be overlooked, nor underestimated. Real-life examples may be drawn upon and used to illustrate or demonstrate various aspects of content covered in undergraduate statistics courses. Historically as teachers of statistics many of us have relied on text book examples to use in teaching undergraduate students. Becoming actively involved in an industrial training programme, trusting and utilising the very statistical skills and tools we have been trying to pass on to students for decades, has considerably enhanced this author's undergraduate teaching programmes. To be able to illustrate or demonstrate various statistical principles and/or procedures using actual case studies from consultancies has created more credibility, and raised students' awareness and appreciation of the statistical content of these courses. From this author's perspective the initial strain of fitting a square peg into a round hole is well worth the time and effort, from both a personal point of view and that of the students.

8 Note

Developed from a paper presented at Sixth Australian Conference on Teaching Statistics, July 2008, Melbourne, Australia.

This chapter is refereed.

References

Allen Consulting Group. (2006). *World class skills for world class industries: Employers' perspectives on skilling in Australia. Report to Australian Industry Group.* Canberra, ACT, Australia: AiGroup.

Azevedo, M. M. (1990). Professional development of teaching assistants: Training versus education. *Association of Departments of Foreign Languages Bulletin, 22*(1), 24–28.

Biggs, J. B., & Moore, P. J. (1993). *Process of learning* (3rd ed.). Sydney, NSW, Australia: Prentice Hall.

Dawe, S., & Nguyen, N. (2007). *Education and training that meets the needs of small business: A systematic review of research.* Adelaide, SA, Australia: National Centre for Vocational Education Research.

Forster, M., Smith, D. P., & Wild, C. J. (2005). Teaching students to write about statistics. In L. Weldon & B. Phillips (Eds.), *Proceedings of the IASE/ISI Satellite conference on statistics education.* CD-ROM and online http://www.stat.auckland.ac.nz/~iase/publications.php

Garfield, J., delMas, R., & Chance, B. (2003). *Assessment resource tools for improving statistical thinking.* Retrieved April 21, 2008, from https://app.gen.umn.edu/artist/articles/AERA_2003.pdf

Gibb, J. (Ed.). (2004). *Generic skills in vocational education and training: Research readings.* Adelaide, SA, Australia: NCVER. Retrieved from http://www.ncver.edu.au/vetcontext/21039.html.

Harris, R., Simons, M., & Moore, J. (2005). *'A huge learning curve': TAFE practitioners' ways of working with private enterprises.* Adelaide, SA, Australia: NCVER. Retrieved from http://www.ncver.edu.au/vetcontext.

Lipson, K. (2007). Assessing understanding in statistics. In B. Phillips & L. Weldon (Eds.), *Proceedings of the IASE/ISI satellite conference on assessing student learning in statistics.* CD-ROM and online at http://www.swinburne.edu.au/lss/statistics/IASE/CD_Assessment

Makar, K., & Confrey, J. (2002). Comparing two distributions: Investigating secondary teachers' statistical thinking. In *Proceedings of the sixth international conference on teaching statistics.* Capetown, South Africa: ISI.

Marr, B., & Hagston, J. (2007). *Thinking beyond numbers: Learning numeracy for the future workplace.* Adelaide, SA, Australia: NCVER. Retrieved from http://www.ncver.edu.au/vetcontext/21039.html.

Martin, P. J. (1997). Adult learning in an industrial setting. In F. Biddulph & K. Carr (Eds.), *People in mathematics education: Proceedings of MERGA 20* (Vol. II, pp. 337–343). Waikato, New Zealand: MERGA.

Martin, P. J. (2005). Enhancing effective communication of statistical analysis to non-statistical audiences. In *Proceedings of IASE satellite conference on statistics education and the communication of statistics*. Sydney: IASE, ISI.

Martin, P. J. (2006). Achieving success in industrial training. In *Working cooperatively in statistics education: Proceedings of seventh international conference on teaching statistics*. Salvador, Brazil: IASE, ISI. Retrieved from http://www.stat.auckland.ac.nz/~iase/publications

Martin, P. J. (2008). Assessment of participants in an industrial training program. In *Proceedings of OZCOTS 2008—Sixth Australian conference on teaching statistics*, Melbourne. Retrieved October 19, 2009, from http://silmaril.math.sci.qut.edu.au/ozcots2008/prelim.html

Mawer, G., & Jackson, E. (2005). *Training of existing workers: Issues, incentives and models*. Adelaide, SA, Australia: NCVER. Retrieved from http://www.ncver.edu.au/vetcontext/21039.html.

McMahon, K., Jaron, D., & Collins, P. (2009). New directions for training packages. Paper presented at *Australian Vocational Education and Training Research Association 12th annual conference*. Sydney: AVETRA. Retrieved from http://avetra.org.au/publications/conference-archives

Misko, J. (2006). *Combining formal, non-formal and informal learning for workforce skill development*. Adelaide, SA, Australia: NCVER. Retrieved from http://www.ncver.edu.au/vetcontext/21039.html.

Mitchell, J. (2008). Capabilities of the emerging 'advanced VET practitioner'. Paper presented at *Australian Vocational Education and Training Research Association 11th annual conference*. Adelaide: AVETRA. Retrieved from http://avetra.org.au/publications/conference-archives

Morgan, C. J., & Deutschmann, P. W. (2003). An evolving model for training and education in resource-poor settings: teaching health workers to fish. *The Medical Journal of Australia, 178*(1), 21–25.

Pretty, J. N., Gujit, I., Scoones, I., & Thompson, J. (1995). *A trainer's guide for participatory learning and action*. London: International Institute for Environment and Development.

Ramsden, P. (1992). *Learning to teach in higher education*. London: Routledge.

Richardson, S. (2004). *Employers' contribution to training*. Adelaide, SA, Australia: NCVER.

Rogers, A. (1986). *Teaching adults*. Milton Keynes, England: Open University Press.

Roth, W. M., & Bowen, M. (2001). Professionals read graphs: A semiotic analysis. *Journal of Research in Mathematics Education, 32*(2), 159–194.

Smith, A., & Oczkowski, E. (2009). Why do Australian companies train their workers? An analysis of 2005 SEUV data. Paper presented at *Australian Vocational Education & Training Research Association 12th annual conference*. Sydney: AVETRA. Retrieved from http://avetra.org.au/publications/

Smith, C., Smith, A., & Smith, E. (2007). *Pedagogical issues for training of mature-aged workers in manufacturing industry*. Sydney, NSW, Australia: Manufacturing Skills Australia.

Strategos Inc. (2005). Lean manufacturing assessment tool. Retrieved October 31, 2012, from http://www.strategosinc.com/downloads/download-central.htm

Svennson, E. (2007). Assessing learning by students' own examination tasks. Experiences from research courses in biostatistics. In B. Phillips & L. Weldon (Eds.), *Proceedings of the IASE/ISI satellite conference on assessing student learning in statistics*. CD-ROM and online at http://www.swinburne.edu.au/lss/statistics/IASE/CD_Assessment

Werner, D., & Bower, W. (1995). *Helping health workers learn*. Berkeley, CA: Hesperian Foundation.

Wild, C., & Pfannkuch, M. (1998). What is statistical thinking? In *Proceedings of the fifth international conference on teaching statistics*. Voorburg, The Netherlands: ISI.

Wild, C., & Pfannkuch, M. (1999). Statistical thinking in empirical enquiry. *International Statistical Review, 67*(3), 223–265.

Raising the Capability of Producers and Users of Official Statistics

S.D. Forbes, J.A. Harraway, James O. Chipperfield, and Siu-Ming Tam

Abstract In both Australia and New Zealand, the National Statistics Offices have developed strong partnerships with academics to raise statistical capability. Both offices recognise the importance of good methodology to underpin official statistics. However, the main target group for Statistics New Zealand (SNZ) has been external users of official statistics, but that for the Australian Bureau of Statistics (ABS) has been its own statistical methodologists (producers) and advancing research. This chapter outlines sets of initiatives from both agencies. SNZ has actively focussed on raising the statistical capability of key groups of users, including schools, small businesses, government, the media and Maori. It has established a network of academics in official statistics whose members are involved in the design, implementation, delivery and assessment of courses for qualification as well as presenting short (1- or 2-day) courses. The ABS places strong emphasis on the recruitment, training and grooming of young methodologists to become leaders in their chosen field of research, and their focus is on collaboration with the university sector and academics to help with this and to foster ABS research. Other initiatives undertaken in both organisations are also briefly mentioned, including the Census AtSchool project.

S.D. Forbes (✉)
Statistics New Zealand and Victoria University of Wellington, Featherston, New Zealand
e-mail: sharleen.forbes@vuw.ac.nz

J.A. Harraway
University of Otago, Dunedin, New Zealand
e-mail: jharraway@maths.otago.ac.nz

J.O. Chipperfield • S.-M. Tam
Australian Bureau of Statistics, Belconnen, ACT, Australia
e-mail: james.chipperfield@abs.gov.au; siu-ming.tam@abs.gov.au

H. MacGillivray et al. (eds.), *Topics from Australian Conferences on Teaching Statistics:*   247
OZCOTS 2008-2012, Springer Proceedings in Mathematics & Statistics 81,
DOI 10.1007/978-1-4939-0603-1_14, © Springer Science+Business Media New York 2014

Keywords Official statistics education • Partnerships between academics and official statisticians • Official statistics training

1 Introduction

Most National Statistics Offices (NSOs) have ongoing significant interest in developing the capabilities of both users and producers of their statistics. Capable producers are required to maintain the high quality, trust and regard in which official statistics are held. Official Statistics need good underpinning methodology so strong emphasis is placed on the recruitment, training and grooming of young methodologists. Capable users are required to ensure that official statistics are used and valued. As Bill McLennan (2005), a past Australian and British Government Statistician said, 'to be useful, official statistics need to be used'. NSOs can add value to the statistics they create by helping key users to attain the skills they need to adequately use their data. In both Australia and New Zealand, the NSOs provide internal training for their own staff and also provide levels of statistical assistance to external users of official statistics. They both have also developed strong partnerships with academics to help raise statistical capability. One of the reasons for having educational providers as partners in raising capability is that while many of the staff in NSOs have expertise in statistics, only a few have expertise in education as well, other than internal training of their staff. Also the primary output purchased by Governments is the production of official statistics not education in statistics. The two organisations discussed here have emphasised different aspects of capability building in their partnerships with universities. The main target group for Statistics New Zealand (SNZ) has been external users, in particular government employees, of official statistics, but that for the Australian Bureau of Statistics (ABS) has been its own statistical methodologists (producers). In both cases, the training of users and the ongoing developing of methodologists, partnerships with academics have helped in developing capabilities of users and producers and also developed long-term strategic relationships. This is not a new idea. For example, in the United Kingdom, there is a Master's course in Official Statistics (M.Sc. Official Statistics) developed specifically for government statisticians by the Office of National Statistics in collaboration with the University of Southampton in 1999 (Brown 2007): http://www.southampton.ac.uk/demography/postgraduate/taught_courses/msc_official_statistics.page.

This chapter comprises two parts, with each part describing initiatives by an NSO in developing productive and successful partnerships with universities. The contrast between the two parts is due to the different but equally important foci of the initiatives and partnerships, with the first focussed on users and the second on producers of official statistics. The first discusses initiatives developed by SNZ and its network of academics in official statistics (NAOS) to raise the statistical capability of key groups of users. SNZ is actively involved in raising the statistical capability of a number of key groups of users, including schools, small businesses, government, the media and Maori. Partnerships with academics have primarily been used with schools, government, new graduates and in the near future, with Maori (indigenous New Zealanders). The second outlines the ABS' collaboration with the university sector and academics in helping to turn young ABS methodologists into research leaders in their field of expertise.

2 Improving Use of Statistical Data Products by Improving Capability of Users

As mentioned the main role of mathematical statisticians within an NSO is to ensure its products and services are informative and relevant to user's needs. Once these products and services are available it does not naturally follow that society is *informed*. This is a concern for all NSOs and this section discusses how SNZ has actively focussed on raising the statistical capability of key groups of users, including schools, small businesses, government, the media and Maori. The goal of this is to improve the use, and therefore value, of SNZ's data products.

2.1 School Students

The focus for NSOs is primarily on making data products including statistical indicators. These products are becoming more numerous, more detailed, more diverse and more readily available. Students, in learning to make their own data products, can use Official Statistics as models of good practice giving them a ready-made and vast resource to draw on. NSOs can also offer access to, or specifically create, data sets for use in all levels of statistics education. Both agencies have a number of products designed to support statistics learning, together with an even larger number designed for public and/or professional audiences. However, to ensure that these data products are accessible, interesting, valued and engaged with, requires that official statistics agencies and statistical literacy educators work together to inform the education community about these products and how to use them effectively in their everyday teaching.

Many NSOs invest either directly, or in collaboration with their Ministries of Education, in developing resources for schools and school students. The aim of this long-term investment is to both produce a future general public that is more statistically literate than the current one and also to potentially increase the future pool of statistics graduates. As Forbes (2010) states 'Today's children are tomorrow's decision makers'. However, it does take substantial work together with a very good understanding of what works in the classroom to turn official statistics into classroom-ready teaching and learning experiences.

For many years, a major focus for SNZ has been collaboration with teachers and academics to provide appropriate statistical literacy resources that can be used in classrooms. In 2008, New Zealand introduced a new, world-leading, Mathematics and Statistics curriculum that covers all levels of schooling (from new entrants to the final year of secondary schooling) and, for the first time, put statistics on an equal footing with mathematics (Ministry of Education 2007). The statistics component of the curriculum has three separate strands running through each level of schooling with the content getting more sophisticated as the level increases. The strands are statistical investigations (using a statistical enquiry cycle), statistical literacy (interpreting statistical and probability statements made by others) and probability. At each of the eight levels of schooling students should learn in 'meaningful contexts' (Ministry of Education 2012). As Wild and Pfannkuch (1999) suggest, experiencing statistics within an inquiry cycle such as PPDAC (Problem, Plan, Data, Analysis, Conclusion) is a fundamental learning experience.

SNZ staff have been directly involved in developing curriculum standards, reviewing assessment materials and giving advice (primarily through their membership of the New Zealand Statistical Association's education subcommittee). SNZ also provides support for teachers via conferences, teacher workshops and Schools Corner (http://www.stats.govt.nz/tools_and_services/services/schools_corner.aspx) on its website. Schools Corner 'contains a variety of teaching resources aimed at both primary and secondary school teachers'. An understanding of statistical concepts is built over many years of learning and ideally should build on students' intuitive reasoning. Using real data, such as selections based on official statistics, can provide practice in statistical analysis, discourse and argumentation. SNZ (free) public releases include 'Hot Off The Press' media releases, Infoshare (time-series data sets), Table Builder, and Census QuickStats about places and subjects can all be used in the school environment.

Products designed specifically by SNZ to support school statistics learning include synthetic unit record files (SURFs) based on real SNZ datasets such as the 2004 Income Survey and the 2006 Census. As stated on the SNZ website 'The New Zealand Income Survey (NZIS) Super SURF is a set of 100 SURFs based on data from the NZIS: June 2003 quarter a largely realistic representation of a portion of the New Zealand population and can be used for teaching and learning purposes (http://www.stats.govt.nz/searchresults.aspx?q=SURFs). More in-depth discussion of SURFs is given in Forbes et al. (2011). All the SNZ products designed for school use are developed either in collaboration with school teachers or by members of staff who have formerly been teachers.

In addition, SNZ jointly funds (together with the Ministry of Education) the New Zealand CensusAtSchools project, http://new.censusatschool.org.nz/, in which students answer surveys on themselves, and can access and work with data samples from the resulting national and international databases. This project is run by Auckland University with annual milestones being agreed collaboratively between SNZ and the university. The ABS has an even greater involvement with CensusAtSchools, funding and running the Australian version directly from their Melbourne office http://www.abs.gov.au/websitedbs/CaSHome.nsf/Home/.

All school teachers would benefit by an understanding of what can be achieved by investigating some of the large national data sets that provide interesting project work for school students as well as ideas for school competitions such as the International Statistics Literacy Project poster competition (http://www.stat.auckland.ac.nz/~iase/islp/competition-second).

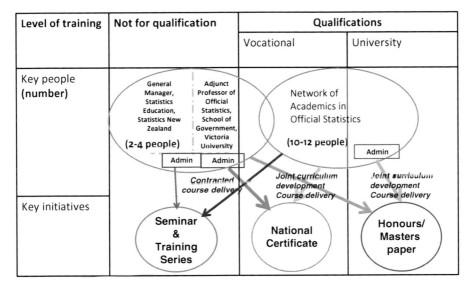

Fig. 1 Key people and initiatives in official statistics training for government employees

2.2 Government Users

The first step that SNZ took to raise the statistical capability of public sector employees was to sponsor a new, half-time position—Professor in Official Statistics in the School of Government at Victoria University of Wellington in late 2009. The aims of this position were increasing the use of official statistics in policy advice and establishing official statistics as a recognised academic sub-discipline within statistics. As shown in Fig. 1, there is only a very small group within SNZ with responsibility for raising statistics capability in the whole external community so in 2006 an NAOS who were interested in the teaching of statistics was established. This network provided SNZ with a pool of people who could be called on to teach courses (whether for qualification or not) and to assess those courses that resulted in a formal qualification. NAOS currently comprises academics from university statistics departments (at least one from each of the seven main New Zealand universities) together with academics from other disciplines (social science, geography, economics, etc.). The strength of such a network is of course dependent on the strength of the relationships between its members. In New Zealand's case, the organisation that has helped these relationships develop over time is the New Zealand Statistical Association that holds regular annual conferences. A number of the NAOS and SNZ members have served on the education subcommittee of this association or been elected members of its executive.

Over time the network moved from being consultative to collaborative to cooperative and now works in direct partnership with SNZ on a number of statistics education initiatives. NAOS members have been involved in the design, implementation, delivery and assessment of courses for qualification as well as presenting short (1- or 2-day) courses for government employees and giving lunch-time seminars. Workshops presented by NAOS members include 'Truth, damned truth and statistics', 'How I learned to stop worrying and love the survey cycle', 'Confidentiality and official statistics' and 'Population Aging'. There are two projects which have been truly cooperative, involving NAOS members in both their design and implementation. The first is a national Certificate in Official Statistics for state sector employees and the second is an honours/Masters course in official statistics for students from a range of disciplines who have at least first-year university statistics. The development and implementation of the latter was done by a small subgroup of NAOS convened by Professor Alan Lee from Auckland University.

3   Courses for Improving the Statistical Capability of Users

3.1   National Certificate of Official Statistics

The National Certificate of Official Statistics was jointly conceived by a 3-way partnership between SNZ, the State Services Commission (agency responsible for cross-departmental state sector issues) and the then Public Sector Industry Training Organisation (responsible for state sector training). From its conception, NAOS members advised on the number of modules or units (called Unit Standards in the New Zealand context) that the qualification would contain and the content of each unit. They also agreed to work collectively with SNZ to deliver and assess these units. The certificate was developed and piloted in 2007 then registered on the New Zealand Qualifications Framework (National Qualifications Framework Project Team 2005). The certificate is at level 5 on this framework (roughly the equivalent of first-year university) and is a 40 credit, vocational qualification comprising five units and assessed on a competency basis. Students do it part-time over an 18-month period. Although the qualification sits outside of the universities system, NAOS members (academics from Auckland, Victoria, Canterbury and Otago universities) currently deliver and/or assess each of the units taught. The following four units are taught in 1- or 2-day workshops:

1. Resolve ethical and legal issues in the collection and use of data in a public sector context.
2. Interpret statistical information to form conclusions for projects in a public sector context.
3. Assess a sample survey and evaluate inferences in a public sector context.
4. Evaluate and use statistical information to make policy recommendations in a public sector context.

A case study approach is used throughout the certificate with two main publications selected each year for use across all the four units for teaching purposes and two different publications for assessment purposes. The four case studies are selected from:

- SNZ's official statistics releases (e.g. Household Labour Force Survey, Retail Trade Survey) and reports (e.g. *Innovation in New Zealand 2005*, Statistics New Zealand 2007)
- Other government agency releases (for example, an evaluation report commissioned by the Ministry of Social Development (Fergusson et al. 2005) and the 2006 Maori Language Survey (Te Puni Kokiri 2007) an exploration of problem gambling using the New Zealand Health Survey (Arnold and Mason 2007))
- Contemporary research reports (e.g. on the benefits of insulating houses (Howden-Chapman et al 2007))

For each unit, students are given a set of written questions that are completed in their own time then submitted to the assessors (who are also contracted academic members of NAOS). The fifth, and largest (16 credit) unit, is a workplace-based statistics project that is verified by the student's manager. All the assessment is competency based with students directed to resit questions until the required standard is reached. The certificate was formally evaluated in 2009 and the results reported in International Association of Statistical Education conferences (e.g. Forbes 2009, 2011). To date, over 100 students from 37 different government and local authority agencies have enrolled in the certificate. Just over 40 % of the students have come from SNZ. The academics involved in the Certificate of Official Statistics report that it gave them exposure to teaching basic and official Statistics in a new environment and to a new type of student (older, more mature, currently in the work force and sometimes in middle management positions).

3.2 Honours/Masters Course in Official Statistics

In its first year, 2011, the honours course in Official Statistics was a joint winner of the Best Cooperative Project in Statistical Literacy award from the International Statistical Literacy Project. From the NSO's (SNZ) perspective, widening the pool of new graduates who have some official statistics knowledge should help recruitment both within the agency and across other government agencies. One of the original aims was to provide a course for new staff that would provide them with an overview of official statistics. It should also help raise official statistics knowledge and capability more generally. From an academic perspective, official statistics provide interesting data in context for motivating student learning in statistics and for developing statistical literacy in society. Although many New Zealand statistics graduates are employed in government departments, local bodies or health authorities investigating large data sets and consequently providing information for policy decisions, New Zealand universities have not previously taught courses with Official Statistics as a specific focus. Some universities view Official Statistics as an optional and less central part of the statistics curriculum, and prefer to concentrate on required material although a few do include some specific official statistics techniques such as index numbers and basic demography tools (e.g. age standardisation and odds ratios) in first-year courses that service many disciplines. Student numbers enrolling in a specialist official statistics course at a single university would be small making it uneconomic to teach so the key to making this a feasible project was the ability to pool resources across universities using an advanced video-conferencing network. Linking across universities allowed experts in key areas of Official Statistics to teach students at all the participating universities in one class from a studio at their own institution.

However, there were some obstacles to be overcome. New Zealand universities have similar but *not* identical teaching terms so some negotiation was required to determine common teaching dates and times when the technology was available at all the sites. New Zealand universities also do *not* all have the same number of credits per paper. The compromise reached was that the course was taught in 2 h weekly slots for 12 weeks. The new course was built into the degree regulations at each university with approved course codes for each institution. Another area of negotiation was prerequisites and it helped that prerequisites for statistics majors were similar at participating universities in New Zealand. Initially, students had to have completed a first-year course in statistics with at least two courses at second year level in quantitative research methods or applied statistics but this was relaxed in 2012 so that allowance could be made for interested social science and public policy students that might not be as well quantitatively prepared. It was also intended that government employees could staircase (move from one to the next) between the National Certificate of Official Statistics to the honours paper and one student has already successfully done so. The national course coordinator was at Victoria University in Wellington where the central website was also hosted. All course material (apart from that with copyright restrictions) was publically available on the website with no login identification required (http://msor.victoria.ac.nz/Courses/STOR481_2013T2). A technician was present at each site to ensure that the technology worked across all sites. All the 2 h lectures were recorded and a video feed from the computer with an inset feed from the camera on the lecturer with the combined audio stream was constructed each week. This was extensively used as revision, in particular by students whose first language was not English.

In the course students are expected to become familiar with:

1. The key aspects of Official Statistics, as distinct from other branches of statistics
2. The legal and ethical constraints on organisations producing Official Statistics
3. The principal methods for data collection, analysis and interpretation of health, social and economic data, including spatial data
4. Methods for presenting and preparing commentaries on Official Statistics

Lectures cover aspects of ethics, sampling, demography, health and other social statistics, index numbers and an overview of national accounts, as well as data

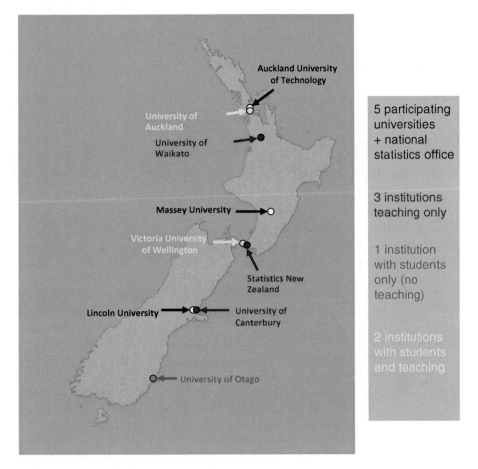

Fig. 2 Geographic distribution of participating institutions in 2012

matching and geospatial techniques such as geo-visualisation with each lecturer being a New Zealand expert in the field they taught. Students were introduced to an extensive set of links to important national data files, including examples from SNZ and other websites which they explored when answering questions on set assignments. There was no final examination, with assessment by five written assignments during the semester, each worth 20 %. Computer packages (such as EXCEL, R or SAS) were used to perform data analyses. Assignments were submitted by email to lecturers teaching the modules in the course and can be viewed on the website. These were marked and returned to students by email. In 2011, 29 students were enrolled at three New Zealand universities as shown in Fig. 2. All these students completed and passed the course. In 2013, 35 students from five universities enrolled and 32 passed.

Some students used this course as a fourth-year honours option, others as a component for a postgraduate diploma in applied statistics and some as a third-year option towards their major in statistics or as a paper contributing to a Diploma for Graduates Endorsed in Statistics, a programme used to build statistics expertise for another major. In future, students are likely from economics, the social sciences, geography, public policy or public health areas. But the course could also extend into education and teacher training.

4 Developing Capability in Producers of Official Statistics Through Partnerships with Academics

The ABS' mission is to 'assist and encourage informed decision making, research and discussion within governments and the community, by leading a high quality, objective and responsive national statistical service, and delivering statistical products of high quality'. Many other NSOs have similar missions. Measuring the quality of an NSO's statistical products and services has been discussed in the literature (see Brackstone 1999). Delivering high quality statistical products and services into the future requires, amongst other things, building and maintaining the expertise or capability of the NSO's mathematical statisticians.

The role of a mathematical statistician (see Linacre 1995) within an NSO is to ensure its products and services have a high level of statistical integrity and are cost-effective. Relevant areas of mathematical statistics applied in an NSO include, but are not limited to, the design and estimation for sample surveys and analysis (e.g. principal component analysis for the construction of socio-economic indexes) of those data typically collected by an NSO, data access and confidentiality methods and time series analysis.

While the mission of NSOs and the role of their mathematical statisticians are expected to remain largely unchanged, the external world is rapidly changing and is becoming more complex. As a result, data intensive methods cutting across different subject matter and professions are increasingly required to solve these problems. There has been an explosion of data (e.g. big data), increasing computing power, budget reductions and increasing sophistication of analysts who demand more timely access and more detailed data (e.g. person-level micro-data). There is also greater competition in the labour market for mathematical statisticians. These external changes heighten the existing challenge for NSOs to build the capability of its mathematical statisticians in order to deliver on its mission.

Section 4.1 describes many ways in which the ABS has engaged with academics to maintain such capability. Section 4.2 specifically discusses the ABS' initiative of building statistical capability through formal agreements with universities. Section 4.3 discusses some indirect, though substantial, ways in which these formal agreements build statistical capability within the ABS and Australia more generally. Section 4.4 outlines some potential barriers to building partnerships with academics.

4.1   Developing Capability in NSOs by Engaging with Academics

Challenges of building capability within NSOs include attracting graduates and training highly capable mathematical statisticians. This section describes how the ABS has historically addressed these challenges through broad-based engagement with academics.

4.1.1   Attracting Graduates to ABS

Like many NSOs, the ABS aims to recruit people with strong analytic, conceptual and communication skills who have demonstrated an aptitude for quantitative methods and have suitable qualifications. Currently there are about 100 mathematical statisticians at ABS, all of whom have a university degree and 25 % have a doctoral or postgraduate degree. Historically, a large component of ABS recruitment is made up of university graduates. In recent times, there has been increasing competition for mathematics and statistics graduates.

ABS puts considerable effort into attracting suitable graduates. Some of the attraction strategies include offering scholarships, attending career fairs, participating in conferences, giving lectures at universities, giving workshops and attending statistical society meetings. ABS also awards scholarships (comprising a modest cash payment) to students who excel in a statistics or mathematics subject to subsidise education costs (fees and books).

The ABS has provided cadetships (comprising a modest salary) to a small number (typically three or less, depending upon quality of the applicants and available funding) of Honours students who are studying mathematical statistics and have high academic achievement. The topic of the Honours research project, a requirement for many Honours courses, is negotiated between ABS, the student and the student's academic supervisor. A condition for accepting a cadetship is that the student agrees to work at the ABS for a minimum period of 6 months. While the success of the cadetship programme has not been formally evaluated, it is widely acknowledged that the programme has resulted in the recruitment of some highly influential statisticians who have made significant contributions to the ABS, both within and outside of the methodology area. Indeed two recent Australian Statisticians started their careers in the ABS as cadets.

4.1.2   Direct Training of ABS Staff

The ABS hosts visits by overseas and local statistical experts for the purpose of training ABS staff. For example, in 2012 the ABS hosted a series of lectures presented by leading academics and experts on the topic 'Disclosure Risk of Linked and Longitudinal Data'. Since 2012, the ABS has funded world experts to deliver short courses and lecture series on Synthetic Data for Disclosure Control, Maximum Likelihood and Bayesian methods. Less recently, short courses covered topics on Model Assisted Methods (MAM) and Variance Estimation for Complex Sample Designs.

4.1.3   Technical Input and Review

Three examples of how academics provide technical input and review include:

- The ABS' Methodology Advisory Committee, comprising leading New Zealand and Australian academics in mathematical statistics, meets twice a year to critique proposed methods or techniques being considered by the ABS.
- Academics from various universities have provided advice about the content of two significant training courses developed and delivered by the ABS (Clark, 2002; Van Halderen and Bhattacharjee 1998). The Survey Methods course is an introduction into sample designs and estimation, with an emphasis on relevant ABS applications. The course is designed to develop the fundamental technical skills required of a mathematical statistician at the ABS. It is a prerequisite for the MAM course. MAM is an introduction to the role of models in sample designs and estimation and is built around the material in Särndal et al. (1992).
- On occasion an academic may collaborate with an ABS statistician on a topic of mutual interest with a view to publication. Recent examples include Chipperfield and Steel (2009, 2011, 2012).
- While the ABS does not have institutionalised arrangements for seconding academics to work in the ABS, in 2011 an Adjunct Professor and Research Program Leader at CSIRO was seconded to ABS for 1 year on a part-time basis. Collaboration during the secondment resulted in the submission of papers on confidentiality to peer-reviewed journals (see, for example, O'Keefe and Chipperfield 2013).

4.1.4   Encouraging Interaction Between Academics and ABS Staff

Examples of the way in which the ABS has encouraged interaction between its staff and academics include:

- Hosting Statistical Society meetings, which allow for formal and informal interaction with academics.
- Funding staff to present at conferences allows for interaction with academics. Annually, the ABS funds an international trip to 'budding' methodologist who has less than 10 years of experience and who has demonstrated strong potential for further development as a professional statistician. The trip typically includes attending conferences, universities and meeting with their counterparts in overseas NSOs.

- Holding honorary academic positions that allow ABS staff to jointly supervise Ph.D. students. It also can mean that universities provide full or part funding for ABS staff to attend conferences or cover other expenses.

4.1.5   Further Study

There are two main ways in which the ABS supports further study. First, ABS staff may apply for up to 6 h per week paid leave to study, provided the study (e.g. course Ph.D. or Masters) is relevant to ABS' business.

Second, the ABS has also provided Ph.D. scholarships to support research that is relevant for official statistics. It is not uncommon for ABS staff to take-up such an opportunity and to return to ABS employment after completing their Ph.D.

4.1.6   Examples of Developing Capability in NSOs Through Funding Agreements with Universities

This section describes a significant collaborative approach undertaken by ABS with one university, and how this has been expanded to other universities, providing an example of a type of partnership that other NSOs might choose to employ.

4.2   University of Wollongong

With a view to enhancing ABS' capability in survey methods and analysis techniques, the ABS has fullyfunded a Professorial Chair in Survey and Statistical Methodology at the University of Wollongong (UoW) in Australia since 2004. A world-class professor with expertise and extensive prior experience with statistical agencies was appointed to the position. Since the appointment, the Chair has improved the expertise of ABS' mathematical statisticians by encouraging innovation, suggesting new ways of approaching methodological problems, and by improving the methodological rigour of ABS research and development.

Of particular note is the assistance the Chair has provided in developing a framework for the analysis of probabilistically linked micro-data that was relevant for the ABS. Also of note is the Chair's work in the development of a framework for the follow-up of non-respondents to ABS surveys. Both these frameworks are being refined and evaluated through ongoing joint work between ABS and the professor.

As part of funding arrangement, an explicit role of UoW, led by the Chair, is to provide direct training to ABS staff, provide technical input and review, and encourage interaction between academics and ABS staff. This specifically involves:

Training courses: The UoW provide three training courses per year. Examples include:

- Principles and Practices of Methodological Research. This provides an introduction to the design-based and model-based frameworks, and their roles in making inference. It provides a summary of the different statistical tools (e.g. Bootstrapping, Multiple Imputation) and how to evaluate the performance of a statistical method (e.g. simulation). It also describes how to manage research and how to disseminate the results (e.g. publication).

- Introduction to Multilevel Modelling for Repeated Measures Data. This course introduces multilevel models and their application within the computer package MLwiN using a mixture of lectures and workshops.
- Methods of Combining Data. This course considers model-based and design-based methods of making inference using more than one set of data (e.g. meta-analysis).

Other courses include: Sample Design in Practice (sampling hard-to-reach populations), Advanced Survey Estimation, Detection and Treatment of Outliers and Logistic Regression. Importantly, the specific topics are based on the capability needs of ABS methodologists.

One-on-one meetings. These meetings typically involve a staff member raising a practical problem with the Chair, who may abstract the problem within an established statistical framework and provide relevant references to published articles in the statistical literature.

Project advice on select projects. Advice generally includes identifying the key statistical issues and references at the beginning of the project, providing ongoing advice during the project, and commenting on the project's final report.

Comment on technical papers. Comments focus on improving the statistical rigour of the papers with a view to making them of publishable quality.

Review articles. A review article provides a broad overview of the literature on a particular statistical problem or method, identifies key references, and explains its relevance and applicability to official statistics.

Symposia: The ABS and UoW organise an annual symposium, where experts and practitioners present talks on a particular theme of statistics. The 2012 symposium was on Analysis of Longitudinal Data and included talks about the Household and Income Labour Dynamics in Australia and the Survey of Australian Children. The 2013 symposium was on Bayesian methods for Official Statistics.

Triennial UoW fellows meeting. A UoW Fellows Meeting includes four presentations and interactive discussion between Fellows and invited guests, which comprise academics and other technical experts. Presenters at these meetings require a high level of technical expertise.

4.3 Australian Technology Network

An emerging objective for many NSOs is to *unlock the statistical value in administrative data* collected by commonwealth government agencies. This presents two major statistical challenges: integrating multiple sources of administrative data, using record linkage or data fusion methods; and facilitating access to analysts, within the community and government, to the data without disclosing information about people or organisations within Australia. There is limited academic expertise and experience in these fields of statistics within Australia.

To attract such expertise, the ABS has committed to part-fund two professor positions, which have been advertised internationally.

The positions are located in the Queensland University of Technology (QUT) and the University of Technology Sydney (UTS), which are members of the Australian Technology Network (ATN).

The ABS signed a memorandum of understanding (MOU) with the ATN in February 2012 to fund these positions. Under the MOU, the ABS Chairs will spend about 20 % of their time helping ABS to build its own capability on data integration and disclosure avoidance techniques by: presenting training courses, collaborating and providing advice on key projects, reviewing the literature, attending and commenting on ABS technical reports.

The ABS, as a foundation partner of the ATN's Industry Doctoral Training Centre (IDTC) opened in 2012, will now have the opportunity to influence what subjects are taught to the Centre's Ph.D. students, and the research topics of the students. In particular, the ABS plans to deliver a semester length course to IDTC Ph.D. students on survey methodology.

In addition, the ABS has agreed to supplement scholarships for up to two Ph.D. students at UTS or QUT who are conducting research that is relevant to the ABS' methodology work programme.

4.4   Indirect Benefits of Funding Agreements

Building capability in an NSO for the long term requires a stable and well-functioning academic community that conducts quality research in problems facing official statistics. Funding academic positions in official statistics can also provide indirect benefits for NSOs. Some examples of where ABS has indirectly benefited from funding the UoW Chair in statistics include:

- The Chair applying for research grants in topics relevant to ABS. These funds were used to recruit research fellows and Ph.D. students in statistics, thereby increasing the pool of statisticians with skills in demand by the ABS, as well as advancing knowledge in these topics.
- The Chair being funded, through consultancies, by other agencies within Australia and internationally to conduct research into topics relevant to ABS.
- Funding the Chair broadly supports the teaching of statistics courses at UoW, again increasing the skills of the recruitment pool for the ABS. UoW currently provides a range of undergraduate courses on survey sampling and data analysis, and provides a postgraduate certificate in statistics. These courses cover areas of statistics that are highly relevant to the ABS. Graduates from the UoW therefore provide a promising pool of future recruits for the ABS.

4.5   Perceived Barriers to Effective Collaborations Between Academics and NSOs

This section comments briefly on factors that can limit the amount and benefit of collaboration between academics and NSOs.

Statistical methods typically applied within the ABS and other NSOs form only a part of the broad field of statistics. As a result, attracting academics with the right expertise and experience with whom to partner can be challenging. An academic with the right expertise may not reside within Australia or New Zealand, given their relatively small population and geographic isolation.

Other barriers include that:

- Academics have restricted access to ABS' micro-data.
- ABS' priorities generally change more rapidly than the research priorities of academics.
- The focus of the ABS is on solving a particular problem while the focus of academics is often on solving a generic problem.

These barriers can be managed, but it requires balancing the interests of both parties.

5   Concluding Comments

The possible strategies for an NSO in aiming to achieve its objectives and responsibilities are very broad and varied. The recent innovations described in this chapter demonstrate that focussing on certain aspects of objectives and certain 'audiences' have produced exciting developments which appear contrasting, but which contribute in different ways to meeting an NSO's overall aims and in ways in which an NSO can partner with universities.

NSOs provide interesting data in context for motivating student learning in statistics and for developing statistical literacy in the wider society. For example, a school teacher with a major in mathematics and statistics would find Official Statistics to be an excellent source of teaching resources. Both the ABS and SNZ have a long history of being involved with and producing products specifically for schools.

SNZ has extended this model with a formal partnership with academics, NAOS, that has helped shape two successful initiatives judging by student achievement, questionnaire responses and informal feedback. These are New Zealand's National Certificate in Official Statistics and Honours Course in Official Statistics. The new honours paper, in particular provides a win–win situation for all the players. There are both tangible and intangible benefits for SNZ in that it makes young graduates aware of the official statistics system as a potential employer and also gives them exposure to official statistics that they will routinely come across in their everyday life as well as possibly in their future employment. Universities have benefitted by being able to provide a specialist course in statistics that would not have been feasible otherwise. But the real beneficiaries are the students who are given the opportunity to investigate a new area of statistics that they might not have been aware of otherwise.

From the perspective of a statistical agency, the world is becoming more complex. The ABS needs to keep its methodologists up to date with contemporary statistical methods, and perhaps develop these methods to tackle complex statistical problems. While the ABS partnerships with academics have been very successful, there are some challenges for the future. The current take-up of statistics in Australian universities presents challenges to the supply of capable graduates with relevant training in the future. To address this, the ABS is beginning to geographically decentralise its staff to its offices across Australia which, given technological advances, is becoming increasingly practical. This means that a preference for living close to family and friends is less of a barrier to working with the ABS than it has in the past. It also means that ABS' mathematical statisticians can be geographically closer to the academic expertise located across Australia. Historically, decentralisation of mathematical statisticians has been resisted to some extent by Australia and other NSOs (e.g. Felligi 2010 discusses this issue from the perspective of Statistics Canada). SNZ, however, has three offices in the major cities of Auckland, Wellington and Christchurch with the largest group of mathematical statisticians being in the head office in Wellington. For the reasons given above, there has also been a sizable group in the Christchurch office. The focus of the Auckland office has been on data collection. While for some time there has been a mathematical statistics presence there, this has only comprised one or two staff.

Forming partnerships with education providers to raise statistical capability has demonstrated benefits for official statistics agencies, the education providers and more importantly the learners of statistics. However, many NSOs are facing tight budgets that will impose interesting challenges for agencies trying to maintain the same level of engagement with academics, let alone expand it. They will need to strike a balance between the imperative of balancing annual budgets and the longer term investment in statistical capability both within and external to the NSO. However, on a positive note, there is a consistently been a rich source of interesting methodological problems statistical problems facing NSOs. This means they are a natural research partner for academics now and into the future. While this has been an area of active collaboration for the ABS for some time it is just beginning to be pursued by SNZ.

Although the possibilities for an NSO to collaborate with academics are extensive and substantial no single NSO is likely to be able to utilise all the possibilities. In discussing two contrasting but complementary foci—on users and on producers of official statistics—this chapter illustrates how the development of partnership arrangements with teachers and academics, whether formal or informal, has had overall long-term benefits for both agencies.

6 Note

Developed from a paper presented at Seventh Australian Conference on Teaching Statistics, July 2010, Perth, Australia.

This chapter is refereed.

References

Arnold, R., & Mason, K. (2007). Problem gambling risk factors and associated behaviours and health status: results from the 2002/03 New Zealand Health Survey. *New Zealand Medical Journal, 1*(1257). Retrieved from http://www.nzma.org.nz/journal/120-1257/2604/

Brackstone, G. (1999). Managing data quality in a statistical agency. *Survey Methodology, 25*, 139–149.

Brown, J. J. (2007). Teaching sampling statistics: experience from the MSc in Official Statistics Programme. *Paper presented to the 56th Session of the International Statistical Institute*, Libson.

Chipperfield, J., & Steel, D. G. (2009). Design and estimation for split questionnaire surveys. *Journal of Official Statistics, 2*, 227–244.

Chipperfield, J., & Steel, D. G. (2011). Efficiency of split questionnaire surveys. *Journal of Statistical Planning and Inference, 141*, 1925–1932.

Chipperfield, J., & Steel, D. G. (2012). Multivariate random effect models with complete and incomplete data. *Journal of Multivariate Analysis, 109*, 146–155.

Clark, R. (2002). Statistical training of statistical services branch staff, Methodology Advisory Committee, catalogue number 1352.0.55.053. Australian Bureau of Statistics, Canberra.

Felligi, I. P. (2010). The organisation of statistical methodology and methodological research in national statistical offices. *Survey Methodology, 36*, 123–130.

Fergusson, D., Horwood, J., Ridder, E., & Grant, H. (2005). *Early start evaluation report, Christchurch: Early Start Project Ltd*. Wellington, New Zealand: Ministry of Social Development.

Forbes, S. (2009). Creation and evaluation of a workplace based Certificate in Official Statistics for government policy makers. *Paper presented at International Association of Statistics Education Conference, Durban, South Africa, included in conference proceedings. Next Steps in Statistics Education*. Retrieved from www.stat.auckland.ac.nz~iase.publications

Forbes, S. (2010). 'Getting better value from official statistics'. *The Information and Decision Support Centre Working Paper Series, No. 3*. Cairo, Egypt: Information and Decision Support Centre.

Forbes, S., Camden, M., Pihama, N., Bucknall, P., & Pfannkuch, M. (2011) 'Official statistics and statistical literacy: They need each other'. In N. Davies & J. Ridgeway (Ed.). *Statistical Journal of the IAOS, 27*, 113–128. IOS press.

Forbes, S. (2011) Collaboration and cooperation: the key to reaching out? Proceedings of the 2011 IASE Satellite Conference-Statistics Education and Outreach. http://www.conkerstatistics.co.uk/iase/proceedings.php

Howden-Chapman, P., Matheson, A., Crane, J., Viggers, H., Cunningham, M., Blakely, T., Cunningham, C., Woodward, A., Saville-Smith, K., O'Dea, D., Kennedy, M., Baker, M., Waipara, N., Chapman, R., & Davie, G. (2007). Effect of insulating existing houses on health inequality: cluster randomised study in the community. *British Medical Journal, 334*, 460–468. doi:10.1136/bmj.39070.573032.80.

Linacre, S. (1995). Planning the methodology work program in a statistical agency. *Journal of Official Statistics, 11*, 41–53.

McLennan, W. (2005). *Verbal presentation to Statistics New Zealand staff* (as past Australian Government Statistician).

Ministry of Education. (2007). *Draft mathematics and statistics curriculum*. Wellington, New Zealand: Author.

Ministry of Education. (2012). Mathematics and statistics achievement objectives—Level 1. *The New Zealand Curriculum Online*. Retrieved Oct 20, 2012, from http://nzcurriculum.tki.org.nz/Curriculum-documents/The-New-Zealand-Curriculum/Learning-areas/Mathematics-and-statistics/Mathematics-and-statistics-curriculum-achievement-objectives.

National Qualifications Framework Project Team. (2005). *The New Zealand National Qualifications Framework, revised paper*. Wellington, New Zealand: New Zealand Qualifications Authority.

O'Keefe, C., & Chipperfield, J. O. (2013). A summary of attack methods and confidentiality protection measures for fully automated remote analysis systems. *International Statistical Review. 81*, 426–455.

Särndal, C. E., Swensson, B., & Wretman, J. (1992). *Model assisted survey sampling*. Berlin: Springer.

Statistics New Zealand. (2007). *Innovation in New Zealand 2005*. Wellington, New Zealand: Author.

Te Puni Kokiri. (2007). *2006 Survey of the health of the Maori language. Final report*. Wellington, New Zealand: Author.

Van Halderen, G., & Bhattacharjee (1998) Training of methodologists in the ABS. Methodology Advisory Committee, catalogue number 1352.0.55.020. Australian Bureau of Statistics, Canberra.

Wild, C., & Pfannkuch, M. (1999). Statistical thinking in empirical enquiry (with discussion). *International Statistical Review, 67*(3), 223–265.

Education for a Workplace Statistician

K.S. Gibbons and Helen MacGillivray

Abstract Transitioning from university student to a statistician working as a collaborative researcher, consultant and workplace educator as well as a data scientist is a very daunting and challenging task. Statistical workplaces can involve at least some aspects of all these components, and some will more easily facilitate learning on the job than others. But all workplace statisticians need proficient statistical thinking, problem-solving and communication skills, as well as a sound statistics foundation for ongoing learning to competently understand, perform and possibly develop statistical analyses. This chapter discusses how early authentic experiential and constructive learning in statistical data analysis and problem-solving courses, combined with experience gained in a developmental and mentored program in tutoring such courses, build the key skills for a workplace statistician. This is illustrated by describing how this has proved invaluable in working as a statistician in a large tertiary hospital that also incorporates a basic science medical research institute.

Keywords Statistical consulting • Communication skills • Data investigation process • Statistical thinking • Tutoring

K.S. Gibbons (✉)
Mater Research Office, Mater Research, Level 2, Aubigny Place,
Raymond Terrace, South Brisbane, QLD 4101, Australia
e-mail: kristen.gibbons@mater.uq.edu.au

H. MacGillivray
School of Mathematical Sciences, Queensland University of Technology,
Gardens Point, GPO Box 2434, Brisbane, QLD 4001, Australia
e-mail: h.macgillivray@qut.edu.au

H. MacGillivray et al. (eds.), *Topics from Australian Conferences on Teaching Statistics:*   267
OZCOTS 2008-2012, Springer Proceedings in Mathematics & Statistics 81,
DOI 10.1007/978-1-4939-0603-1_15, © Springer Science+Business Media New York 2014

1 Introduction

Preparing for the workplace after completing an undergraduate degree in any field is a daunting prospect. This is made easier in many disciplines, knowing that you will be communicating with like-minded colleagues, who have had similar training and on a day-to-day basis are performing comparable tasks. However, the possible employment opportunities after completing a statistics degree are endless, and as a result, very diverse, particularly if partnered with computational studies, and even more so, an information technology degree. Although all disciplines need communication and problem-solving skills in their graduate capabilities, they are especially important for statistics graduates because of both the nature of the discipline and the very diverse workplaces and careers open to such graduates. Graduate statisticians are employed in a wide range of workplaces, with a varied statistical support base, and as such require the skills to not only be able to communicate with other statisticians, but more importantly, with clients and professionals in their work environment who face complex quantitative problems and/or may not be comfortable with statistical concepts and methodology. In all work situations and careers, statisticians need to understand, model and tackle problems in contexts that can be complex, unfamiliar or both, and to plan, implement, analyze and interpret investigations in collaboration with others, often with those who are familiar with the contexts of interest. Communication capabilities, in the broad sense that encompasses discourse, interaction and joint learning as well as specific proficiencies, are integral to all these skills, are complementary to a sound statistical methodological understanding which prepares for ongoing learning, and should be a vital part of any statistics degree.

Statistical consulting is demanding of such skills, and working as a statistician in a large complex of hospitals that include a medical research environment is an excellent example of just how demanding. The statistician is working as a collaborative researcher, consultant and educator of doctors, nurses, scientists and allied health staff. The training of statisticians for such roles must incorporate the skills required for not only performing statistical analyses but also for consulting with clients and researchers with diverse statistical knowledge and understanding, effectively teaching both statistical concepts and their use, including the use of statistical software. Additional pressures can include ensuring that sound statistical advice is not lost in the power struggle of competition for funds or in research in other disciplines.

In many ways, statistical consulting is at the heart of the discipline, science and profession of statistics. Much has been written on the nature of statistical consulting and its core role in the practice of, and developments in, statistics and the whole discipline of statistics. Such writings have also included advice on teaching statistical consulting, and some have described courses specifically designed to teach statistical consulting. The literature is briefly explored in Sect. 2, with emphasis on the core role of statistical consulting in all statistical workplaces.

What opportunities are there for undergraduate students pursuing statistical studies to learn the skills necessary to be able to work collaboratively and communicate with non-statisticians in the "real world"? Seeing and working through case studies can motivate and provide interesting examples. Solving problems by closely following exemplar solutions and prescribed paths can demonstrate theory and show how others have solved problems. Opportunities such as working in groups to complete assessment tasks and addressing staff and fellow students can develop teamwork and presentation skills. All of these approaches contribute to learning, but all tend to be orchestrated and teacher-centred. They tend to lack authentic experiential learning, and do little to foster the skills required to investigate real contexts collaboratively with client disciplines and to communicate concepts to others with little or no statistical experience. It is the need for these combinations of skills that have led many statisticians to warn against statistics education pedagogies that prepare students for analysis but not for statistical consulting (Kenett and Thyregod 2006).

It is in, and from, the writings of statisticians, particularly those involved in the practice of statistics, that we find the emphasis on the whole statistical data investigation process and the need for experiential learning of its statistical thinking. Advocacy of these has increasingly featured in statistical education in the past two decades, although such advocacy has not necessarily resulted in implementation of truly experiential learning. What is said about statistical consulting and the learning of it resonates again and again with teaching that involves authentic learning of statistical thinking for all students, whether mainstream or in client disciplines. The word "mainstream" refers to courses for students majoring in statistics. In addition, comments such as McLachlin's (2000) that distinguish between "meeting client expectations and addressing client needs" precisely describe one of the many challenges faced by university statisticians in working genuinely collaboratively with client disciplines on design of statistical curriculum for their students. The statistical consultancy elements underpinning statistical education advocacy are discussed in Sect. 3. There has been less emphasis, but almost as much need for such, on the importance of experiential and constructivist approaches to learning the statistical thinking of probability and the whole processes of modelling stochastically and statistically.

As an undergraduate, the first author describes herself as fortunate to have had early authentic learning experiences consistent with those advocated by statistical consultants and statistical educators. These were introduced in her university in the early 1990s, developed for students in both mainstream and "service" statistics courses, and have received extensive and consistent feedback from graduates, peers and accrediting bodies that they provided excellent foundations for statistics across workplaces ranging from research to highly applied. These experiences are briefly summarized in Sect. 3.

However, a further facet that has received little attention in statistics education has proven to be significantly advantageous. The experience gained by the first author during her undergraduate degree in a developmental and mentored program in tutoring statistics has proven to be invaluable in subsequent employment in a consultative medical research environment, dominated by professionals with little or no statistical experience. Through tutoring different aspects of statistics, including the experiential learning by other students of data investigations and problem solving, as well as to students in a range of disciplines, the principles of communicating statistics were learnt and developed over a number of years. This program was specifically aimed for those learning to teach tertiary statistics and mathematics, but was consistently given unanimously positive feedback in evaluations received from participants and graduates who went on to either academic or non-academic careers. The report from the first author was even more emphatic—namely that without these tools, the role of the statistician in a workplace where the majority of staff have a minimal statistics background "would have been near impossible". It was the unexpected vigor of this reaction that led to the research, analysis and reflection discussed here. Section 4 describes the program, the tutoring experiences, some of the skills learnt, and some of the synergies between statistical consulting and facilitating student learning in an experiential learning environment. The last is a critically important point—it is not just any teaching experience, but the types of teaching that involves working with students in authentic experiential learning as advocated from statistical consulting.

The requirements of the statistician in a workplace with diverse statistical demands are outlined in Sect. 5, with examples and reflections on workplace components illustrating the training provided by the developmental and mentored tutoring program. Section 6 briefly considers similarities and comparisons between different types of statistical workplaces which involve statistical consulting and in which statistical communication is of paramount importance. The chapter concludes with summary analogies to explicit training in statistical consulting, and with recommendations for programs in training statisticians.

2 Views of Statistical Consulting from Consultants

Kirk (1991) describes statistical consulting as "the collaboration of a statistician with another professional for the purpose of devising solutions to research problems". Bisgaard and Bisgaard (2005) outline three different consulting roles: a pair of hands, the expert, the catalyst/collaborator/coach. Many authors portray statistical consulting much more broadly. Zahn and Boroto (1989) depict successful statistical consulting as "increasing the usefulness of knowledge dissemination in a client's field". Barnett (1986) comments "we see, tied up together, the role of the statistician as consultant, consultancy as the stimulus for research in statistics, and consultancy as the basis for teaching statistics". Gullion and Berman (2006) describe it as help with any statistical aspect of an empirical problem, with a very broad view of what statistical aspect means, closely aligned with Chambers (1993) "greater statistics". Boen and Zahn (1982) comment that statistical consulting could be viewed as collaborative or straight (fulfilling a specialist role) consulting. These are the "entwined collaboration" and "serial collaboration" of Cameron (2009). However it is described, the compass and magnitude of the importance of consulting to statistics is evidenced by the extent and diversity of forums for it, publications and presentations on what statistical consulting is, and the skills and qualities required to do it and do it well. The subset of references here cannot do full justice to the extent of publications and the thought given to the core of statistics—its practice.

All such publications emphasize the importance of communication, again in a broad and deep way involving information flow, discourse, interchange of ideas, critical thinking and problem solving. Barnett (1976) describes statisticians as needing to be translators and communicators. In a long list of advice and skills needed, Joiner (2005) includes the ability to "listen carefully and ask probing questions". Many authors describe the diverse skills and personal attributes needed, including Derr (2000), Russell (2001), Cabrera and McDougall (2002) and Kenett and Thyregod (2006), while others analyze what makes a successful consulting session including "the need to solve the right problem" (Zahn and Boroto 1989), and even assisting with descriptions of types of clients (Moolman 2010).

With so much emphasis on the importance of statistical consulting and the many and diverse skills required, it is not surprising that many publications also emphasize the need for training, but there are fewer publications describing training courses. For example, after large sessions in 1990 at ICOTS3 (Third International Conference on Teaching Statistics) and in 1998 at ICOTS5, specifically on training students for statistical consulting, papers at subsequent ICOTS on this topic appear to be both fewer and more dispersed. Absence of preparation for consulting is lamented (for example, Russell 2001; Kenett and Thyregod 2006) but there are also excellent examples of courses in conjunction with statistical consulting centres (for example, Smyth 1991), in capstone courses (Smith and Walker 2010), in preparing students for industry placements (Rangecroft and Wallace 1998), and, most significantly, in earlier undergraduate years using role play (Taplin 2003, 2007) and community-based projects (Jersky 2002; Thorne and Root 2002). Many authors comment directly or indirectly on how consulting projects and case studies can influence statistics teaching at all levels in both content and pedagogy, but the authors of courses for training in consulting express a requirement for small classes, quoting numbers such as 10–20 students. It is of critical importance to note that the emphasis is on hands-on experience: of experiential learning of the whole process of statistical consulting, of student ownership of learning and of learning through mistakes in a safe educational environment.

3   Environments and Strategies for Learning Statistical Practice, Thinking and Problem-Tackling

3.1   The Statistical Data Investigation Process from Statistical Consultants and Educators

Much has been written in the past few decades on facilitating the learning of statistical thinking and reasoning, and its importance for all students across all disciplines and for educating statisticians-to-be. Approaches include data-driven learning, more emphasis on data production and the measuring and modelling of variability (Moore 1997), real data and contexts, and generally a more holistic approach that reflects the practice of statistics. The emphasis in creating environments for learning—in statistics and in general—has been on active learning, hands-on experience, authentic contexts and problem solving. As for statistical consulting, it is the broad meaning of statistical thinking: making sense of information—including data—in which variation and/or uncertainty is present. It is thus inclusive of chance and data which should be regarded as intertwining and interacting elements of statistical thinking.

In considering the desirable key components of university-based training of a statistician to prepare for a research consulting career in an organization involving multiple disciplines and with objectives to create economic, environmental or community value, Cameron (2009) comments that such training is an appropriate foundation for most statisticians wherever they may be employed. Cameron consults what many "wide and experienced" statisticians have written, including Box's (1976) emphasis on statisticians as scientists and the "learning process" of scientific method, and Chambers' (1993) view of statistics as "learning from data" and "greater statistics" compared with "lesser statistics" which "tends to be exclusive, oriented to mathematical techniques, less frequently collaborative with other disciplines".

In adding first and fifth components to Chambers' (1993) description of how statisticians practice "greater statistics", Cameron (2009) identifies the following:

- Formulating a problem so that it can be tackled statistically
- Preparing data (including planning, collecting, organizing and validating)
- Analyzing data
- Presenting information from data
- Researching the interplay of observation, experiment and theory

Kenett and Thyregod (2006) identify five steps of statistical consulting:

- Problem elicitation
- Data collection and/or aggregation
- Data analysis using statistical methods
- Formulation of findings, their consequences and derived conclusions
- Presentation of findings and conclusions/recommendations

Such descriptions are articulations of statistical investigation, consulting and problem solving that date back to the 1970s (Hunter 1977) and earlier (see Woodward and Schucany 1977). Statistics educators will instantly recognize similarities to the Plan, Collect, Process, Discuss (PCPD) and the Problem, Plan, Data, Analysis, Conclusion (PPDAC) expressions of the data investigative cycle. Cameron's (2009) description is based on descriptions by professional statisticians. Wild and Pfannkuch's (1999) popularization of PPDAC is adapted from MacKay and Oldfield (1994) and reflects the statistical process (see for example, Shewhart and Deming 1986). The description of the data-handling cycle that featured in the UK National School Curriculum in the mid-1970s (Holmes 1997) has become the PCPD cycle that is at the heart of extensive pedagogies and resources (Marriott et al. 2009).

Cameron's (2009) model of what professional statisticians do in the practice of statistics is not only consistent with dimension one (the investigative cycle) of Wild and Pfannkuch's (1999) model of statistical thinking but also reflects dimension two of their model in the types of thinking fundamental to statistics. These include: recognition of the need for data, changing the representation to assist understanding and problem solving, investigating variation, reasoning with statistical models and incorporating statistics and context. The authors comment that they are "not concerned with finding some neat encapsulation of statistical thinking"—their focus is on what professional statisticians do in solving real problems involving the need for modelling and analyzing context information, uncertainty and data.

Thus, the descriptions and advocacy of the data investigation process and statistical thinking in the education of statisticians comes from statisticians and statistical practice. It is imperative that statistics education retain the knowledge and awareness of this, and that the whole statistics community also become more aware of developments in statistics education and the synergies with statistical practice. As the statistical education literature tends to reflect developments and changes that have already been implemented and evaluated, it is also important to note that the advances consistent with the advocacy from statistical consulting started across the world over the past two decades. However, the literature from statistical consultants, such as Kenett and Thyregod (2006), indicates that there is a lack of penetration and implementation. More attention is needed from academic communities to the combined advocacy of statistical consultants and statistical educators.

3.2 Experiential Learning of the Statistical Data Investigation Process

It is very clear from both the statistical consulting and the statistical education literature that it is authentic experiential learning that is vital. In criticizing university teaching that does not include sufficient focus on the first two and last two steps of their statistical consulting cycle, Kenett and Thyregod (2006) comment that

> to be able to respond properly to the needs of the client, it is important to take part in the collection of data, or at least have the opportunity to watch data being collected or generated.

and that not being involved in collecting data

> has led some graduates to be of the opinion that taking part in the collection of data is a waste of the statistician's precious time, and it even implies the risk of getting dirt on your hands.

Both statistical consultants and statistical educators champion the constructivist and experiential learning approaches. Garfield (1995) comments that

> students will understand material only after they have constructed their own meaning for what they are learning.

We add to that with emphasis that students understand the whole data investigation process only when they have lived it. And that they understand probabilistic thinking and its modelling only when they construct meaning that links with everyday contexts and with data, and builds on previous experiences.

Experiential learning of the problem elicitation and data collection and/or aggregation steps of Kenett and Thyregod (2006) were introduced in the first course in mainstream statistics in the authors' university in the early 1990s. The reasons were similar to those of Kenett and Thyregod (2006) and Garfield (1995), namely no matter how real the data and context given to students, the lecturer telling the story of the context, the problem and the data collection was no substitute for students' experiencing steps 1 and 2 for themselves. As a group project, students proposed their own topic of interest to them, planned and implemented their data collection, explored their data and presented their exploration in a report and an oral presentation. It was also intended as foundation for a second year statistics course, which included students designing and carrying out their own experiment (Mackisack 1994). The first course strategy was so popular with the mainstream mathematics and statistics students that it was trialled in 1994 with the large class (approximately 400 students) engineering statistics course. The engineering students identified so much with "their" data and "their" questions that they were not content to just explore their data, but wanted to analyze it using statistical techniques as they met them during the semester. Coincidentally, Jersky (2002), in reporting on the development of a second year consulting course, comments that in 1994 they noted that (mainstream) students from a first statistics course wanted to use their knowledge in an investigation meaningful to them. In another coincidence, 1994 was the first year that Kirkup (see, for example, Kirkup et al. 1998, 2010; Kirkup 2013) introduced student-centred enquiry into first year science laboratory courses. This approach in science is now called inquiry oriented learning (IOL). Anderson and Sungur (2002) use the term inquiry-based learning in describing teaching strategies in statistics, as ranging from practice/drilling up to inquiry, with the intermediate steps being (in order after practice/drilling): activity based, guided discovery, discovery, guided inquiry. Similar descriptions are used in the science literature. Words used in such descriptions must always be clearly defined—as is done in Anderson and Sungur's paper. For example, the word "discovery" might be used by some as representing a similar level of learning to Anderson and Sungur's (2002) "inquiry". However, if well-defined, such metrics are a useful reference to gauge the mix and progress of learning components in a course.

The 1994 trial developed into a semester-long free-choice full data investigation as an integral part of introductory statistics courses in all engineering, science and mainstream mathematics and statistics programs. Some of the growth of developments can be found in MacGillivray (1998a, b, 2005) and Forster and MacGillivray (2010). Progress evolved based on feedback from students, observation and analysis of student work, ongoing collaboration with tutors, lecturers and external peers, and was facilitated by national grants which enabled evaluation and helped build resources, models and pedagogies. As in courses described by Thorne and Root (2002) and Halvorsen (2010), students identify their topic and issues of interest and start designing their data collection early in the semester, but receive feedback and ongoing assistance throughout the semester as individual groups and in the whole-class sessions. Similar to Kuiper (2010), early software use helps students discover the power of statistics in investigating the many-variabled situations that are typical of real world contexts. Overall course structures, teaching approaches and assessment packages were developed from what was learnt from observing and helping students in their data investigations, moderated by the restrictions and requirements of individual courses. The thousands of datasets and topics produced by students also became extraordinarily valuable as teaching resources. Some of the approaches that evolved from the many years of experience with a vast diversity of students' motivations, interests and capabilities can be found in MacGillivray et al. (2014) and even in applications such as learning support for postgraduates across disciplines (MacGillivray 2011).

As well as evaluations from students during the courses, feedback from graduates and input from client disciplines and accrediting bodies have been consistently positive, as well as providing valuable constructive feedback. Graduates from other disciplines often introduce themselves as "I was in the group that did the project on ...". Statistical graduates, whether they went into non-academic or academic research-based workplaces, have described it as

the best foundation for learning to think statistically
a wonderful foundation for learning to conduct statistical research.

Much of what was learnt from working with so many students in inquiry-based learning resonates with the observations and advocacy from statistical consultants. As described above, genuine experiential learning of the whole process of data investigations from first thoughts to reporting is the forerunner of learning statistical consulting. Like Taplin (2003) and Kuiper (2010), it was found that students can start such experiential learning before, or as, gaining the knowledge that a traditional statistics teaching approach assumes must be a prerequisite. The authors' experiences over many years are similar to Taplin (2003) who comments that

it is the practical experience of performing consultations that students are motivated by and learn from

and that

If students learn best from making their own mistakes, then a dominant role of educators is to place students in positions where they make these mistakes safely.

However, a major point of contrast between the authors' courses and the courses referenced above is the class sizes, namely hundreds compared with sizes of 10–20 or 40–50. As discussed in Sect. 5, tutoring in such large courses has provided invaluable experience in communicating with diverse interests, disciplines, motivations and capabilities.

It is of interest to note that Halvorsen (2010) is reporting on learning strategies that started in 1989. How can the whole statistical community gain more knowledge and become more aware of innovative teaching practices that are not necessarily reported as "research"? How can academic statisticians and statistical educators become more aware of the world of statistical practice that is not necessarily reported as statistical "research" or in educational research literature? This is why there must be forums (including both conferences and publications) on teaching statistics for the whole statistical community; such forums must be far broader than pure research forums and must constantly strive to reach out to new participants who can bring expanding richness and knowledge to the complex tapestry of teaching statistics.

3.3 Learning Probabilistic Thinking

The need for more attention to be given to reform of the teaching of probability within statistics has been more recently recognized and is gradually gaining attention. As with the development of inquiry-based and discovery learning in introductory statistics, observation of students, analysis of student work, feedback and evaluation, and collaboration with tutors and other lecturers have enabled the development of an introductory course in probability and distributions which constantly relates to data and everyday experiences, and embodies constructivist principles. Students appreciate how it uses and builds on students' prior experiences and concepts, as represented by a student comment:

> Using what we already knew to learn other stuff was really good and helped us learn other stuff.

A wide variety of contexts to facilitate learning is used in line with cognitive theory (Garfield 1995), and the approach is student centred and active problem tackling in a problem-solving environment as described by Gal et al. (1997),

> an emotionally and cognitively supportive atmosphere where students feel safe to explore, comfortable with temporary confusion, belief in their ability and motivation to navigate stages.

Although there is little direct reference by consultant statisticians to the learning and teaching of probabilistic thinking and its role in learning to think statistically, the importance of modelling and thinking stochastically underpins much in the literature and plays a prominent role also in statistical education (Wild and Pfannkuch 1999).

Some information on the course and its development can be found in MacGillivray (2006, 2007). To date, seven different staff with varying backgrounds have enthusiastically embraced its principles, teaching approaches, materials, resources and structure. The students who take this course are in mathematics, statistics or education programs with some in science programs, but it has also contributed to components of courses across engineering. Although it links with some introductory calculus, its approach is statistical with emphasis on language, communication, everyday concepts, and visual tools including simulations.

The course has included a smaller free-choice inquiry project in which students collect, investigate and report on data, in two contexts of their choosing, that could be Poisson (discrete variable) or exponential (continuous variable). However, from the point of view of tutoring experience as foundation for statistical workplaces, it is working with students in the active learning, problem-tackling environment that provides key experiential learning of communication and facilitating the learning by others.

4 The Tutor Development Program and Skills Learnt

Teaching as the facilitation of learning is especially relevant for statistics, and it is not surprising that many expositions on developing the skills of statistical consulting have parallels in learning to teach statistics. Barnett (1986) comments that statistical consulting is the basis for teaching and research. One of Joiner's (2005) items on a list of 22 skills for effective statistical consulting is

> Be a good teacher—much success in consulting depends on being able to help others understand statistical tools, and their strengths and weaknesses.

Becoming a good teacher of statistics is a complex, never-ending process. Some clues come again from another of Joiner's (2005) tips for effective consulting,

> Be able to listen carefully and to ask probing questions.

and a comment from an applied statistician with a long history of excellent student evaluations of his teaching

> I learnt understanding of statistics through teaching service courses.

This section describes a developmental and mentored tutoring program, and the skills learnt from it and from tutoring in the introductory and large classes described above, that are proving of such value for a workplace/consulting statistician. The term "tutoring" here could consist of:

- Providing assistance on a 1-1 basis to students
- Providing assistance to a class of students on work as designated by lecturer/course coordinator, either alone or with another tutor or lecturer
- Providing assistance in a class situation in a computer laboratory or a problem-solving situation, where students might be working in groups or individually
- May include some demonstration/delivery to a class in a less formal situation than lectures

4.1 The Tutor Development Program

The first part of the developmental tutoring program is working as a mentored volunteer duty tutor in the drop-in facility on one campus of the university-wide centre providing learning support in mathematics and statistics across all disciplines. This centre, called the QUTMAC (QUT Maths Access Centre) and now part of a larger learning support facility, was established in 2004 and has developed into an essential component of the overall learning support services of the university (MacGillivray and Croft 2011). The drop-in facilities are collaborative student spaces with resources and duty staff (volunteer or QUTMAC staff) on a roster that covers approximately 30 h per week over three campuses. Undergraduates with an appropriate balance of academic background and level of achievement apply for the volunteer positions. For example, a student who has completed the first year mathematics and statistics courses requires a high grade point average, but this may not be as necessary for a student with third year mathematics and statistics. Personal criteria relevant to tutoring are also used, but high achieving and committed students to whom such volunteer work appeals are generally given every encouragement. Volunteer drop-in duty tutors generally work for an hour a week, interact with fellow duty tutors and QUTMAC staff, consult QUTMAC staff for advice and assistance, and attend specialized training sessions for the duty tutors.

Mentored volunteer drop-in duty tutoring provides an excellent introduction to the development of communication and tutoring skills, with mostly one-to-one assistance in an informal friendly environment with no time pressures or formal obligations, and with any concerns or difficulties referred to the QUTMAC director. Since 2004, the concept has been consolidated and developed with more integrated and systematic preparation, mentoring and feedback. Written anonymous evaluations are obtained from all of the program's preparation, training and reflection sessions, and these formal evaluations, combined with informal feedback from tutors during and subsequent to their participation, are used to evaluate the service. Formal and informal feedback has been increasingly and consistently highly favourable, with almost all QUTMAC volunteer duty tutors praising the scheme as invaluable, with comments such as

Fantastic! Love it.
Brilliant for learning.
1-1 conversations to learn ways of explaining.
Love being able to help with no pressure.

The volunteers most suited for tutoring greatly enjoy the experience, and some love it so much that they continue as volunteer duty tutors after they are appointed as sessional paid tutors in formal tutorials, and even continuing as volunteers as postgraduate students. A few discover that they do not enjoy it and are able to withdraw. Applications for the volunteer positions have remained high and students are proud of their work, emphasizing it on their CV's. Pre- and post-semester get-togethers gradually developed into interactive training sessions and reflective forums. Informal preparation has now grown to a training day with input from other learning support components and professionals. Inclusion of presentations about peer mentoring have been trialled, but with limited success as the QUTMAC volunteers found such general programs of little relevance to their work. They have always insisted that they are trainee tutors, and do not like the term "peer tutor".

The first author was a drop-in volunteer duty tutor in her second year. After completing at least one semester of QUTMAC duty tutoring, she, along with the other volunteer tutors, were invited to participate in a 2-day tutor training program run by the QUTMAC director and mathematics and statistics staff. They were then included on the list of potential sessional tutors in mathematics and/or statistics.

The 2 days of tutor training had the following general program:

Day 1:

- The nature of the tutorial
- Duties and responsibilities
- Working with the lecturer/discipline coordinator
- Planning and preparation
- Connecting with students
- Understanding students' school and other backgrounds
- Overview of ways people learn mathematics and statistics
- Consideration of examples to select for trainees' demonstrations
- Group discussion about reflection and articulation of good tutorial practice

Day 2:

- Presentations of examples and subsequent feedback from peers and staff, with discussion
- Administration matters
- Training in marking using marking schemes and exemplars

For the majority of students completing the QUTMAC duty tutoring and tutor training, this was their first exposure to being on the other side of the teaching environment; teaching, not being taught. Although all students had achieved highly in their own studies and were very confident in both the details and the roles of the material, it was important—as recognized and acknowledged by all participants—to learn additional skills to communicate with students and to explain in ways that would assist students to understand and gain confidence in the statistics being taught. The training days proved to be daunting at times because no matter how well one know how to answer or work through a question, it is still nerve-wracking doing it in front of one's peers who are pretending to be students! But at the close of such training programs the formal evaluations and informal feedback demonstrated that participants felt they were all armed with a greater comprehension of the challenges of helping others to understand.

After completing the above program, the first author was appointed as a sessional tutor in her third year of study, and continued as a sessional tutor in the third and fourth year of undergraduate studies in a double degree program in mathematics and information technology, during the honours year, and even a small amount during the early part of her full-time employment after graduation. Tutoring was in three different statistics courses to cohorts ranging over all of science, engineering, mathematics, together with some education, surveying, health and other disciplines. These courses were the introductory courses discussed in Sect. 3 above, two of which had been taken by the first author as a student. All tutoring involved some preparation of material for computer laboratory sessions or problem-solving tutorials, some presentations, working with students in small groups and one-on-one during the class sessions, and marking subsequent assessment. Some classes were taken in tandem with another tutor, and some not, depending on the student demand for the timeslot, but classes were regularly visited by the lecturer/course coordinator who also acted in a tutorial capacity as needed. Full assistance was provided to tutors by the lecturer/course coordinator, with discussions either in tutor/lecturer meetings or individually, with ongoing information and tips provided electronically.

4.2 The Tutoring Experience and Skills Learnt

A wide range of skills were learnt over the course of the QUTMAC volunteer duty tutoring, the 2-day training program, and the extended experience of mentored tutoring, and some have proved to be particularly relevant to life as a consultant statistician. Communication is a key factor, but not only is this far more than merely being knowledgeable, articulate and able to present and explain clearly, it is also an integral part of the full spectrum of skills gained from tutoring, provided, of course, that the tutor is fully engaged in, and dedicated to, facilitating learning. A good tutor learns to listen and observe, and thus, through reflection and discussion with the lecturer and other tutors, learns skills in communicating and working with a wide range of

- Different groups of people
- Skills and knowledge background
- Motivations and personalities
- Learning and working styles

Listening to students, observing their work, participating in discussion of their investigation projects, helping them tackle problems, and explaining concepts, techniques and their applications provides great depth to the tutor's own understanding of statistics and statistical thinking. For example, the feedback from graduates asked to contemplate how to prepare for lecturing included comments such as

> Can't imagine how to lecture without tutoring experience.
> How could anyone not tutor before lecturing?
> Either tutor before or academic workload should include tutoring.
> Irresponsible not to include tutoring in workload.
> Even interactive lectures do not substitute for 1-1 for teacher learning.
> I took it for granted that people tutor before lecturing.
> It would be the maddest thing to let researchers loose on lecturing [without tutoring experience].

Graduates who went on to a range of statistical workplaces have attested to the value of learning to communicate and work collaboratively with other disciplines on often complex applied problems. The analogies to learning to be an effective statistical consultant are numerous and obvious.

However, it must be emphasized that the value in the tutoring experience was because of the nature of the tutorials, the pedagogies of the courses and the collaborative mentoring of the lecturer/course coordinator. The focus of the tutorials was on working with the students, facilitating their learning, and not on merely standing in front of a class providing instruction. The pedagogical approaches of the courses were similar, providing experiential learning of the full data investigation cycle and problem solving in probability and distributional modelling, in an environment of collaborative and supportive learning. Tutoring that involves no more than instruction by the tutor in courses with out-of-date instructional dictatorial pedagogy provides no more opportunities to learn to teach and communicate with students or clients than reading a book by oneself.

One of the key components of the QUTMAC duty tutor is to work one-on-one with students to help them discover how to progress through specific problems. In particular, working through the cycle of understanding the problem, explaining the necessary theory and assisting the student with applying the theory is closely related to statistical consultative work in graduate life. A particularly useful aspect of the sessional tutoring is the interaction with groups of students throughout the semester on their free-choice projects exploring the data investigation cycle. This includes helping with forming an interesting question, developing methods to collect the data, determining the appropriate data analysis, and assisting with analysis and interpretation of data after collection. This is very closely related to work as a consulting statistician. The tutors who experienced such investigations as students are more confident going into tutoring statistics, but all tutors gain invaluable skills through this learning to facilitate and assist with authentic data investigations, providing wonderful experience for future graduate roles.

5   A Professional Setting

It is essential that statistics educators and curriculum designers, in courses for future statisticians and for future statistical clients, find out as much as possible about what is asked of graduate statisticians in workplaces. Unfortunately, there are few forums and little support for such discovery. Although the description below is of just one example of a statistician at work, the requirements of the position are particularly broad, encompassing a diversity of roles that are representative for many graduate statisticians. Over the 7 years of employment as a consultant biostatistician, there have been many times when the skills described in Sect. 4 have been utilized. This section therefore describes the various roles and the importance in those roles of the communication, teaching, knowledge of the data investigation process, statistical thinking and facilitation skills acquired through the managed tutor development program and the types of courses tutored. Both the consultative and research aspects of the roles have lead to significant educational developments, reinforcing the statistical consulting literature emphasizing the nexus between statistical consulting and teaching. The educational, consultative and research aspects of the roles are discussed.

5.1   The Overall Setting

Since graduating, the first author has been employed in the research support services division of Queensland's Mater Health Services/Mater Research. Research at Mater is extremely diverse, with seven hospitals (both public and private, across two campuses), a research institute and additional clinical support services (such as pathology, pharmacy and radiology) on the South Brisbane Campus.

The Mater Research Office (MRO) aims to support high-quality research activities on campus across all disciplines and hospitals and for any staff member. In addition to research projects, quality assurance activities are also supported. Services offered include statistical advice and analysis, assistance with data management and clinical research coordination, ethics and governance review and determination of research funding opportunities. The Office assists with over 200 projects a year, ranging from simple audits through to complex clinical trials. The role has also grown to encompass the supervision of research-related data management activities. The formation of the Data Management and Analysis (DMA) Team (of which the first author is Team Leader), within the MRO, has allowed for significant expansion of these services.

Specifically, the following duties are undertaken by the DMA team:

- Statistics education activities
- Randomization services
- Statistical advice
- Statistical analysis
- Development of research databases
- Review reporting of statistical results
- Review statistical software
- Provide statistical service to the Data and Safety Monitoring Committees
- Develop templates for electronic survey processing

Both the clientele and the projects supported by the DMA team are extremely varied. Staff from a wide range of disciplines carry out research projects. While many have undergone some element of statistical education in their training (particularly medical staff), the range of statistical experience is enormous. A substantial portion of the support services provided are to novice researchers, many who are required to complete a research project to fulfil their discipline-specific training. The majority of these clinicians are willing to learn basic research techniques such as literature searching and study design; however, there is generally a lack of desire to both understand and carry out the statistical aspects of their projects. At the other end of the spectrum are senior clinicians, many of whom have exemplary research track records and are both willing and eager to further educate themselves on the statistical techniques available and how they can be applied to their research. A growing number of requests are coming from allied health, nursing and midwifery staff, many of whom have no formal statistical education, and most have no research training either, but have chosen to undertake postgraduate study incorporating a research project. Given this diverse clientele base, the ability to communicate statistical concepts to a wide range of experience levels is vital.

The primary role of statistical consultation was supplemented by an additional focus on education of staff. The delivery of statistical education has changed and expanded over the past 7 years. Feedback from researchers and observations by the DMA staff have influenced changes, but the most powerful agent for change and expansion has been the success of the educational programs, developed and delivered by the first author. These have succeeded in providing a foundation for consulting and collaborative research, as well as improving the confidence and skills of staff with a wide range of backgrounds and with highly variable prior contact with statistics. Teaching statistics in the workplace has its own challenges, and it might be thought to be very different to tutoring introductory undergraduate statistics, but it turned out that the skills and confidence developed as described in Sect. 4 above were exactly right for developing, implementing and evolving workplace educational programs to support clients, researchers and the consulting work. In addition, as included in the discussion below, the success of the evolving educational programs has significantly increased the appreciation of statistics and the demand for statistical consulting.

5.2 Development of the Educational Programs

5.2.1 The Introductory Course

In earlier years, a 2-day introductory statistics course was run for hospital staff, aimed at clinicians who had little to no background in statistics. This was a small group course (maximum of 30 participants) addressing both the underlying theory and the interpretation of statistical concepts commonly used in clinical research. There were a number of components to the course; a website was developed that included lecture notes and presentations, as well as exercises to be completed to ensure the learning was understood. Java applets were embedded in the web pages to perform some of the statistical computing functionality that was being taught, but the emphasis was on the participants working through scenarios without primarily using statistical computing software. It was made known to the participants prior to registering for the workshop that there was an expectation that some pre-reading had been completed to ensure that the face-to-face component of the course was as useful as possible, but this proved unrealistic for the majority of participants due to their work commitments.

One of the challenges in presenting this course was the wide range of backgrounds from which the participants came. Midwives, nurses, junior doctors and professors all attended, with some having no background in statistics, to others just looking to refresh their skills (for example, they may have already completed Masters in Public Health). Individual goals also varied; many were currently undertaking a research project of their own, whereas others were interested in being able to more adequately assess the evidence to help inform their clinical decision making.

Despite such diversity, the feedback on evaluations from participants regarding the content and presentation of the courses for the 3 years it was run was favourable, with general feedback including:

- Over 90 % indicating the information on the website was sufficient for all topics covered
- All respondents agreeing the presentations were interesting and relevant, and that the presentations complemented the teaching material
- Between 10 and 20 % (depending on the topic) indicating more exercises would be useful
- For the topics covering types of errors and sample size, prediction, correlation and regression, the web-based material could be further simplified and the content reduced
- A practical aspect with a commonly used statistical software package would be beneficial

The format and content of the material changed slightly over the 3 years in response to such feedback. Initially the course was delivered on two consecutive weekdays, but the requirement for clinicians to commit to 2 days away from clinical duty was often unrealistic. The course delivery was subsequently changed to two consecutive Saturdays. However, this also proved difficult, and it was decided that a 2-day course was not the ideal format for delivery of statistical education to health staff, particularly clinicians. Although it was felt that the course was very beneficial in its final iteration, the resource requirements to run such a course, the limited cost recovery involved and the difficulty for clinicians to attend led to investigation of other educational strategies. These have been trialled over the past few years, with the two formats discussed below proving to be ultimately successful.

This 2-day introductory statistics course, of which the course materials are still regularly used in other educational settings, was aimed at novice researchers on campus, and was the first author's initial educational experience in the workplace. Given the target audience, the course was designed to provide participants with the basic tools necessary to be able to read a scientific article and critically appraise the statistical methods used. As such, the course involved many examples, learning to interpret statistical output from computer programs as well as understanding statistical methods and results, and the way they were expressed, in journal articles.

The style of teaching is not dissimilar to that experienced as an undergraduate tutor, particularly of service courses, as the participants have little to no background in statistics, and are often only participating to obtain relevant professional development accreditation. At the end of each course, an evaluation survey was conducted to ascertain whether the participants found the course useful and whether the style of presentation should be changed. It was noted on evaluations that the approachability of the presenter, use of practical examples and information on other aspects of research projects (for example, data collection and management) were all strengths of the course, demonstrating the practical application of skills learnt from the tutoring experience. The course was a success from an educational viewpoint, and was changed to other forms for logistical reasons. The confidence in teaching brought to the workplace contributed to the success of, and the increasing demand for, educational services.

5.2.2   Mini Workshops Using SPSS

Three-hour mini workshops whose primary focus is an introduction to the software package SPSS (in response to the feedback about the need for more hands-on learning) have proved so popular that further expansion into other areas is envisaged. Workshops are held in a small group environment but still with diverse backgrounds and goals. The workshop is designed to provide an opportunity to perform common data management and relevant statistical techniques in SPSS under the guidance of a tutor. A booklet is provided and real datasets used. Each workshop contains two staff members from the DMA team; one demonstrating the material, and the other helping participants who may require extra assistance. It is made clear to the participants both prior to and at the beginning of the workshop that the statistical background is not covered. However, if participants need a refresher on introductory statistics they are directed to the web pages used for the previous 2-day course.

These workshops have been incredibly popular. Additional workshops to those originally scheduled have been staged in an attempt to meet the demand, and workshops for individual work groups have been presented when requested. Due to the overwhelming response, the DMA team has recently developed an online enrolment system, with features such as wait-listing and automatic email reminders, in an attempt to reduce the administration overhead that had resulted from the overwhelming response. Course materials (including the booklet developed for the workshop and the example datasets) have also been uploaded onto the DMA team website, and are frequently emailed to staff who are unable to attend a workshop.

Following each workshop, participants are emailed a link to an anonymous online survey to provide feedback about both the courses and other forms of statistical support that they would find useful. Included in the survey are also questions regarding the level of knowledge of SPSS prior to the workshop, and the goals of the individual. Over half of all workshop participants had never used SPSS prior to the workshop, with a further 30 % having used SPSS in the past but not recently, and the remainder having some experience of using SPSS for basic data analysis. Encouragingly, nearly 80 % of participants indicated that they had either a basic or adequate knowledge of statistics, indicating that the workshop was reaching the intended audience—staff with some prior statistical exposure but with a lack of practical experience. The individual goals noted by participants reflected this; many participants indicated that they wanted to learn the practical aspects of SPSS, as well as formal statistical analysis. They wanted also to learn data management methods, including the transfer of data from other formats (such as Microsoft Excel) to SPSS, correct assignment of variable attributes, basic data entry, cleaning and transforming data.

The free text comments entered by participants reflected that their needs had been met, and the course structure, including materials, timing and course size, was appropriate for the intended audience. Comments from participants who had no prior exposure to SPSS included

> I did find the training a lot less scary than I had thought it would be
> I was delighted that I could keep up; great having a presenter and an assistant to pitch in and help me out when I missed a step

> As a novice I felt well catered for, really enjoyed it and came away inspired rather than
> overwhelmed as can often occur.

It was also commonly noted that the participants' goals were met.

> The workshop was excellent!!! I thoroughly enjoyed it and I feel that my goals were well
> and truly met.
>   I learnt a lot and will feel more confident in using the program. It met my objectives
> about data management and filing.

Participants found the workbook very useful, for example

> The workbook was great pre-reading, workshop and take home tool.
>   Great presentation, easy to understand information and excellent supporting
> documentation.

However, it was noted on a number of occasions that participants would still find it useful to have a complementary course on basic statistics, of a similar length and style, which could be combined with this workshop.

5.2.3   Smaller Focused Sessions for Training Projects

To partially address the removal of the statistical course, and the need for this type of training on the campus, smaller focused sessions have been provided for individual groups of staff. These 1 h sessions were integrated into a series of research workshops that took place once a month and designed to assist trainees to develop a research protocol for their training project. Initially it was limited to only the obstetrics and gynaecology registrars, however, following the success with this group, the program has been formalized and is run twice a year (with a session every fortnight) with registrants from all across campus. Participants are required to have a research supervisor, and at the end of the course present their research protocol in a 3 min presentation, which is judged by senior members of the research institute. Initially the statistical aspect of the research protocol was allocated one session only, and in early iterations of this program a mini lecture was given on basic statistical concepts. However, through both direct feedback from participants and indirect feedback through observation, it was found that this did not suit the audience. Instead, the preferred format is now a short 20-min presentation which outlines why it is important to have a solid statistical analysis plan and robust sample size considerations in a research protocol, as well as outlining the various support services on campus available to researchers to assist in the design process. Following this presentation participants are allocated to break-out groups of approximately eight people with a member of the DMA team and various aspects of their projects are discussed and preliminary advice given. This small group work has proven extremely beneficial, as researchers not only receive statistical and data management advice, but they also receive feedback from other clinicians on the potential clinical implications of their research. The success of this new format can be easily gauged by the number of follow-up requests the DMA team receives from this program. Over three-quarters of participants contact a DMA team member or the program facilitator for additional assistance.

5.3   Consulting Component

The primary function of the DMA team leader is to consult with researchers from across the hospital campus to assist with their quality assurance and/or research projects. The level of consultation can range from a brief chat to reassuring a researcher that the statistical methods they have employed are appropriate, to formulating a research plan and educating the researcher on what statistical methods are available and should be used for their project. The confidence gained through tutoring has translated to the ability to spend time educating the researchers and encouraging them to attempt to write their own statistical methods and perform their own analyses, rather than writing the relevant sections of a statistical plan and conducting the analyses for them. While this method does not appeal to all researchers (particularly trainees who have no interest in their research project apart from completing it to satisfy training requirements), numerous comments have been made that the research clients appreciate being taught the statistics, and being able to converse with a statistician, compared with previous negative experiences of consultant statisticians who had few communication skills and would not attempt to educate the researchers.

5.4   Research Component

For the majority of research projects the aim is to be involved from beginning to end; to ensure that an appropriate sample size is calculated, the data are collected in a fashion suitable for data analysis, analyses performed properly and results interpreted correctly. The experience of learning and then tutoring the full data investigation process has played an integral role for the statistical consultant in assisting researchers to design their research projects and be able to offer advice not only for statistics but potential pitfalls in data collection as well as the appropriate steps to take when developing the research protocol. The usefulness of the input from the DMA team leader is evidenced by the increasing number of referrals to the service through existing researchers, as well as clients who have previously only used the services of the statistician at the end of their project, now asking for statistician input at the beginning.

6   Workplace Statistician or Statistical Consultant?

The professional setting outlined in Sect. 5 is of a position described as a statistical consultant, with the primary requirement being statistical consulting work in both senses of statistical services and collaboration. Given the many comments in the extensive literature from statistical consultants on the importance of communication skills, deep knowledge and authentic experience of the data investigation process, it is therefore not surprising to find how valuable the managed tutoring program, and the type of tutoring, are in preparation for statistical consulting work. What has been enlightening, however, is not only the extent of the value of this preparation, but the way in which it has led to significant educational developments in the workplace, and to greatly increased demands for the statistical consulting and appreciation of the importance of data and sound statistical analysis and interpretation.

However, this could be seen as being within one workplace, whereas a "statistical consultant" is often thought of as providing statistical services and/or injecting collaboration into several workplaces, or at least in an assortment of disparate contexts. What are some of the possible differences between a consulting statistician in one workplace and an external "statistical consultant"? Clearly, there is much overlap in terms of the sorts of activities that must be performed, and the kinds of skills each needs. Both need a strong foundation for statistical analyses and, importantly, statistical *thinking*. Both also need to be able to communicate, bi-directionally, to understand the problem being posed in a sensible context, and then to be able to communicate the way to handle the problem back to the person asking the question. In some workplaces the roles may be very distinct (perhaps in smaller or more research-oriented workplaces), while in others the roles may be very similar (for example in a large business-related setting), but nevertheless the twin features of the importance of having highly developed statistical thinking *and* the communications skills to convey that thinking are likely to be persistent needs irrespective of the workplace.

Given the diversity of workplaces in which statisticians are needed, any differences between the roles are swamped by differences across workplaces. However, it is worth reflecting on possible differences in order to place the professional setting of Sect. 5 into a broader context. Workplace statisticians might typically handle a less diverse set of problems than external statistical consultants, in the sense that the kind of problem to be solved will tend to come from a smaller set of directions. Of course, the range of statistical methods that will be useful remains dauntingly broad, but the need to "get your head around" the underlying setting may possibly be a slightly easier task in some workplaces.

Another potential difference relates to workplace culture. External statistical consultants may not really even need to deal with a workplace culture, as such, as they may operate at a different, external level. The effect of this difference might be twofold: statistical consultants might be less familiar to their clients than someone who shares their workplace, and this will almost certainly change the nature of communication. They will, typically, be less "close", less prone to the "drop-in" style of question, and, thus, probably able to more formally prepare for the contact they will have with their clients. Also, external consultants may feel more removed from the problems their clients have. Workplace culture is also a key issue. Section 5 reports a dichotomy, for example, between young medical researchers who seem disengaged from the statistical methods, at least initially, while there is also a cohort of senior researchers who do wish to engage in statistical thinking for their own learning. This dichotomy, and the ability of the workplace statistician to know which type of person they are dealing with, sets a dynamic between the statistician and the researcher that may not be so common in a typical external statistical consulting role. Because the workplace statistician works in the same place as the people needing statistical expertise, there may be a shared purpose about their activities that impacts on the communications between them. Obviously, this feature would also depend critically on the scale of the organization, with large organizations probably making the roles of workplace statistician and external statistical consultant less distinct.

Consideration of the professional setting described in Sect. 5 reveals both similarities and contrasts between the roles of the DMA Team (and its leader) and an external statistical consultant. Because the statistician's role is across seven hospitals, a research institute and clinical services, there are effectively a number of workplaces and cultures involved. Although all the workplaces are in health, there is a great diversity of people needing statistical assistance, ranging over midwives, nurses, registrars, researchers and professors, and hence of types of problems from simple to complex, and from data collection and management to highly sophisticated scientific research.

Possibly a major difference is the opportunity for education in the workplace, but this raises the question of how many workplace statisticians are involved in providing statistical education in their workplaces, and how many external statistical consultants are called upon to provide statistical education within workplaces? Should learning to teach statistics be a part of the statistics curriculum for all future statisticians?

Whatever the workplace or employment roles, it is clear that the benefit of the approach discussed here is that it covers all of the bases that are needed as the transition to employment is made. The keys are the critical acquisition of communication and teaching skills that the managed tutoring program offers, as well as the authentic knowledge and experience of the data investigation process for oneself and in assisting others. There may be other paths for acquiring these qualities, but it is clear from both the literature and experience that these are the qualities to acquire along with sound understanding of statistical methodology and its applications.

7   Conclusion

Over 7 years of employment as a consultant biostatistician, there have been many times when the skills described in Sect. 4 have been utilized. A number of examples have been described, highlighting the use of communication skills learnt through tutoring, knowledge of the data investigation process and experience in consulting that continued the skill development started in the tutoring program. As discussed, education has become a key component of the services provided. There have been many opportunities across a very large and diverse workplace to educate staff about various aspects of statistics, ranging from spontaneous unstructured sessions (generally consultations), short seminars (either at a trainees' session, campus wide grand rounds or invited presentations at research meetings) to a structured 2-day education course. The wide range of experiences from the tutoring program have helped in all aspects of the three roles: consulting, collaboration and education.

Sound knowledge of statistical concepts and methodology and high achievement in formal assessments are no longer enough for a graduate statistician to succeed in a consultative environment interacting with professionals with little or no statistical knowledge. Preparation of graduate statisticians for life after university should include the opportunity during their degree to foster the skills required not only to communicate statistical concepts but also to work with professionals to develop and implement projects. A course specifically on statistical consulting, if well-designed and implemented, is clearly of value to those who choose, or are able, to take it. But, as advocated by the many outstanding statistical practitioners quoted here, the skills needed for statistical consulting and statistical workplaces, in general, must be recognized, acknowledged and embedded in curricula aiming to develop statisticians. This is the first key message of this chapter.

The second key message is that such skills can be developed and enhanced by universities by implementing an undergraduate mentored tutoring program, including a training program, where students are provided with the opportunity to assist and tutor other students in a safe mentored environment with integrated training and sound staff support. It is important to note that simply tutoring a class or participating in a training program is not enough. The courses experienced by the tutors as students, and then tutored by them, must genuinely provide experiential learning of statistics that reflects how statisticians work in the workplace and how they tackle problems. Then the tutoring itself must also be authentic experiential learning of teaching, so that the skills enhanced through tutoring are those of listening, observing and helping a wide range of personalities, capabilities, backgrounds and motivations—all quintessential communication skills of core importance in statistical consulting. The tutoring experience must involve authentic interaction with students (as opposed to merely demonstrating), regular meetings and involvement with the lecture/course coordinator (to foster the mentoring relationship) and tutoring courses that have a large component of experiential learning to ensure the most is gained out of this fantastic opportunity.

The integration of these aspects into an undergraduate statistics degree will result in a new generation of statisticians, capable of smoothly transitioning into working life, with the skills necessary to consult, collaborate with, and educate a wide range of professionals with little or no statistical knowledge.

8   Note

Developed from a paper presented at Seventh Australian Conference on Teaching Statistics, July 2010, Perth, Australia.

This chapter is refereed.

References

Anderson, J., & Sungur, E. (2002). Enriching introductory statistics courses through community awareness. 6th International Conference on Teaching Statistics. http://icots6.haifa.ac.il/PAPERS/3A4_ANDE.PDF ISI: Voorburg, Netherlands

Barnett, V. (1976). The Statistician; jack of all trades, master of one. *The Statistician, 25*(4), 261–279.

Barnett, V. (1986). Statistical consultancy—A basis for teaching and research. In R. Davidson (Ed.), *The Proceedings IASE/ISI 2nd International Conference on Teaching Statistics.* Voorburg, The Netherlands: Vancouver, ISI. Retrieved from http://iase-web.org/documents/papers/icots2/Barnett.pdf

Bisgaard, S., & Bisgaard, S.-E. (2005). *ENBIS workshop on statistical consulting and change management,* Newcastle, UK. Retrieved from http://www.enbis.org/Workinggroups/consultancy/ws_consultancy_05.html

Boen, J. R., & Zahn, D. A. (1982). *The human side of statistical consulting.* Belmont, CA: Lifetime Learning.

Box, G. E. P. (1976). Science and statistics. *Journal of the American Statistical Association, 71,* 791–799.

Cabrera, J., & McDougall, A. (2002). *Statistical consulting.* New York: Springer.

Cameron, M. (2009). Training statisticians for a research organisation. In *Proceedings of the International Statistical Institute 57th Session.* Durban, South Africa: ISI.

Chambers, J. M. (1993). Greater or lesser statistics: A choice for future research. *Statistics and Computing, 3,* 182–184.

Derr, J. (2000). *Statistical consulting: A guide to effective communication.* Pacific Grove, CA: Brooks/Cole.

Forster, M., & MacGillivray, H. L. (2010). Student discovery projects in data analysis. In C. Reading (Ed.), *The Proceedings IASE/ISI 8th International Conference on Teaching Statistics,* Ljubljana. Voorburg, The Netherlands: ISI. Retrieved from http://icots.net/8/cd/pdfs/invited/ICOTS8_4G2_FORSTER.pdf

Gal, I., Ginsberg, L., & Schau, C. (1997). Monitoring attitudes and beliefs in statistics education. In I. Gal & J. Garfield (Eds.), *The assessment challenge in statistics education* (pp. 37–54), Amsterdam: IOS Press.

Garfield, J. (1995). How students learn statistics. *International Statistical Review, 63,* 25–34.

Gullion, C. M., & Berman, N. (2006). What statistical consultants do: Report of a survey. *American Statistician, 60*(2), 130–138.

Halvorsen, K. (2010). Formulating statistical questions and implementing statistics projects in an introductory applied statistics course. In C. Reading (Ed.), *The Proceedings IASE/ISI 8th International Conference on Teaching Statistics,* Ljubljana. Voorburg, The Netherlands: ISI. Retrieved from http://icots.net/8/cd/pdfs/invited/ICOTS8_4G3_HALVORSEN.pdf

Holmes, P. (1997). Assessing project work by external examiners. In I. Gal & J. Garfield (Eds.), *The assessment challenge in statistics education* (pp. 153–164). Amsterdam: Ios Press.

Hunter, W. G. (1977). Some ideas about teaching design of experiments, with 2sup5 examples of experiments conducted by students. *The American Statistician, 31,* 12–17.

Jersky, B. (2002). Statistical consulting with undergraduates—A community outreach approach. *The Proceedings IASE/ISI 6th International Conference on Teaching Statistics,* Capetown, ISI, Voorburg, The Netherlands. Retrieved from http://iase-web.org/documents/papers/icots6/3a1_jers.pdf

Joiner, B. (2005). Statistical consulting. In S. Kotz, N. Balakrishnan, C. Read, & D. Vidakovic (Eds.), *Encyclopedia of statistical sciences* (2nd ed.). New York: Wiley.

Kenett, R., & Thyregod, P. (2006). Aspects of statistical consulting not taught by academia. *Statistica Neerlandica, 60*(3), 396–411.

Kirk, R. E. (1991). Statistical consulting in a university: Dealing with people and other challenges. *The American Statistician, 45*(1), 28–34.

Kirkup, L. (2013). Inquiry-oriented learning in science: transforming practice through forging new partnerships and perspectives. *Australian National Teaching Fellowship Final Report*, Office of Learning and Teaching, Australian Government. Retrieved from http://www.olt.gov.au/altc-national-teaching-fellow-les-kirkup

Kirkup, L., Johnson, S., Hazel, E., Cheary, R. W., Green, D. C., Swift, P., & Holliday, W. (1998). Designing a new physics laboratory programme for first year engineering students. *Physics Education, 33*, 258–265.

Kirkup, L., Pizzica, J., Waite, K. M., & Srinivasan, L. (2010). Realizing a framework for enhancing the laboratory experiences of non-physics majors: From pilot to large-scale implementation. *European Journal of Physics, 31*(5), 1061–1070.

Kuiper, S. (2010). Incorporating a research experience into an early undergraduate statistics course. In C. Reading (Ed.), *The Proceedings IASE/ISI 8th International Conference on Teaching Statistics*, Ljubljana. Voorburg, The Netherlands: ISI. Retrieved from http://icots.net/8/cd/pdfs/invited/ICOTS8_4G1_KUIPER.pdf

MacGillivray, H. L. (1998a). Developing and synthesizing statistical skills for real situations through student projects. In *The Proceedings IASE/ISI 5th International Conference on Teaching Statistics*, Singapore (pp. 1149–1155). ISI: Voorburg, The Netherlands. Retrieved from http://iase-web.org/documents/papers/icots5/Topic8n.pdf

MacGillivray, H. L. (1998b). Statistically empowering engineering students. In *Proceedings of the 3rd Engineering Maths and Applications Conference* (pp. 335–338), The Institution of Engineers, Australia.

MacGillivray, H. L. (2005). Helping students find their statistical voices. In L. Weldon & B. Phillips (Eds.), *The Proceedings of the ISI/IASE Satellite on Statistics Education and the Communication of Statistics*, Sydney. Voorburg, The Netherlands: ISI. Retrieved from http://iase-web.org/documents/papers/sat2005/macgillivray.pdf

MacGillivray, H. L. (2006). Using data, student experiences and collaboration in developing probabilistic reasoning at the introductory tertiary level. In A. Rossman & B. Chance (Eds.), *The Proceedings IASE/ISI 7th International Conference on Teaching Statistics*, Brazil. Voorburg, The Netherlands: ISI. Retrieved from http://iase-web.org/documents/papers/icots7/6B4_MACG.pdf

MacGillivray, H. L. (2007). Weaving assessment for student learning in probabilistic reasoning at the introductory tertiary level. In B. Chance & B. Phillips (Eds.), *Proceedings IASE/ISI Conference on Assessing Student Learning in Statistics*, Portugal. Voorburg, The Netherlands: ISI. Retrieved from http://iase-web.org/documents/papers/sat2007/Macgillivray.pdf

MacGillivray, H.L. (2011). Statistical thinking foundations for postgraduates across disciplines, Proc IASE Satellite Statistics Education and Outreach, Malahide, Ireland, August http://iase-web.org/documents/papers/sat2011/IASE2011Powerpoint5A.3MacGillivray.pdf

MacGillivray, H. L., & Croft, A. C. (2011). Understanding evaluation of learning support in mathematics and statistics. *International Journal of Mathematical Education in Science and Technology, 42*(2), 189–212.

MacGillivray, H.L., Utts, J., & Heckard, R. (2014). Mind on Statistics (Australia and New Zealand 2nd edn). Cengage Learning Australia (650pp).

MacKay, R. J., & Oldfield, W. (1994). *Stat 231 course notes fall 1994*. Waterloo, Ontario, Canada: University of Waterloo.

Mackisack, M. (1994). What is the use of experiments conducted by statistics students? *Journal of Statistics Education, 2*(1). Retrieved from http://www.amstat.org/publications/jse/v2n1/mackisack.html

Marriott, J., Davies, N., & Gibson, L. (2009). Teaching, Learning and assessing statistical problem solving. *Journal of Statistics Education, 17*(1). Retrieved from www.amstat.org/publications/jse/v17n1/marriott.html

McLachlin, R. D. (2000). Service quality in consulting: What is engagement success? *Managing Service Quality, 10*, 239–247.

Moolman, W. H. (2010). Communication in statistical consultation. In C. Reading (Ed.), *The Proceedings of the 8th International Conference on Teaching Statistics*, Ljubljana. Voorburg, The Netherlands: ISI. Retrieved from http://iase-web.org/documents/papers/icots8/ ICOTS8_4H2_MOOLMAN.pdf

Moore, D. S. (1997). New pedagogy and new content: The case of statistics. *International Statistical Review, 65,* 123–165.

Rangecroft, M., & Wallace, W. (1998). Group consultancy, as easy as falling off a bicycle? In L. Pereira-Mendoza, L. S. Kea, T. W. Kee, & W. Wong (Eds.), *The Proceedings of the 5th International Conference on Teaching Statistics* (pp. 359–364), Singapore. Voorburg, The Netherlands: ISI.

Russell, K. G. (2001). The teaching of statistical consulting. *Journal of Applied Probability, 38A,* 20–26.

Shewhart, W., & Deming, W. (Eds.) (1986). *Statistical method from the viewpoint of quality control.* New York: Dover Publications (original work published 1939).

Smith, H., & Walker, J. (2010). Experiences with research teams comprised of graduate students, faculty researchers, and a statistical consulting team. In C. Reading (Ed.), *The Proceedings IASE/ISI 8th International Conference on Teaching Statistics*, Ljubljana. Voorburg, The Netherlands: ISI. http://iase-web.org/documents/papers/icots8/ICOTS8_4H1_SMITH.pdf

Smyth, G. K. (1991). Experiences in training students in statistical consulting and data analysis. In *The Proceedings of the 3rd International Conference on Teaching Statistics* (pp. 446–450), Dunedin, ISI. Voorburg, The Netherlands: International Statistical Institute, Voorburg.

Taplin, R. H. (2003). Teaching statistical consulting before statistical methodology. *Australian & New Zealand Journal of Statistics, 45,* 141–152.

Taplin, R. (2007). Enhancing statistical education by using role-plays of consultations. Journal of the Royal Statistical Society: Series A 170 (2), 267-300.

Thorne, T., & Root, R. (2002). Community-based learning: motivating encounters with real-world statistics. In *The Proceedings IASE/ISI 6th International Conference on Teaching Statistics*, Capetown. Voorburg, The Netherlands: ISI. Retrieved from http://iase-web.org/documents/ papers/icots6/3a3_root.pdf

Wild, C. J., & Pfannkuch, M. (1999). Statistical thinking in empirical enquiry (with discussion). *International Statistical Review, 67*(3), 223–265.

Woodward, W. A., & Schucany, W. R. (1977). Bibliography for statistical consulting. *Biometrics, 33,* 564–565.

Zahn, D. A., & Boroto, D. R. (1989). The wanted and needed conversation: a tool to enhance consulting effectiveness. *ASA Proceedings of the Statistical Education Section, 1989,* 66–67.

Improving Teachers' Professional Statistical Literacy

Robyn Pierce, Helen Chick, and Roger Wander

Abstract Given the deluge of data that school principals and teachers receive as a result of student assessment, it has become essential for them to have statistical literacy skills and understanding. Earlier work with primary and secondary teachers in Victoria revealed that, although most saw school statistical reports as valuable for planning and thought that they could adequately interpret them, their confidence was often not well founded, with some fundamental misconceptions evident in their statistical understanding. Based on these results, a workshop was developed to target key aspects of statistical literacy particularly relevant to the education context. The workshop incorporated simple hands-on activities to develop understanding of box plot representations, critiquing descriptions of distributions and applying the newly learned principles to participants' own school reports. Although principals and teachers responded favourably to the activities, delayed post-testing indicated limited retention of the relevant aspects of statistical literacy. These results suggest that when teachers are dealing with data on only one or two occasions in a year, it may be important to provide timely and efficient access to reminders of basic concepts.

Keywords Teachers' statistical literacy • Interpreting data • Box plots • Attitudes to data use

R. Pierce • R. Wander
Melbourne Graduate School of Education, University of Melbourne,
Parkville, VIC 3010, Australia
e-mail: r.pierce@unimelb.edu.au; rdwander@unimelb.edu.au

H. Chick (✉)
Faculty of Education, University of Tasmania,
Private Bag 66, Hobart, TAS 7001, Australia
e-mail: Helen.Chick@utas.edu.au

H. MacGillivray et al. (eds.), *Topics from Australian Conferences on Teaching Statistics:* 295
OZCOTS 2008-2012, Springer Proceedings in Mathematics & Statistics 81,
DOI 10.1007/978-1-4939-0603-1_16, © Springer Science+Business Media New York 2014

1 Introduction

Statistical literacy is typically considered to involve the ability to read and interpret statistical information, often in everyday contexts, and draws on understanding of numeracy, statistics, general literacy and data presentation (see, e.g. Gal 2002). The importance of statistical literacy for all has been highlighted by statisticians (e.g. Wallman 1993), and its scope within the school curriculum and appropriate pedagogical approaches have received attention (e.g. Watson 2006). What is being referred to there is knowledge sufficient to make sense of data and statistical information that are likely to be encountered in the broad community (e.g. in the media). However, there may be specific aspects of statistics and reporting styles that are specific to particular sectors of the workforce. We use the term *professional statistical literacy* to refer to this slightly more specialised knowledge.

An increased push for measuring educational outcomes and a growth in institutional capacity to generate and analyse large data sets have resulted in schools having to deal with extensive statistical information about student outcomes and similar data. The government expectation is that teachers will use this information to inform decisions regarding school planning and teaching practice, and "develop a more objective view about the performance of their students compared to those in other schools and in relation to state-wide standards" (Ministerial Council on Education, Employment, Training and Youth Affairs, n.d.). Clearly a necessary requirement for successful data-driven choices is that school principals and teachers have sufficient statistical literacy to interpret such data. However, a Statistical Society of Australia Inc. (2005) report pointed out that statistics has a poor image and profile among students, parents and the general public. Indeed, negativity towards statistical information and lack of confidence in analysing statistical data may discourage education personnel from other than cursory engagement with such information. Tom Alegounaris, a board member of the Australian Curriculum and Assessment Authority and president of the NSW Board of Studies, commented that "teachers lack the expertise to analyse student results" and "were seen to resist using data", arguing further that "teachers had to discard their 'phobia' of data" (Ferrari 2011). Michelle Bruniges, Director-General of Education in NSW, similarly suggests that "school improvement is being held back because many teachers lack confidence and skills to analyse National Assessment Program for Literacy and Numeracy (NAPLAN) student test data" (Milburn 2012).

This chapter examines the issues associated with statistical literacy in the education workplace and reports on the trial of a workshop for teachers. This workshop was developed in order to address previously identified barriers and misconceptions associated with analysing and interpreting system reports of student assessment (hereafter referred to as SRSA). The chapter draws attention to conceptual and attitudinal issues that need to be addressed in any statistics courses designed for pre-service or practicing teachers.

2 Evidence Base for Teachers' Statistical Literacy Workshop

In researching professional statistical literacy in the education sector, we chose to focus on the system reports associated with student assessment in Victoria, specifically reports from the National Assessment Program for Literacy and Numeracy (NAPLAN) and the Victorian Certificate of Education (VCE) data service. Prior to the development of a workshop for teachers, data were collected from 938 Victorian primary and secondary teachers as summarised in Table 1. These data were intended to give us an indication of both attitudinal factors and statistical literacy issues that might impact on teachers' work with SRSA.

2.1 Underpinning Frameworks and Results

The surveys mentioned in Table 1 had items investigating teachers' attitudes towards data, and their understanding of how to interpret SRSA. The background for and results from these sections are discussed in what follows; they were particularly relevant to informing the design of the professional statistical literacy workshop.

2.1.1 Theory of Planned Behaviour

Gal (2002) specifically highlighted the role of dispositions in contributing to whether or not individuals will choose to "activate" their statistical knowledge. A series of items framed by Ajzen's (1991) *Theory of Planned Behaviour (TPB)* probed teachers' attitudes, subjective norms and perceived behavioural controls that may act as enablers or barriers for their intention to engage with SRSA (see Pierce and Chick 2011a). Ajzen (1991) proposed the TPB as a framework for

Table 1 Data collected to provide an evidence base for the design of the workshop for teachers

Group (size)	Data collected	Sample type
1 ($n = 84$)	Pilot survey targeted attitudes and perceptions affecting engagement with SRSA	Convenience sample of secondary mathematics and English teachers from non-government schools
2 ($n = 150$)	Paper-based survey followed by focus group. Survey targeted demographics, access to SRSA, attitudes, perceptions and statistical literacy with respect to SRSA	Cluster sample: five school regions then one network from each region, then two primary and two secondary schools from each network, then seven teachers and principal or nominee from each government school
3 ($n = 704$)	Online survey using simplified version of previous survey with items rephrased or modified in the light of focus groups and paper-based survey	Random sample of 104 primary and secondary government schools, with expectation of 60 % school agreement and 50 % within-school teacher response rate

Table 2 Likert items probing teachers' attitudes and perceptions towards SRSA and planning

	Statement	SD (%)	D (%)	N (%)	A (%)	SA (%)
2.1 AT	SRSA tell me things about my students that I had not realised	4	16	25	48	7
2.2 AT	SRSA are helpful for grouping students according to ability	3	10	21	57	10
2.3 AT	SRSA are useful for identifying topics in the curriculum that need attention in our school	2	5	12	60	22
2.4 AT	SRSA are useful for identifying an individual student's knowledge	2	12	21	56	9
2.5 AT	SRSA are helpful for planning my lessons	4	10	28	50	8
2.6 AT	SRSA are useful to inform whole school planning	2	4	15	58	22
2.7 BC	I don't feel I can adequately interpret the SRSA I receive at our school	21	42	22	13	2
2.8 BC	Practical constraints mean that it is not possible to change teaching in my area in response to SRSA	10	42	34	13	2
2.9 BC	I find that most SRSA are not relevant to my teaching	10	46	31	10	3

studying intention to change behaviour, and identified three key components that determine that intention:

> The first is the *attitude* toward the behaviour and refers to the degree to which a person has a favourable or unfavourable evaluation or appraisal of the behaviour in question. The second predictor is a social factor termed *subjective norm*; it refers to the perceived social pressure to perform or not to perform the behaviour. The third antecedent of intention is the degree of *perceived behavioural control* which … refers to the perceived ease or difficulty of performing the behaviour and it is assumed to reflect past experience as well as anticipated impediments and obstacles. (p. 188, emphases added)

He argued that favourable attitudes and subjective norms, together with greater perceived behavioural control, would result in a stronger intention to perform the associated behaviour. This has been supported by research (see, e.g. Armitage and Conner 2001), showing that the three components are strongly predictive of behavioural intent, which, in turn, can account for a considerable proportion of variance in actual behaviour.

Table 2 shows the results from half of the TPB questions used in the survey conducted with Group 3, focusing on attitudes, signified by AT, and behavioural controls, signified by BC. These indicated that most teachers saw SRSA as valuable for planning at a school, curriculum and lesson level (2.3, 2.5, 2.6), and that they did not think that they had a problem interpreting the SRSA (2.7). Although these results were generally positive, there were significant numbers of neutral and disagreeing responses.

2.1.2 Framework for Professional Statistical Literacy

A second series of items on the Group 3 survey focused on teachers' professional statistical literacy. Using a framework proposed by Pierce and Chick (2011b)—building on earlier work by Curcio (1987), Gal (2002) and Watson (2006)—these questions assessed teachers' ability to interpret data in a professional situation. This

Fig. 1 A framework for considering professional statistical literacy (Pierce and Chick 2011b, p. 633)

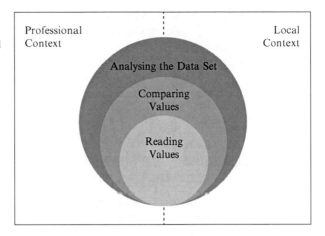

framework (shown in Fig. 1) acknowledges that effective data interpretation requires attention at multiple levels in a hierarchy. The lowest level, *reading values*, involves understanding features such as keys, scale and graph type, together with the capacity to read specific data points on the graph or table. The second level, *comparing values*, requires attention across multiple facets of a graph or across one or more representations (graphs or tables). Finally, the third level, *analysing the data set*, involves considering the data as a whole entity, including interpreting variation, and attending to the significance of results. The framework also acknowledges the role of context, in association with the three levels of technical facility. *Professional context* concerns information relevant to the profession and needed to interpret the data set (e.g. meaning of specialist terms such as "band"). The second, *local context*, comprises knowledge about the situation or context that gave rise to the data that is not evident in the data set alone (e.g. knowledge of the local school situation that may have affected test results). These two context components may overlap, hence the dashed line between them in Fig. 1. The structure of this framework for professional statistical literacy was verified in Pierce, Chick, Watson, Les and Dalton (2014).

It should be noted that in Victoria the most common graphic used in SRSA is a box plot, with whiskers extending only to the 10th and 90th percentiles because testing is considered unreliable at the extremes. Since the graphic does not represent the full distribution of results, there is potential for misinterpretation; furthermore, research (see, e.g. Pfannkuch 2006) tells us that students commonly exhibit confusion between the frequency and density of data points in a box plot. In the survey items focusing on these aspects, misconceptions were found to be prevalent, implying that teachers' confidence in their capacity to interpret SRSA—as revealed by their responses to 2.7 in Table 2—was not well founded (see Pierce and Chick 2013, for further details). Specifically, whereas most teachers could correctly read values from tables and identify a school's weakest area from a graphic, more than 70 % misinterpreted certain aspects of box plots. Furthermore, data collected through focus groups and surveys (from Groups 2 and 3 in Table 1) suggested a need to check and address participants' conceptual understanding of data and graphics before focusing on interpretation and consequent workplace decision making. It was also clear from Group 2 responses and focus group discussions that a lack of appropriate vocabulary hindered teachers' ability to describe and compare distributions of results.

3 The Teachers' Statistical Literacy Workshop

3.1 Learning Objectives for the Workshop

The information obtained about teachers' attitudes and statistical literacy informed our objectives—in terms of the messages to be emphasised and the choice of content to be targeted—in the half-day professional learning workshop that we designed for teachers to improve attitudes and statistical literacy. The data suggested that a significant minority of teachers did not feel that the data were valuable nor that they provided useful information about their students (items 2.1 and 2.4, Table 2), and that a majority did not understand the fundamental construction of a box plot (a typical set of SRSA boxplots is shown in Fig. 2). These results implied that any professional learning workshop clearly needed to address both attitudes and competence.

The structure and nature of the program developed paid attention to previous research on the elements of successful professional learning programs. Ingvarson et al. (2005), for example, examined four studies on teacher professional

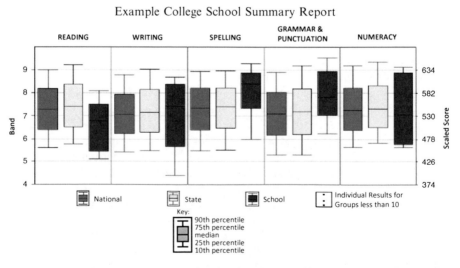

Fig. 2 The hypothetical school summary report showing the results of the hypothetical class's results (the *right-most box plot* in each group of three) compared with national and state results (*leftmost* and *middle*, respectively)

development undertaken by the Australian Government Quality Teacher Programme. In total this encompassed 80 professional development activities and 3,250 teachers. They found that "consistent significant direct effects were found across the four studies for the impact of content focus, active learning, and follow-up on knowledge and professional community" (p. 1). These findings—together with the recommendations of Martin (2008, 2010) based on his experience of providing statistical training in a variety of industries—informed the development of the "Making sense of data" workshop for teachers described below. Martin (2008, 2010) drew attention to the deep learning that can take place when statistical knowledge is set meaningfully in the participants' workplace context so that the new knowledge and the skills being taught relate to real workplace needs. For these reasons the workshop focused on the detail and format of the statistical reports most commonly sent to Victorian schools from the various government education authorities.

Based on our research results and the above professional learning principles—a focus on content, active learning and attention to participants' practice—a workshop was designed to actively engage teachers with the statistical content of SRSA. The learning objectives for the workshop were that teachers should be able to:

1. Demonstrate understanding that the box plots in the NAPLAN reports provided to their schools only represent the middle 80 % of the cohort since the whiskers extend only to the 10th and 90th percentiles, and, as a consequence, interpret box plots taking into account that the weakest and strongest students are not represented by this particular graphic.
2. Demonstrate understanding that the "fences" of a box plot divide the cohort into quarters so that the length of each section gives an indication of density not frequency, and, as a consequence, interpret box plots appropriately avoiding such misconceptions as "there are too many students in the tail".
3. Engage with the data because they realise it tells them something about their students.
4. Make use of the data to inform planning for teaching by identifying patterns in students' strengths and weaknesses.

3.2 Workshop Structure and Tasks

The activities were designed to actively involve teachers with relevant scenarios that targeted key concepts. At the beginning of the session teachers were introduced to a hypothetical class of 30 students with a School Summary Report (see Fig. 2) and individual SRSA results. Each student in the class was depicted on a separate narrow card as an image with individual assessment data (one such student is shown in Fig. 3, together with his NAPLAN data). The class size of 30 was chosen not only for its realistic estimate of Victorian class size, but also to allow simple determination of the top and bottom 10 % of the cohort together with the location of the median.

Fig. 3 One of the set of 30 hypothetical students used as a data set for the professional learning workshop with individual NAPLAN data results. Data recorded in the text below the student figure were James' NAPLAN results — Reading 523.4, Writing: 582.4, Spelling: 594.8, Grammar & Punctuation: 557.2, Numeracy: 541.1

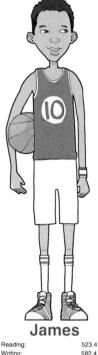

James

Reading:	523.4
Writing:	582.4
Spelling:	594.8
Grammar & Punctuation:	557.2
Numeracy:	541.1

3.2.1 Task 1: From Students' Scores to Box Plots

The teachers, working in groups of three or four, used the image cards and their data (Fig. 3) to plot the distribution of Reading scores on a large number line. The cards themselves were used as data points (see Fig. 4). Using this plot, they then built a NAPLAN-like box plot by (1) dividing the group into quarters, (2) placing a red box over the middle 50 %, (3) turning over the top and bottom 10 % of images to hide those students and their data (while still leaving a place holder in the plot) and (4) extending whiskers to the furthest visible students (see Fig. 4). It should be noted that dividing the class into quarters was mildly problematic for some teachers because the class size was 30. The statistical solution to this was discussed briefly, but the main emphasis was on building the idea that each quarter contains equal numbers of students. This exercise used images of students, not just points, in order to reinforce the message that this data provide information about students, thus linking the abstract box plot graphic image to concrete information. The box plot construction exercise emphasised, first, the density rather than frequency of students represented by the sections of the plot and, second, the importance of paying attention to the key, which on Victorian SRSA notes that the whiskers extend to the 10th and 90th percentiles.

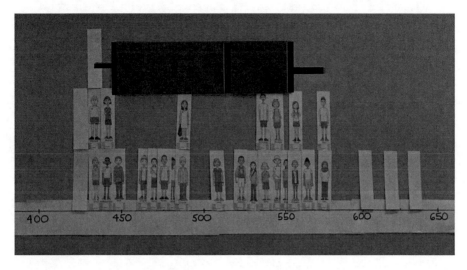

Fig. 4 The distribution of reading scores for the set of 30 hypothetical students used as a data set for the professional learning workshop, with the associated boxplot

3.2.2 Task 2: From Box Plots to Students' Scores

Given that the background research data had suggested teachers were confused regarding the density vs. frequency issue, the workshop included an activity that was essentially the reverse of Task 1, in which it was highlighted that a given box plot could represent any number of possible distributions. Each group was given a new box plot and asked to plot student image cards (with no scores given) in a possible distribution to fit the box plot. The variety of distributions suggested from the different groups allowed discussion of possible and impossible solutions. The activity also afforded explicit consideration of possible ranges, densities, and skewness.

3.2.3 Task 3: Interpreting and Discussing Box Plots

In the third task participants were asked to consider a series of statements describing Example College's Numeracy results. The wordings used in the descriptions—good or otherwise—were based on actual descriptions given by teachers to an open-response survey item in the evidence-base phase described earlier. These descriptions, together with our observations in the early phase of data collection with Group 2, revealed that many teachers appeared to have difficulty using appropriate language to describe graphical information and statistical results. The workshop task was intended to address these difficulties and focus on descriptive precision and appropriate vocabulary. In the workshop each group of teachers was given ten statements to consider (see Fig. 5). The task was to (1) identify and discard incorrect statements; (2) rank the remaining statements in terms of helpfulness and then (3) write their own summary of one of the SRSA box plot reports.

1. The middle 50% of students have scaled scores between about 460 and 630.

2. The distribution is positively skewed with half of the students scoring below approx 530

3. There is a narrower range of ability in those students between the 10th and 50th percentile than between the 50th and 90th percentile.

4. The weakest 10% of students scored below 460.

5. The distribution for numeracy is centred around 530.

6. Most of the scores are in the 50 to 75 percentile.

7. The bulk of the students are above the 50th percentile.

8. Graph down the bottom mainly.

9. Fewer children scored in the lower range than did in the upper.

10. Numeracy has a low mean.

Fig. 5 Teacher statements describing Example College's Numeracy results

3.2.4   Task 4: Analysis of Real Data

After working on box plots and allowing teachers to articulate global impressions of the Example College group's abilities (in relation to the State across five tests), the focus of the workshop shifted to having teachers examine students' responses to individual NAPLAN test items via what is known as an Item Analysis Report. This report provides statistics about school, state and national performance on specific questions used in student tests. This activity was introduced through examples discussing the absolute and relative differences between results (e.g. determining whether or not it is significant that a school has only 56 % of its students getting a question correct compared to 61 % of the students in the whole state), with a focus on practical local measures of similarity and difference. This led to the final half of the workshop, which provided an opportunity for each school group to work on interpreting its own school SRSA.

4   Outcomes

4.1   Participants and Method

The workshop participants were a subset of Group 2 in Table 1, with the same teachers who had provided the survey and focus group data, but chosen from just one of each of the primary and secondary schools in the networks previously involved (this choice of school was random, leading to a selection of participants from 10 schools out of the original 20). Five workshops were conducted, one for each of the five pairs of primary and secondary schools. At least 4 weeks after taking part in their single half-day workshop, the participants—together with the other Group 2 teachers from those schools who had not attended the workshop—completed an online survey targeting attitudes, perceptions and statistical literacy with respect to SRSA, similar to the initial paper-based survey they had taken to provide the evidence base for the workshop design. Allowing for teacher attrition over time and non-participation in the follow-up survey, 82 of the original 150 teachers completed the survey (noting that only 123 were still available to participate), and 45 of these attended the workshop.

4.2 Workshop Outcomes

The outcomes of the workshop were noted through the workshop observations of the authors, and via more formal feedback from participants via the follow-up survey. Those involved in presenting and organising the sessions (the authors, along with representatives from the project's partners) were aware of the impact of the workshops as they progressed. After each presentation the project team noted the strong evidence for the teachers' clear need for conceptual activities related to box plots. In one group, the assumed data expert from the school was observed to be confidently making erroneous statements about some aspects of the data set. This helped confirm the researchers' decision to include a workshop activity where the validity of statements made about data could be examined.

There was informal but audible appreciation from the participants as they came to understand the box plot representation, with many indicating that they had not had such understanding prior to the workshop. This feedback indicated that some participants realised for the first time, for example, that a "long tail" on a box plot indicates diversity of student results rather than large numbers of students; that the box does not represent the whole class; and that bounds at the 10th and 90th percentiles mean that the character and number of students *not* represented within the displayed box plots is almost certainly of practical significance for teaching. Participants also commented that working with images of students prompted mental links to their own students, and that they now had a greater perception that such data could be relevant to their practice as teachers. The observers also noted that participants were able to use this understanding to critique descriptions of distributions, and could appropriately apply the principles learnt to an analysis of their own school reports.

The researchers also noted a high level of engagement from participating teachers throughout each of the workshop sessions. Working in groups of three to four, the participants were observed to make group decisions about all aspects of the activity where a box plot was built from the data on the figures' cards. In response to a final comments question in the online post-workshop survey mentioned above, many commented favourably on the value of that exercise:

> I thought the hands on method (using the actual number lines & cards) was excellent & increased my understanding of box & whisker graphs and data analysis.

I found the explanations of how to produce and read box plots really helpful. The manipulatives were great.

Teachers' general perceptions of the workshops were overwhelmingly positive. Some comments mentioned increased confidence, greater understanding of box plots, and use of (or intention to use) the materials and ideas with other staff in participants' schools. Indeed, some participants requested access to the materials so they could replicate the box plot activities at future staff meetings. The comments also indicate that principals and teachers at both primary and secondary levels responded favourably to the simple hands-on learning activities creating box plots. Typical of the comments are the following:

Excellent opportunity to finally learn how to properly analyse box and whisker plots (hopefully)—we dedicated a 2-hour staff meeting to sharing this information with the rest of our staff. Feel much more confident when asked what comments I can make in response to data when presented to me.
It was good also to work with secondary teachers which gave a slightly different perspective (although we used our own school data).
Was really good to discuss and interpret the data, and have practical tasks to complete relating to data analysis. Made you think! Some good ideas to take back to staff on how to analyse data more accurately.

Some of the comments indicated why changes to statistical literacy understanding may have been limited for participants. Three comments in particular indicated that even a focused, hands-on workshop may be insufficient to ensure on-going understanding.

It was excellent. However, you tend to forget some of the stuff because you are only exposed to it for a short period of time. You need to be exposed to it for a few more sessions to not only become confident with it and ask questions, but then become so familiar with it that it becomes second nature.
I think I would need a week of intense professional development to be able to fully understand how to read these graphs. While I get exposure to this and I am getting a better understanding as years pass I still struggle on some aspects.
I find data interpretation to be quite challenging, and would like to attend further PDs on this subject if they should arise.

4.3   TPB Factors and Statistical Literacy

The delayed post-workshop survey allowed some comparisons between workshop participants and non-workshop participants. On average those who had attended the workshop had higher scores on attitude items and lower scores for perceived barriers than those who had not attended the workshop, indicating the likelihood of improved attitudes and perceptions among participants. The statistical literacy questions on the survey were scored, using a simple partial credit scale, and it was found that the statistical literacy scores for those who attended the workshop (W) were higher than for those who had not had this experience (NW) but this difference is not statistically significant: $\left(\bar{x}_W = 26.4, \ s_w = 7.5; \ \bar{x}_{NW} = 24.5, \ s_{NW} = 9.1; \ t_{df=79} = 0.975, \ p = 0.3327 \right)$.

Table 3 Percentage of teachers choosing the truth or otherwise of given statements in the post-workshop survey, from those who had (W) and had not (NW) attended the workshop

		Definitely true		Definitely false		Not enough information		I don't know	
		W	NW	W	NW	W	NW	W	NW
3.1	The spread of the 50th–90th percentile results is wider in writing than in spelling	82*	84*	8	9	2	2	8	4
3.2	In the writing results, fewer students were between the 25th percentile and the median than were between the median and the 75th percentile	27	59	50*	14*	11	14	11	14
3.3	Victoria College's writing results have a greater range than the state results	86	86	7*	7*	4	4	2	2

Note: Items relate to a report similar to that shown in Fig. 2. The correct responses are indicated with an asterisk

However, the results in Table 3 suggest that although the frequency density misconception had been corrected for some—but by no means all—teachers (see the data in Row 3.2), there remained confusion about some aspects of the presentation of box plots. This was caused by not reading the key and difficulties interpreting the consequences of a "reduced" box plot (with 10 %/90 % whiskers) (see Row 3.3). It may be that the image of the box plot is too powerful and hinders individuals' capacity to (a) remember that some data are not represented by components of the plot and (b) recognise the possible range of such "invisible" data.

5 Conclusions and Implications

The workshop described here certainly appears to have had a positive effect on attitudes and perceptions, as well as having improved some aspects of teachers' professional statistical literacy. Unfortunately, at least one significant misconception persisted despite the attention given to what is and is not shown in box plots, and one teacher claimed to be "still baffled" after the program. Another comment reflected the mixed success of the program: "[The activities] were worthwhile, interactive, hands on, well paced, informative and I learned lots (but obviously not enough to pass this 'test'!)". For many teachers, reading box plots is not yet "second nature", although it is acknowledged that the peculiarities of the non-standard format used in Victoria's particular box plot representations may exacerbate the problem. Nevertheless, the teachers acknowledged the effectiveness of the materials and activities used in the workshop for developing understanding.

For those with statistical experience and expertise, many of the concepts associated with professional statistical literacy for teachers and highlighted above are relatively elementary. On the other hand, for those who deal with data only a few times a year and with limited prior knowledge, the outcomes from the evidence-based research and the post-workshop survey suggest that there are some areas that many teachers do not understand sufficiently. In any case, it appears important to provide timely access to reminders of what others may regard as basic concepts. For these teachers, extended programs and/or refresher programs may be of benefit. Following the qualified success of the workshop, the researchers are interested in examining ways of providing targeted statistical literacy professional learning for teachers in a timely and accessible way, without having to formally timetable a face-to-face presentation for groups. The authors have used their experiences with the workshops to create a series of short online packages, each requiring just a few minutes of interaction. The packages have been designed to emulate some of the practical activities, but are accessible for teachers' use at their convenience. These have been trialled with some groups of teachers, with initially encouraging signs, although the formal research into their effectiveness is not yet complete.

In order for teachers to use data effectively in their planning, it is essential that they acquire the kind of fluency that allows them to understand how the abstract data representations depict their very real class or school. In addition, they need to understand where each student is or could be located within the data. This means that the critical misconceptions explored here—the density/frequency issue and the all-the-data-are-within-the-plot assumption—need to be overcome. Equally importantly, a shift in attitudes, confidence and perceptions about the data's value and comprehensibility is also likely to lead to greater and better use being made of such school assessment data.

6 Note

Developed from a paper presented at Eighth Australian Conference on Teaching Statistics, July 2012, Adelaide, Australia.

This chapter is refereed.

Acknowledgments This study is supported by The Australian Research Council (LP100100388) with the Victorian Department of Education and Early Childhood Development (DEECD) and the Victorian Curriculum and Assessment Authority. Other members of the research team are Ian Gordon, Sue Helme (University of Melbourne), Jane Watson (University of Tasmania), Michael Dalton, Magdalena Les (VCAA) and Sue Buckley (DEECD). Some of this work was conducted while the second author was with the University of Melbourne. We thank all teachers who have provided data but especially those from the 20 focus Victorian DEECD schools. An earlier version of this chapter was presented at the 2012 OZCOTS conference.

References

Ajzen, I. (1991). The theory of planned behavior. *Organizational Behavior and Human Decision Processes, 50*, 197–211.

Armitage, C., & Conner, M. (2001). Efficacy of the theory of planned behaviour: A meta-analytic review. *British Journal of Social Psychology, 40*, 471–499.

Curcio, F. (1987). Comprehension of mathematical relationships expressed in graphs. *Journal for Research in Mathematics Education, 18*, 382–393.

Ferrari, J. (2011, July 15). Teachers are 'phobic' over student test data. *The Australian*, p. 7.

Gal, I. (2002). Adults' statistical literacy: Meanings, components, responsibilities. *International Statistical Review, 70*, 1–51.

Ingvarson, L., Meiers, M., & Beavis, A. (2005). Factors affecting the impact of professional development programs on teachers' knowledge, practice, student outcomes and efficacy. *Education Policy Analysis Archives, 13*(10), 1–26.

Martin, P. J. (2008). Assessment of participants in an industrial training program. In H. L. MacGillivray & M. A. Martin (Eds.), *Proceedings of OZCOTS 2008—6th Australian Conference on Teaching Statistics* (pp. 136–142). Retrieved April 28, 2014, from http://iase-web.org/documents/anzcots/OZCOTS_2008_Proceedings.pdf

Martin, P. J. (2010). Evaluating statistics education in vocational education and training. In *Data and context in statistical education: Towards an evidence based society (Proceedings of 8th International Conference on Teaching Statistics)*. Ljubljana, Slovenia: IASE, ISI. Retrieved from http://www.stat.auckland.ac.nz/~iase/publications

Milburn, C. (2012, August 27). Teachers 'failing' on test data. *The Age*. Retrieved from http://www.theage.com.au

Ministerial Council on Education, Employment, Training and Youth Affairs. (n.d.). *Benefits of participating in national assessments*. Retrieved April 29, 2009, from http://www.curriculum.edu.au/verve/_resources/Benefits_of_participation_in_national_assessments1.pdf

Pfannkuch, M. (2006). Comparing box-plot distributions: A teacher's reasoning. *Statistics Education Research Journal, 5*(2), 27–45. Retrieved August 2, 2009, www.stat.auckland.ac.nz/serj

Pierce, R., & Chick, H. (2011a). Teachers' intentions to use national literacy and numeracy assessment data: A pilot study. *Australian Education Researcher, 38*, 433–447.

Pierce, R., & Chick, H. (2011b). Reacting to quantitative data: Teachers' perceptions of student achievement reports. In J. Clark, B. Kissane, J. Mousley, T. Spencer, & S. Thornton (Eds.), *Mathematics: Traditions and [New] Practices (Proceedings of the 34th Annual Conference of the Mathematics Education Research Group of Australasia)* (pp. 631–639). Adelaide, SA, Australia: AAMT.

Pierce, R., & Chick, H. L. (2013). Workplace statistical literacy: Teachers interpreting box plots. *Mathematics Education Research Journal, 25*, 189–205. doi:10.1007/s13394-012-0046-3

Pierce, R., Chick, H., Watson, J., Les, M., & Dalton, M. (2014). A statistical literacy hierarchy for interpreting educational system data. *Australian Journal of Education*. Advance online publication. doi:10.1177/0004944114530067

Statistical Society of Australia Inc. (2005). *Statistics at Australian universities: A SSAI-sponsored review*. Braddon, ACT, Australia: Author.

Wallman, K. K. (1993). Enhancing statistical literacy: Enriching our society. *Journal of the American Statistical Association, 88*(421), 1–8.

Watson, J. M. (2006). *Statistical literacy at school*. Mahwah, NJ: Lawrence Erlbaum.

Statistical Training in the Workplace

Ian Westbrooke and Maheswaran Rohan

Abstract The workplace provides a distinctive context for statistical education in contrast with other statistical training settings, such as through educational institutions, short courses and online course. We focus on our experience in delivering statistical training to staff at our workplace, where we have developed courses for people who are not statistics graduates and who work in an operational rather than a research organisation. We review the literature, describe our workplace and the statistical needs of its staff, and discuss the important but often ignored area of data handling. We then outline the courses we have developed, on study design and statistical modelling courses, explaining our choice of R and R Commander software. We review some differences between training in the workplace and in the education sector, and conclude with some experiences from outside our workplace and final comments.

Keywords Study design • Data handling • Statistical modelling • R • R Commander • R Commander

I. Westbrooke (✉)
Department of Conservation, Christchurch, New Zealand
e-mail: iwestbrooke@doc.govt.nz

M. Rohan
Department of Conservation, Hamilton, New Zealand
e-mail: mrohan@doc.govt.nz

H. MacGillivray et al. (eds.), *Topics from Australian Conferences on Teaching Statistics:*
OZCOTS 2008-2012, Springer Proceedings in Mathematics & Statistics 81,
DOI 10.1007/978-1-4939-0603-1_17, © Springer Science+Business Media New York 2014

1   Introduction

The workplace provides a distinctive context for statistical education, with focused subject areas, ongoing relationships between participants and trainers, and typically an applied emphasis. This contrasts with other statistical training settings, such as through educational institutions, short courses and online courses, where participants typically come from a wide range of backgrounds, timeframes are often limited, and applications are likely to be more wide ranging. In this chapter, we focus on our experience in delivering statistical training to staff at our workplace, where we have developed courses for people who are not statistics graduates and who work in an operational rather than a research organisation. After reviewing the literature, in Sect. 2, we describe our workplace and the statistical needs of its staff in Sect. 3, and in Sect. 4 we discuss the important but often ignored area of data handling. Our study design course is outlined in Sect. 5, while Sect. 6 describes our modelling courses. Section 7 explains our choice of R and R Commander software. In Sect. 8, we review some differences between training in the workplace and in the education sector. Finally, we relate some experiences from outside our workplace in Sect. 9 and make concluding comments in Sect. 10.

2   Literature on Statistical Training in the Workplace

The statistics education literature focuses, understandably, on the formal education sector, particularly at school and tertiary levels. The core of publications relating to the workplace comes from the proceedings of the 4-yearly International Conferences on Teaching Statistics (ICOTS). ICOTS conferences have included workplaces in their ambit, and since at least 1998 have the workplace as one of about nine topic areas in their programmes. Relevant ICOTS papers deal primarily with how the education sector can, should or does relate to the workplace or industry. Reports from within workplaces especially about training non-statisticians are very limited. Hamilton (2010) provides a perspective on in-house training for non-statisticians in a national statistical office, while Forbes et al. (2010) look at training non-statisticians in the state sector from a combined workplace and university perspective. Most other relevant literature comes from a more general perspective. Barnett (1990) asks how different organisations can meet statistical needs. He suggests either employing professional statisticians as employees or consultants; or developing skills in staff not trained as statisticians. He then explains how consultancies or tertiary organisations are providing either open or in-house statistics courses to increase skills of non-statisticians in the workplace. In a keynote address to ICOTS 5, Scheaffer (1998) addressed "Bridging the gaps among school, college and the workplace", looking to "expand the use of statistics in industry while producing a statistics curriculum in schools and colleges that can be defended and sustained".

A theme in a number of ICOTS papers is the importance of context for workplace training, and the particular importance of including modern teaching approaches such as emphasising real data, statistical concepts rather than mathematical derivations, and using projects and hands-on computing (e.g. Stephenson 2002; Francis and Lipson 2010).

3   Our Workplace Context: The Department of Conservation (DOC)

New Zealand's Department of Conservation (DOC) is the central government organisation charged with promoting and implementing the conservation of the country's natural and historic heritage. Thus, DOC is responsible for managing approximately one-third of New Zealand's land area, along with a number of marine reserves; protecting and managing much of the country's indigenous biodiversity, including many unique ecosystems and species; promoting recreation; and facilitating tourism. The 1,800 staff members include several hundred science graduates undertaking science and technical work at national, regional and local levels.

DOC needs evidence-based information to carry out effective management. Typical questions posed by managers include:

- What are the trends in abundance and health for native species and ecosystems, and how can management make a difference?
- How are visitors using parks and conservation lands and facilities, and what issues need to be managed?

To answer these questions adequately, managers need to move beyond the broad, qualitative assessments that have often underpinned decision-making to an evidence-based approach, which demands quantitative assessments that are based on data. As in many environmental and social arenas, there is plenty of variability involved in conservation, so statistics become essential.

Science and technical staff, and others involved in research and monitoring, are generally graduates in various fields, whose qualifications range from first degrees to Ph.D.s. Increasingly, these staff members are expected to perform duties that require a competence in study design and statistical analysis. In addition, a very wide range of staff require basic skills in effective data entry, management and exploration, including effective graphing. A number of staffs require training in statistical modelling skills, starting from the linear model and up through its extensions, including mixed models for repeated measures. In addition, smaller numbers have specialist subject requirements, such as estimating animal abundance or survival analysis, including mark-recapture models. Since only two DOC staff members are appointed primarily as statisticians, the provision of statistical training for other staff is of vital importance.

We initially assessed statistical training needs through analysing the requests made to us, and by talking to staff and managers. Key areas we found for development were data handling and exploration; modelling and study design.

4 Effective Data Entry, Management and Exploration

Practical data handling and exploration are essential prerequisites for the successful application of statistics in the workplace, but are often insufficiently covered in training. In particular, data entry and preparation is an important but often neglected area of statistical practice, and is essential for the key tasks of data exploration and analysis. In fact, there is a key phase during which data preparation and exploration need to interact to ensure that the data are in a suitable state for analysis. Data errors, e.g. as a result of incorrect data entry, are very common and can lead to serious biases and incorrect inferences if left uncorrected.

We have found that Microsoft Excel is a good general tool for data handling. Legitimising the use of Excel for data entry, storage and initial exploration has helped to facilitate moving data from pieces of papers and into computers for analysis (see www.reading.ac.uk/SSC/publications/guides/topsde.html for additional information).

At DOC, more than 300 staff members have taken part in a 1-day course named *Data Handling in Excel—Entering, Managing and Exploring Data*. This course not only introduces tools that assist with data entry, such as freeze panes, protecting data and data validation, but also covers the exploration of data using tables and, if time permits, the production of graphs. One of the key emphases is on the importance of standard data formatting: each observation has its own row; each variable is entered into a column with a meaningful name; and only raw data are entered into data sheets, with no blank rows—analyses and summaries go elsewhere. Fortunately, this layout works not only when using Excel's excellent cross-tabulation tool, the Pivot Table, but also when data are transferred to dedicated statistical packages. When staff see the advantages of using this layout, particularly through the quick and easy creation of summary tables using Pivot Tables, this approach is readily adopted.

4.1 Creating Effective Graphs

To facilitate data exploration and improve the quality of presentation, we developed another 1-day course on graphs, which has an accompanying manual (Kelly et al. 2005). This course draws heavily on Tufte (2001) and Cleveland (1994). We developed exercises for this course, including one (Fig. 1) that allows participants to learn for themselves about Cleveland's recommended order of visual perception (Cleveland and McGill 1985). Other exercises include demonstrations of how easily

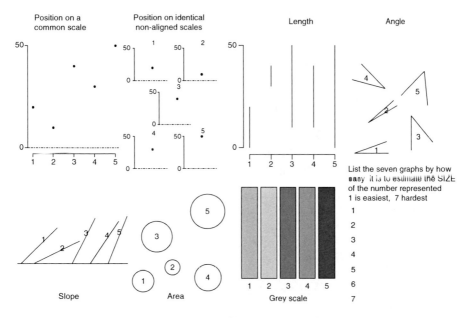

Fig. 1 An exercise from our graphs course. Participants are asked to order the seven graphs according to how easily the size of the number represented can be estimated. This exercise allows participants to learn for themselves about the options for presenting quantitative data in graphs and leads into considering the accuracy of visual perception of different approaches

default Excel graphs can be improved, and an example showing the inadequacies of pie graphs based on an excellent book by Robbins (2005). We plan to add a module on graphing in R, using *ggplot2*.

5 Study Design Course

Our initial decision to emphasise the handling and analysis of data in our courses was supported by the findings from a study of biodiversity monitoring projects we carried out in 2008, which showed that data analysis was the area that required the most strengthening (see Westbrooke 2010; Fig. 1). However, it also revealed that attention needed to be given to study design.

Tertiary statistical courses for non-statisticians generally equip graduates almost exclusively for carrying out experiments and statistical significance testing. However, DOC staff members are mostly involved in observational studies rather than experimental research, and management decisions are generally much better informed by an emphasis on effect sizes rather than hypothesis testing. Therefore, we developed a 3-day course in practical study design for applied conservation ecology, with a focus on observational studies.

A major challenge has been ensuring that staff understand the basics of randomisation and replication, and why they matter. Another important aspect has been clarifying the differences between experiments and observational studies, particularly in terms of strength of inference, and providing guidance on when and how to implement different types of study. We focus on two main areas:

- The four Ws—Why, What, Where and When—with particular emphasis on why, which is the key question for setting clear and realistic objectives. The other three Ws refer to what measure is to be used, and the effective use of replication in time and in space to achieve the objectives.
- The three Rs—Randomisation, Replication, and stRatification.

Participants each bring along an example of a study they are currently involved in designing. They introduce their study on the first morning and we then use these as examples throughout, coming back at the end with the group to evaluate what the topics covered during the course mean for the development of the design for these studies. The use of participants' own examples, and other relevant examples from the workplace, makes the course more effective in gaining involvement and ensuring that they can relate the lessons to their work both in the course and when they return to their job.

6 Modelling Courses

We were often asked early on, "How do I fit these data into an ANOVA", because that was often the only statistical model to which graduates in other subjects had been exposed. Another theme was "What test do I apply to these data", as statistics was equated with hypothesis testing. However, in our courses, we prefer to emphasise model building and the estimation of effect sizes, which are especially important in a management organisation, where there is likely to be much more interest in estimating an effect and a confidence interval than on whether or not it is different from zero.

In our 3-day introductory statistical modelling course, we revise the linear model, with sessions on ANOVA and multiple linear regression, including ANCOVA. We then extend this to the generalised linear model (*glm*), with Poisson and binomial errors for count and binary response variables, respectively. When time has allowed, we have added tree-based models and/or generalised additive models.

We teach participants to follow five steps when modelling:

Step 1: Investigate the data

Identify the response and explanatory variables, and their data types (continuous, nominal, ordinal), and construct graphs and tables to explore the distribution of the variables and the relationships between variables.

In class, this step provides the opening for teaching the use of statistical software for data exploration and visualisation.

Step 2: Fit the model

Choose an appropriate error structure based on the design and information from Step 1, and use software to fit the model.

This step provides the opportunity for us to explain error structure and the assumptions underlying the modelling.

Step 3: Analyse the model

Examine the model output from the software, and apply model selection criteria such as likelihood ratio approaches or the Akaike Information Criterion (AIC) for model comparison and variable selection.

This step allows us to explain how to interpret model output, and to discuss issues around model and variable selection, and multi-colinearity.

Step 4: Assess the model

Examine the assumptions defined in Step 2 graphically and numerically. If Step 4 fails, go to Steps 1 and 2 and try alternative models until a satisfactory model is obtained.

Step 5: Interpret the model

Interpret the results in the context of the overall problem, with estimates of relevant effects, including confidence intervals.

Many DOC staff members have limited mathematical skills, and participants struggle to write down expressions to predict the values from statistical model outputs, or to back-transform onto the original scale. Therefore, we avoid teaching more abstract statistical concepts, aiming to explain the technical aspects that are needed using outputs from the computation of statistical models. Attendees do not learn how to compute parameter estimates and we do not expose them to more than very basic equations; for example, normal equations of the general form $\hat{\beta} = \left(\mathbf{X'WX} \right)^{-1} \mathbf{X'Wy}$ are too complex for almost all of those who attend. This means that:

(i) Participants are not exposed to the design matrix $\mathbf{X}$. This makes it difficult to provide a full explanation of the need for a reference category for factor variables. Instead, we provide an informal explanation of the need for a reference category.

(ii) We avoid talking about $\mathbf{W}$ and iteratively reweighted least squares (IRLS). When the *glm* output shows the number of iterations, we explain that the computation is repeated a number of times to converge on the estimated value for the parameter.

(iii) We only touch in passing on the relationship between maximum likelihood estimation and least squares estimation.

(iv) The concept of starting values for parameter computation is ignored.

Our other main modelling course involves the analysis of repeated measures, with either continuous or discrete responses. DOC carries out many large and small monitoring projects throughout the country, which typically involve collecting information from the same sampling unit (subjects) over several years. We provide an introduction to analysing such data using mixed models.

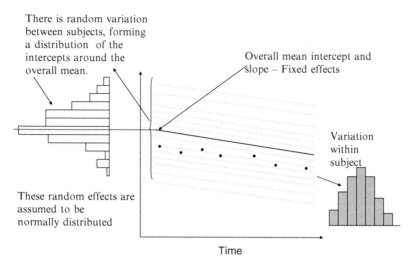

There is random variation
between subjects, forming
a distribution of the
intercepts around the
overall mean.

Overall mean intercept and
slope – Fixed effects

Variation
within
subject

These random effects are
assumed to be
normally distributed

Time

Fig. 2 Random intercept model

First, participants learn from a real example that the standard classical approach is
not suitable for analysing this type of data, illustrating the violations of the classical
assumptions using residual plots for each subject. We then explain the need to modify
the model by allowing for random variation between the subjects; for example, we
introduce the random intercept model using the graphical approach shown in Fig. 2.

We then explain that variation can be divided into two types:

(i) Stochastic variation within a subject, similar to our usual classical model errors.
(ii) Variation between subjects around the overall mean, known as random effects.

Thus, we end up with the response as a function of fixed effects and random
effects, which we present as:

$$\text{Response} \sim \text{Fixed effects} + \text{Random effects} + \text{Error}$$

and illustrate using Fig. 2.

Although serial correlation may feature in some of our datasets, time constraints have so far prevented us from addressing these during the courses.
Therefore, we advise participants that input from a statistician is needed to analyse
these types of data.

As in the introductory course, a number of technical statistical issues are often
discussed in simple language during the course. This approach is well received by
the participants, who are happy to accept our word on the more technical aspects
and are generally more interested in understanding the application rather than the
theory of statistics. Applied topics such as model selection, testing assumptions and
interpreting the results are of particular interest to those attending these courses.

7 The Benefits of Using R and R Commander Software

We use R (R Development Core Team 2012) as the software for our statistical training. With its free access, enormous flexibility and the availability of almost all statistical techniques, the usage of R has increased exponentially worldwide. R comes with base libraries and recommended packages, as well as more than 2,500 contributed packages.

R was chosen as the statistical software for use in DOC because of its power and its free availability. This reduces the cost to the organisation, and ensures a ready access to the software and portability of skills learnt. Initially, we taught our introductory statistical modelling course using standard R, with participants typing and submitting R code. However, in our experience, it is a challenge to teach both statistical methods and the R language for our biology-oriented group of participants— they often ended up with syntax errors, even though we provided R code on the screen and explained how to write the code, and users found it hard to understand and correct these errors. Aside from syntax errors, other difficulties in using standard R included:

(i) A number of R-functions and their options need to be understood. For example, to create good graphs in base R, users need to learn available graphical options such as *lwd*, *lty*, *cex*, *pch*, *type* and *legend*.
(ii) Users find it difficult to understand where to use the appropriate brackets, such as normal brackets (), square brackets [] and curly brackets {}.
(iii) Users need to learn about the availability and location of each function. For example, in order to avoid mathematical details in the binomial model, we use the *ilogit* function in the *faraway* library to compute predictions of the binomial parameter p in a logistic regression for various values of the covariates. When we present such a function, participants ask how they can learn about the availability of such things in R. We explain it is not always easy, and that we learn about them by browsing help files and books, searching the internet, talking to colleagues, and asking questions on appropriate forums or mailing lists.
(iv) R error messages are often far from friendly to casual or first-time users. For example, what is the meaning of the error message "*object ilogit not found*"? We would suggest typing *??ilogit* in the R console as the first step to solving this problem, and following it up with the options mentioned in (iii).

Thus, participants can feel daunted or overwhelmed by the computing aspects of R that are needed to complete their data analysis. Instead, the majority of participants prefer to start by using a menu-based approach, as this is an environment with which they are more familiar and reduces their learning load.

We found three menu-based packages in R: R Commander, R Excel and Deducer. After some comparative evaluation, we found that R Commander was well suited to DOC's needs.

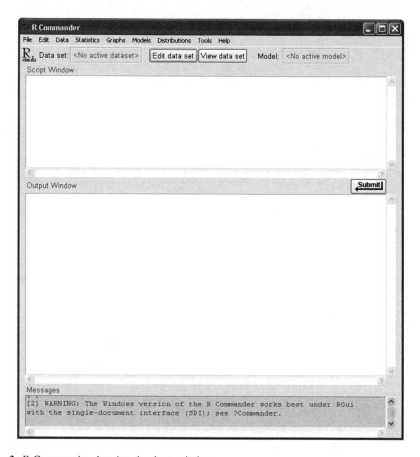

Fig. 3 R Commander showing the three windows

7.1   *R Commander*

R Commander (Fox 2005), which is available in the R package *Rcmdr,* provides a simple point and click interface to R, including linear and generalised linear models, and some graphing capacity. This free, menu-based statistical package within R creates an easy way to learn statistical modelling with R.

R Commander comes with a menu at the top plus three windows (see Fig. 3):

(i)	Script window	–	R script is automatically written here when a menu is clicked. The user can also create or modify code here
(ii)	Output window	–	This operates much like the R console, echoing script as it is submitted and showing outputs
(iii)	Messages window	–	Information is provided about the active dataset and error messages

We converted most of our statistical modelling course materials so that participants could use R Commander instead of typing R code. We found that this made the course more relaxed for both the trainers and participants, allowing more time to learn about statistical methods rather than coding aspects. For example, instead of writing R code such as *xyplot(fc~year.measured | tag, data = FBI, auto.key =TRUE)*, which participants often needed help with, now we just ask them to click *Graphs >> XY conditioning plot* and complete the dialogue. This graph allows visualisation of the response for a given condition and often makes it easier to recognise data entry errors.

Advantages of using R Commander include:

(i) New and infrequent users can start using R without confusion or panic.
(ii) R Commander provides standard R code for each operation, which can assist with learning R programming commands and allows us to modify the script when needed. We sometimes give instructions on how to modify the code; for example, to add a regression line to a plot created by *xyplot()* using the menus, we ask participants to add, *type = c("p", "r")* anywhere within brackets of the *xyplot* code, highlight the whole of this piece of code and click the *submit* button. Code from R Commander can be saved and later run in R directly, with very minimal modification.
(iii) Participants start to learn about R packages, which are automatically loaded in R Commander. For example the *lattice* package is used without their knowledge when *XY conditioning plot* is called.

Issues with R Commander include:

(i) Modelling capability in R Commander focuses on the linear and generalised linear model. It does not provide access to more complex statistical modelling and graphing techniques, such as generalised additive models, regression trees and mixed models. Thus, participants who require functionality that is not available in R Commander need to learn to use R code directly.
(ii) Some stability issues were occasionally encountered through our teaching experience, but these were largely overcome by ensuring that we used the *sdi* (Single Document Interface) rather than the default *mdi* (Multiple Document Interface) interface to R under Windows, and by using an up-to-date version of *Rcmdr*.

Four main R Commander menus—*Data, Statistics, Graphs* and *Models*—are used in our statistical modelling course. A number of data manipulation and statistical operations can be carried out under these menus. All possible operations can be viewed in the manual or by clicking appropriate menus. Some examples are:

(i) *Data* menu:
    Data can be imported to R Commander from various formats, such as text, csv files, SPSS, Minitab and STATA when *Import data* is clicked. The *Manage variables in active data set* submenu allows data manipulation such as transformations, converting numerical variables to factors and recoding variables.

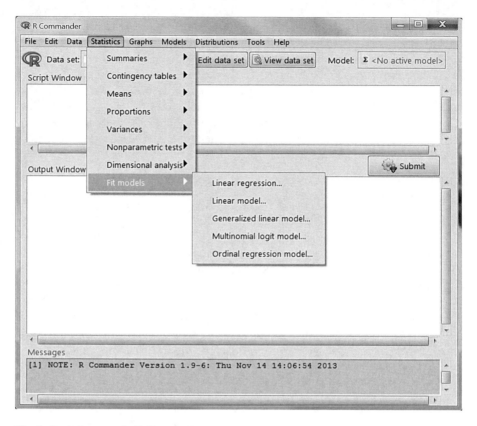

Fig. 4 Statistics menu in R Commander

(ii) *Statistics* menu:

Most statistical analyses that are required at undergraduate level can be carried out using this menu (Fig. 4). ANOVAs and *t*-tests are available under the *Means* option, while the *Fit models* options allow for the linear model, generalised linear model, the multinomial logistic model and ordinal regression model.

(iii) *Graphs* menu:

Tools for graphs in R Commander are excellent, but not totally comprehensive. A number of graphs ranging from histograms to 3D graphs including conditional plots can be generated. Figure 5 shows the range of graphs available.

(iv) *Models* menu:

This menu (Fig. 6) provides diagnostics for a model, both graphically and numerically, including making available tests for multi-colinearity and autocorrelation.

R Commander provides a limited range of statistical tools, but has plug-in packages that can be manually added using the *Tools* menu. Plug-in packages include one for survival analysis and one that provides an introduction to the sophisticated graphics package *ggplot2*.

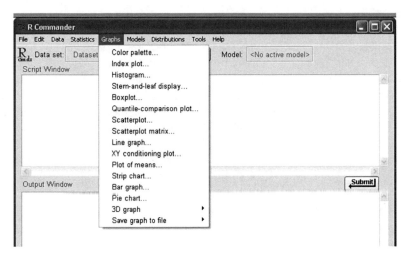

Fig. 5 Graphs in R Commander

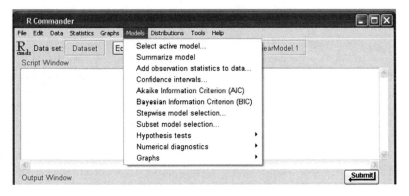

Fig. 6 Assessing models in R Commander

We carried out an informal survey amongst a group of our course participants about R Commander. More than 70 % of the 21 participants surveyed agreed that R Commander is useful for their own work, while none disagreed. More than 70 % also agreed that they did not need to memorise the R functions as much, while 10 % disagreed.

Experience at DOC has shown that R Commander significantly eases the steep learning curve for R. In our statistical modelling course, we have found that it allows both participants and trainers to concentrate more on the statistical content. A number of participants have also found this approach to be a useful bridge to writing code in R, and are now programming in R independently. This ability to use code in R provides a good basis for those progressing to our repeated measures course, which involves mixed modelling, where we put R Commander aside and use R directly. Some staff members, especially those who only use statistical tools occasionally, have found that the R Commander environment is adequate and convenient for their needs, and feel no need to progress to writing R code.

While software packages such as Minitab and SPSS might have some advantages over R Commander for some teaching purposes, we have found that R Commander works very well for our first modelling course, and functions very effectively as a bridge to our main statistical software, R, as the author (Fox 2005) intended. We chose R over our previous software SPSS because it meets our needs in a statistical package best, and for its free availability. R can be used simultaneously by a large number of people in the workplace or anywhere else without any licence issues.

8   Differences Between Training in the Workplace and in the Education Sector

While the courses we deliver to DOC staff have similar content to those taught in educational institutions, the workplace context has led to some distinct features. First, we emphasise practical applications and examples using real data, with a basic outline of the theoretical background; formulae and mathematical notation are kept to a minimum, with no derivations or proofs. Second, we teach intensive block courses (typically 1 or 3 days long) rather than multiple sessions over a longer period such as a quarter or semester; we have found that it is much easier for staff members who are faced with many competing priorities to commit to attending short courses, particularly since participants are dispersed across New Zealand, as conservation management is often carried out in remote areas. Third, we work with small classes, up to a maximum of about 12, and have a high trainer to participant ratio; one trainer can cope with up to five or six participants, so we usually aim for two trainers per course. Finally, we do not carry out formal assessment of participants, as it would take up precious classroom time, and there is less need for formal qualifications in the workplace context; instead, we ask the participants to assess the course and its applicability to their work.

For the statistical modelling course, respondents have assessed six statements on a 5 point scale (from "strongly disagree" to "strongly agree"). The statements covered whether the overall content was relevant to my work; whether the explanations and practical computing increased understanding of statistical models and R and whether the 3-day programme met overall expectations. For 30 respondents from 2009, 2011 and 2012, 177 (98 %) of 180 responses were evenly split between "agree" or "strongly agree". An open question that asked *what worked well* revealed two themes, with 11 mentioning practical exercises or examples, and one of these and six others mentioning the small class or the availability of two tutors. In response to *what could be improved*, there were no obvious common themes except that 11 gave no response (as against five for *worked well*) and five stated little, not much or nothing.

9 Statistical Training Beyond DOC

Each of the authors has had a very good response internationally when presenting workshops or seminars involving R Commander. Recently, one author (M.R.) carried out a 1-day workshop prior to an international conference in Sri Lanka, which used R Commander and briefly covered material that was similar to our 3-day statistical modelling course. This was well received by the 26 participants from seven countries. One author (M.R.) was also asked on the spot to deliver an informal seminar for fisheries scientists at the Secretariat of the Pacific Community (SPC), New Caledonia, because there was a thirst for learning more about R and R Commander. When shown the graph in Fig. 7, participants could not believe it could be so easy to create such a graph either in R Commander or using a package such as *gplots*; previously, they had taken hours to make a similar graph by writing R code.

Similarly, there was such wide interest on accessing R using R Commander that one author (I.W.) added an unscheduled seminar to his presentations at the 2011 International Conference on Health Statistics in the Pacific in Suva, Fiji. He also found that R Commander worked very well in a 2-day workshop on

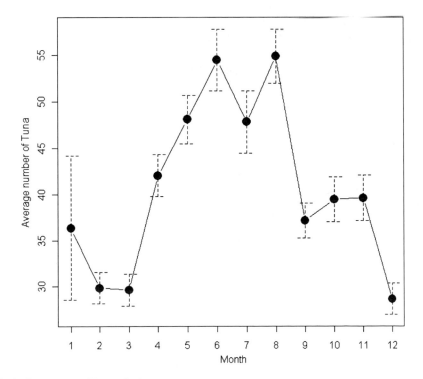

Fig. 7 Tuna count with standard errors

statistical modelling at the University of Queensland, which was aimed at non-statisticians. We have noticed that R Commander is generally becoming popular in the Pacific region and beyond, with workshops using R Commander appearing more commonly.

10   Conclusions

Our experiences from training DOC staff members and presenting seminars more widely have shown us that:

- Training that allows observational data to be distinguished from experimental data and which provides modelling skills that are applicable to different types of data is critical. There needs to be an emphasis on the estimation of effect sizes rather than hypothesis testing.
- Effective data management and exploration (especially graphing) skills are needed, to provide the basis for data analysis.
- R Commander works very well for introducing statistical modelling, especially for graduates in non-statistical disciplines. It allows trainees and workshop participants to concentrate on the concepts and application of models to the data, rather than the mechanics of the computations involved.
- The workplace context means that courses work best as intensive block courses, rather than as a series of shorter sessions, with a practical rather than theoretical emphasis. The use of real examples and datasets that attendees readily understand is also critical, with hands-on computing an integral part of all sessions. The evaluation of how well courses meet workplace objectives is more important than evaluating individuals.

The key to a statistician making a difference in a large workplace is to have a strong training and advocacy role. One or two statisticians can make a difference, in our case to help protect New Zealand's unique biodiversity and protected areas. We receive great support from the wider statistical community through consultation, the receipt and provision of specialist training, and the availability of resources such as R and more specialist software. To make academic statistical training of biological, ecological and social science students more applicable to the workplace, our experience shows that there is a need for a stronger statistical modelling approach, with less emphasis on hypothesis testing.

11   Note

Developed from a paper presented at the Eighth Australian Conference on Teaching Statistics, July 2012, Adelaide, Australia.

This chapter is refereed.

Acknowledgements We would like to thank the many statisticians and others who have helped with the development of training at DOC, especially Neil Cox, Jennifer Brown, Richard Duncan and Tim Robinson who have played major roles in developing some of the courses. We wish to acknowledge colleagues and the chapter referees who assisted us by providing feedback and comments on this chapter and the ICOTS and OZCOTS papers that preceded it.

References

Barnett, V. (1990). Statistical trends in industry and in the social sector. http://iase-web.org/documents/papers/icots3/BOOK1/C8-3.pdf

Cleveland, W. S. (1994). *The element of graphing data*. Summit, NJ: Hobart Press.

Cleveland, W. S., & McGill, R. (1985). Graphical perception and graphical methods for analyzing scientific data. *Science, 229*, 828–833.

Forbes, S., Bucknall, P., & Pihama, N. (2010). Helping make government policy analysts statistically literate. In C. Reading (Ed.), *Data and context in statistics education: Towards an evidence-based society. Proceedings of the Eighth International Conference on Teaching Statistics (ICOTS8, July, 2010), Ljubljana, Slovenia*. Voorburg, The Netherlands: International Statistical Institute. http://iase-web.org/Conference_Proceedings.php?p=ICOTS_8_2010

Fox, J. (2005). The R Commander: A basic-statistics graphical user interface to R. *Journal of Statistical Software, 19*(9), 1–42.

Francis, G., & Lipson, K. (2010). The importance of teaching statistics in a professional context. In C. Reading (Ed.), *Data and context in statistics education: Towards an evidence-based society. Proceedings of the Eighth International Conference on Teaching Statistics (ICOTS8, July, 2010), Ljubljana, Slovenia*. Voorburg, The Netherlands: International Statistical Institute. http://iase-web.org/Conference_Proceedings.php?p=ICOTS_8_2010

Hamilton, G. (2010). Statistical training for non-statistical staff at the office for national statistics. In C. Reading (Ed.), *Data and context in statistics education: Towards an evidence-based society. Proceedings of the Eighth International Conference on Teaching Statistics (ICOTS8, July, 2010), Ljubljana, Slovenia*. Voorburg, The Netherlands: International Statistical Institute. http://iase-web.org/Conference_Proceedings.php?p=ICOTS_8_2010

Kelly, D., Jasperse, J., & Westbrooke, I. (2005). Designing science graphs for data analysis and presentation: The bad, the good and the better. Department of Conservation Technical Series 32. Wellington. Department of Conservation. www.doc.govt.nz/upload/documents/science-and-technical/docts32.pdf

R Development Core Team. (2012). R: A language and environment for statistical computing. R Foundation for Statistical Computing, Vienna, Austria, ISBN 3-900051-07-0. www.R-project.org/

Robbins, N. B. (2005). *Creating more effective graphs*. Hoboken, NJ: Wiley.

Scheaffer, R. L. (1998). Statistics education—bridging the gaps among school, college and the workplace. http://iase-web.org/documents/papers/icots5/Keynote3.pdf

Stephenson, W. R. (2002). Experiencing statistics at a distance. http://iase-web.org/documents/papers/icots6/4d3_step.pdf

Tufte, E. (2001). *The visual display of quantitative information*. Cheshire, CT: Graphics Press.

Westbrooke, I. (2010). Statistics education in a conservation organisation—Towards evidence based management. In C. Reading (Ed.), *Data and context in statistics education: Towards an evidence-based society. Proceedings of the Eighth International Conference on Teaching Statistics (ICOTS8, July, 2010), Ljubljana, Slovenia*. Voorburg, The Netherlands: International Statistical Institute. http://iase-web.org/Conference_Proceedings.php?p=ICOTS_8_2010

Part E
Postgraduate Learning

Engaging Research Students in Online Statistics Courses

Glenys Bishop

Abstract This paper outlines the experiences of the Statistical Consulting Unit at the Australian National University (ANU) in providing an online short course in introductory statistics. This course is now in its fourth iteration, and a second online course in experimental design and analysis was added in 2012, with a third one under development. These courses use the CAST e-books, developed by Dr. Doug Stirling, for the main statistical content, and the Moodle delivery platform. They are non-compulsory and not for credit, but free to ANU research students and staff.

This paper considers various indicators to measure the success of an online course. A number of lessons have been learnt about engaging students in an online environment, assessing the time required to complete the course, offering incentives and receiving feedback.

Keywords Online learning • Student engagement • Postgraduate

1 Introduction

The introductory course described in this paper is offered to research students and staff at the Australian National University (ANU) free of charge. It has some characteristics which set it apart from regular university courses. Enrolment can occur at any time throughout the year and there is no final deadline. Students can request their enrolment be renewed the following year. There are no set start and end dates.

G. Bishop (✉)
Statistical Consulting Unit, Australian National University, Canberra, ACT 0200, Australia
e-mail: glenys.bishop@anu.edu.au

H. MacGillivray et al. (eds.), *Topics from Australian Conferences on Teaching Statistics:* 331
OZCOTS 2008-2012, Springer Proceedings in Mathematics & Statistics 81,
DOI 10.1007/978-1-4939-0603-1_18, © Springer Science+Business Media New York 2014

The course is neither compulsory, nor does it count for credit in any program offered by ANU. It fits the definition of learning support provided by MacGillivray and Croft (2011), in that it is an enabling programme, designed to support postgraduate students to acquire statistical knowledge but without any associated credit towards their course.

The challenge is to find ways to engage students given all this flexibility.

Much has been written about engaging students in the Statistics classroom using problem-based activities, cooperative learning and alternative methods of assessment. For example, Enders and Diener-West (2006) and Da Silva Nascimento and Dos Santos Vaz Martins (2007) have investigated teaching and assessment methods for introductory courses aimed at graduate students. Garfield and Ben-Zvi (2007) provide an overview of research into Statistics teaching and learning methods for students at all levels. My need is to consider ways of engaging students in an online environment.

Everson and Garfield (2008) focus on teaching Statistics in an online environment with one of their test groups consisting of beginning graduate students with no prior statistical knowledge. This group is very similar to the target audience for the course described here. Their emphasis is on encouraging discussion among students.

More generally, Salmon (2002) introduced the idea of e-tivities which she describes as "frameworks for enhancing active and participative online learning by individuals and groups". An e-tivity involves at least two people working together in some way. Among other things they are based on interactions among course participants mainly through written messages. They are designed and led by an e-moderator.

Tudor (2006) emphasised the importance of interaction with the professor (lecturer) to engage students and she achieved this through small group discussions, feedback on exams and weekly online communications.

These authors (Da Silva Nascimento and Dos Santos Vaz Martins 2007; Salmon 2002; Tudor 2006) all assume that all participants will start and finish the course at the same time. Thus they can be put into discussion groups or e-tivities at the start and given a due date for completion. This is difficult to accomplish when there are no fixed start and end dates.

Section 2 of this paper explains why this course was developed, the introductory course is described in Sect. 3, while Sect. 4 summarises measures that have been used to evaluate and improve the course. Conclusions and lessons learned are discussed in Sect. 5.

2   Background

The Statistical Consulting Unit (SCU) at the ANU provides statistical support to Honours and graduate research students, as well as research staff as part of the university's research endeavour. Of the more than 300 clients per year that the SCU sees, about 80 % are research students.

The support mainly takes the form of free consultations about any aspect of the statistical cycle, including assistance with forming the research question, design of experiments, survey design, data entry and methods of analysis, with an emphasis on collaboration and skilling the client. Like Finch and Gordon (2010), we have found that the range and variety of these interactions with clients have helped us to observe important aspects of statistical thinking that arise in the research environment.

In addition, the seven consultants (4.7 full-time equivalents) in the unit also serve on research students' supervisory panels, when there is a major statistical component in the thesis. They have built strong relationships with some of the research groups in the university and are called on to participate in statistical workshops for groups of research students within a discipline.

In the past, the SCU used to present four 3-day short courses each year: Introductory Data Analysis, Experimental Design and Analysis (EDA), Regression and General Linear Models and either Survey Design and Analysis or Multivariate Analysis. The aims of these courses were to promote awareness of appropriate statistical thinking, enable participants to recognise when they needed to ask for statistical help, facilitate communications between clients and applied statisticians, and enable researchers to do some of their own data analysis. After each course there was always a surge of new clients seeking consulting appointments.

However, these courses required a lot of resources, not only in preparation and presentation of the material but also in administration. Furthermore, the number of consulting clients continues to grow, thus reducing the amount of time available for presenting short courses. These courses were not able to satisfy demand since the total number of places in a year was limited to 78; there was always a waiting list. Students would often enrol in any course, regardless of its suitability to their needs at the time just to study some Statistics. The SCU sought an alternative that would allow students access throughout the year, that would be less resource intensive and that would not limit the number of participants and would start at an introductory level.

At the same time there was a strong demand in the Colleges of Science for more Statistics education for their students.

So in 2010 work began on the development of an online introductory Statistics course. That year, a pilot course was developed to be delivered online and named Introductory Statistics Online (ISO). At the time of writing, in 2013, this course is in its fourth iteration. In addition, a second online course, EDA, has been added. A further online course, Introduction to Modelling, is under development.

3   Course Description

3.1   Overview

Using the American Statistical Association guidelines (GAISE) (2005) for an introductory course, I thought it was important to emphasise data and concepts rather than theory and recipes. While the guidelines emphasise the importance of including

- Subject matter is conceptual not formula-driven;
- Facilities used are already available in ANU where possible;
- Relevance is ensured by using examples from ANU research;
- Supported by research supervisors;
- Minimum maintenance is required by the SCU staff;

Fig. 1 Desirable properties of the online Statistics course

On completion of the course you will be able to:
- Describe the role that statistics plays in a research project;
- Apply statistical principles when setting up your research project;
- Organise your data in a manner suitable for statistical analysis;
- Interpret statistical package outputs for hypothesis tests;
- Recognise when you need to ask for statistical help;
- Communicate your research questions to an applied statistician;
- Conduct some of your own data analysis.

Fig. 2 Learning outcomes for Introductory Statistics Online (ISO)

data, supervisory staff within the university went further and recommended the use of data collected by other ANU researchers to illustrate relevance. The demand from the scientific supervisors at the time indicated an urgent need for a course and that they should be happy with it. The SCU itself wanted a course that was not going to be as resource intensive to develop and maintain as the suite of 3-day courses had been. Thus the desirable properties were as listed in Fig. 1.

An extra property was that if the course was to be useful, it should engage the students. This property is of course the subject of this paper.

At the time that ISO was being developed, ANU had established Web Access To Teaching and Learning Environments (Wattle) (2013), an online learning environment. The main underpinning platform is Moodle (2012), and this has been augmented with other online learning facilities such as Digital Lecture Delivery (DLD) (2013), Wimba Voice Tools (2013) for audio and voice recording and Adobe Connect (2013), a video conferencing tool. So it was decided to use this delivery platform.

By referring to Bloom's taxonomy of educational objectives (1956–1964), the learning outcomes, shown in Fig. 2, were determined. They tend to come from the declarative, procedural and conditional knowledge dimensions. In other words, it is not our aim to train practising statisticians but rather to develop an awareness and appreciation of how statistical thinking fits into a research project from another discipline.

The statistical content is provided by Computer Assisted Statistics Texts (CAST) developed by Stirling (2000–2012). CAST consists of a suite of electronic books dealing with introductory and more advanced statistical concepts, exercises and notes for lecturers. The advantages of CAST are its dynamic illustrations of concepts, its interactivity offering students the opportunity to try things, its very limited use of formulae and its exercises for much of the introductory material. In addition theory

Topic 1: the role of sound statistical thinking in a research project, introduction to data structures, guidance on setting up a data file, the importance of variation in statistical methodology
Topic 2: summarising and presenting a data set, using descriptive statistics, charts and tables
Topic 3: fundamental concepts of survey design,
Topic 4: fundamental concepts of experimental design
Topic 5: introduction to statistical inference, including the sampling distribution of the mean
Topic 6: attaching a precision measure to estimates
Topic 7: fundamental concepts of hypothesis tests
Topic 8: using statistical models to compare two groups

Fig. 3 Topics in the 2012 and 2013 version of ISO

and exercises use interesting international data sets, and the student can choose which of several categories of data they want to see in examples. Categories include Biometric, Business, Climatic and Official Statistics, although not in every example. Finally, under a creative commons licence, CAST can be downloaded to a local website.

With help from Stirling [personal communication], we have created a customised e-book for use in this course. To supplement this material, we have used data provided by ANU research students to create some meaningful practical examples. They are described in more detail in Sect. 3.

In its current form, ISO consists of eight topics, summarised in Fig. 3. This differs from the first two iterations of the course. They included correlation and regression but did not include survey design or experimental design. These two topics were inserted early in the course because a large proportion of our clients have design issues in their data collection. This is commensurate with the findings of Finch and Gordon (2010).

Each topic contains a link to its description and objectives. There are instructions on which sections of the CAST e-book to work through, which exercises to try and links to any additional material. Each topic also has a quiz for self-assessment, with questions that are usually either multiple choice or matching. Each topic also includes a request to the students to indicate approximately how long they spent on the topic. Links take the student directly to the relevant sections of CAST. As an example, topic 8 is shown in Fig. 4. The GenStat solutions are visible because the student has viewed the GenStat Exercise file. The SPSS solutions are not visible, indicating the student has not viewed that file.

Three levels of certificates are used in ISO: bronze for completion of the first four topics, silver for completion of the first six topics and gold for completion of all eight topics. These certificates give students a goal and documentation of achievement. In the first two iterations of the course, students could attain Platinum standard by completing an assignment involving analysis and interpretation. This was discontinued in 2012 because of low uptake and other issues explained in Sect. 4. The numbers of certificates awarded so far are shown in Table 2 of Sect. 4.

8 # Models for Comparing Groups   □

This topic shows how statistical models are developed.

Instructions

Read the description and objectives of this topic.

Read Chapter 8 of the CAST anu e-book, paying particular attention to section 1, and the dynamic illustrations in sections 2, 3, 4 and 5.

Try a few hypothesis tests in CAST exercises 7.3

Save the file New Worm data to your computer.

(Optional) Use the saved data file to carry out either the Topic 8 GenStat Exercise or the Topic 8 SPSS Exercise. You may find it useful to read about counts of worms in native rats in the case studies again (see Topic 1).

Complete the Topic 8 Quiz

Please **answer** the choice question 'Time for Topic 8'.

☑ Topic 8 Quiz
❓ Time for Topic 8
❓ Course evaluation
🗒 Topic 8 GenStat Exercise   ☑
🗒 Topic 8 SPSS Exercise   ☑
🗒 Topic 8 GenStat Exercise Solutions   ☐

Fig. 4 How topic 8 appears in the online delivery platform

Table 1 Association between attendance at the initial meeting and completion of pre-course and topic 1 quizzes January to April 2011

Initial meeting	No pre-course quiz	Did pre-course quiz	Total	Test for association
Did not attend	32	23	55	
Attended	12	29	41	$\chi^2 = 7.91$
Total	44	52	96	P-value 0.005
	No topic 1 quiz	Did topic 1 quiz	Total	Test for association
Did not attend	38	17	55	
Attended	18	23	41	$\chi^2 = 6.13$
Total	56	40	96	P-value 0.013

3.2 Student Support

As Tudor (2006) noted, students need to feel that there is a real lecturer presence behind the course and a number of ways to achieve this have been used.

In 2010 and 2011, students who registered for the course in any given month were invited to an introductory lecture in which I introduced myself and invited them to contact me if they had any questions about the course. During this lecture I also described the key features of the course and CAST. Attendance at these meetings was poor but those students who attended were more likely to complete at least one quiz than those who did not. See Table 1.

I was concerned that students who could not attend the introductory lecture were missing out on information that would help them with the course. To replace this live lecture, I made a 9-min MP4 recording, showing various features of CAST with my voice describing them. While this is probably not as good as the live lecture attended by some students, it does reach all students.

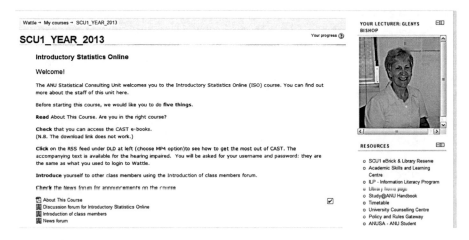

Fig. 5 The Welcome page for ISO in 2013

Each monthly group in 2010 and 2011 was given a 10-week deadline for completion with exhortations after 6 or 7 weeks. This usually resulted in a plethora of emails, requesting extensions. In 2012 a deliberate decision was made to have no set start and end dates, to allow maximum flexibility for students. In retrospect, it might have been useful to try e-tivities (Salmon 2002), such as discussion starters (Tudor 2006), with each monthly group constituting a discussion group. The flexibility built into the current version means e-tivities of this type are no longer possible.

Other ways to demonstrate the lecturer's presence is through the use of forums. There are three in this course: a news forum, an introductory forum and a discussion forum. I use the news forum to make announcements and issue general messages of encouragement. Students cannot initiate news items. The introduction forum is for students to introduce themselves to each other and to me. Each year I demonstrate how to use this forum by introducing myself and providing a link to the Staff page of the SCU. Only about 25 % of students introduce themselves. It has been my practice to make some response to each student who does so. The discussion forum was set up for students to ask questions about any aspect of the course. Many are reluctant to do this but will instead email their question directly to me. When their questions are of general interest I enter them with my answers into the forum, while preserving the student's anonymity.

Sometimes I email individual students who have been making some progress but stopped, to offer help. This requires regular monitoring of the course.

Finally, I have now included a photograph of myself in the course so that students can see a real person, rather than a committee of faceless people, is involved. See Fig. 5.

We have found that an effective way to engage students is to engage supervisory staff support for ISO. We have used a number of contexts to raise the profile of the course. One is to demonstrate the course in an information seminar for supervisory staff; it is a good idea to make sure there are some people present who can already see the course's benefits. Occasionally research students bring their supervisors to statistical consultation meetings and that is a good time to draw attention to the course and how the student would benefit. One of our consultants has tapped into the research student workshops for a large discipline; through this liaison, the research student coordinator has urged all of her charges to enrol in the course. The provision of helpful statistical advice to a colleague from another discipline can trigger enrolments from that colleague's discipline. Finally, the SCU has an Advisory Committee of stakeholders chaired by the Pro-Vice Chancellor for Research and Research Training; progress and developments in the online courses are reported at every meeting.

To date, no School or College within ANU has made the course compulsory for their research students. However, some supervisors and the research student coordinators of some schools within the university have emphasised its usefulness. Their students are more likely to enrol in and persist with the course than other students.

3.3 Case Studies

One of the desirable properties of the course was to include data collected by ANU researchers. Consultants approached a number of research students and sought their permission to use their data. Working with those students I produced descriptive summaries. The three data sets selected were not very large, had categorical and numerical variables and could be used to illustrate regressions, transformations and graphical methods.

A fourth study does not include the raw data but is a report, available on the internet, and illustrating aspects of the statistical cycle in research (See Mallee fire ecology 2013).

Initially the case studies with data were used in exercises and an assignment for students to practise regression analysis in a statistical package. However, it became clear that these exercises were very time consuming and many students skipped them. Only two students completed the assignment. An additional difficulty was that students were asked, at the start of the course, to nominate which statistical package they planned to use during the course, as steps in the analysis were included in several topics before the final assignment. This caused a lot of angst for some students and, I think, caused some of them to give up.

In version three of the course, the case studies were incorporated fairly superficially into quiz questions. Their real benefit has been to use them in the follow-up courses, EDA and Introduction to Modelling. In version four (2013), some simpler local data sets such as Canberra annual rainfall have been used in worked examples. Using the Moodle conditional facility, students cannot access the solution file to an example until they have, at least, viewed the initial file which sets out the data, the research question and GenStat and SPSS instructions.

In general, most of the data sets presented to the SCU have levels of complexity that do not lend themselves well to an introductory course analysis.

4 Evaluation

MacGillivray and Croft (2011) have provided a framework for evaluating learning support. Their framework includes both qualitative and quantitative data on usage and on effect. Both types of data for usage of this course have been compiled and there is a small amount of qualitative data on effect but, unfortunately, nothing quantitative.

4.1 Quantitative Usage

The main quantitative measure is the number of students who enrol. Clearly the word has passed around, as can be seen from the numbers enrolling shown in Table 2.

It could be argued that, since so few of those who enrol actually obtain certificates, the course is not successful. However, what does not show up in the figures is the number of students who are accessing the course material but not doing the quizzes. While Wattle does not keep records of this sort of activity, weekly browsing by the lecturer reveals that many students have accessed the course without attempting quizzes. They are most likely looking at the CAST material. The number of researchers being exposed to statistical material is far greater than the 76 per year who attended the SCU short courses in the past.

Table 2 Numbers of new enrolments in ISO and number of certificates awarded in each year that the course has been offered

Year	2010[a]	2011	2012
Enrolments	34	163	148[b]
Bronze	2	2	6
Silver	–	5	4
Gold	–	11	7
Platinum[c]	1	1	–
Total certificates	3	19	17

[a]In 2010, the course was open for 2 months and only to Science students
[b]Some 2011 students re-enrolled the following year bringing the total enrolments in 2012 to 165
[c]Platinum level withdrawn at the end of 2011

Table 3 Formal evaluation feedback for the second and third iterations of the course

Year	2011	2012	2012a
Has your statistical knowledge and understanding improved as a result of taking this course?			
Yes I have learnt a lot	6	6	4
Yes I have learnt a few things	4	2	7
It has consolidated what I already knew	5		1
It has had no effect on my knowledge and understanding of Statistics			
No, I feel very confused by the material presented			
Do you think the material covered in this course will be useful in your research work?			
Yes, I have already thought of ways to use what I have learnt	7	4	
Although I have not used much of it yet, I can see that it will be useful	4	4	
Possibly	3		
It did not go far enough for me	1		
It was too confusing to be useful			
Would you recommend this online course to your fellow students?			
Yes definitely	10	8	8
Yes maybe	4		4
Not sure	0		
Probably not	1		
Definitely not			
Please take 1 min to comment on any aspect of this course.			

2012a is for feedback halfway through the course

4.2 Qualitative Usage

Moodle includes an anonymous feedback facility. We developed our own simple feedback form, and responses are summarised in Table 3. The feedback is anonymous but it is clear from the number of responses and the comments made that feedback is usually given by those who have completed the course. Because so few students were completing the whole course, a feedback form was added at the end of topic 4 in 2012 in an effort to capture some interim opinions. That form omitted the second question. The results are shown as 2012a in Table 3. Generally, responses demonstrated overall satisfaction with the course.

Excerpts from the 1-min comments in the 2011 iteration are shown in Table 4. Some are about CAST in particular, some are about the quizzes and some are about other aspects of the course. Using the comments about quizzes and examining the results of quiz questions enabled me to make improvements to quizzes in 2012. Questions routinely badly answered were reworded. Quizzes for the later topics and those requiring calculations to be performed using statistics software were made more relevant by asking for interpretation of computer output provided.

The comment about objectives led to their revision. Each quiz question was examined to ensure that it was in line with the topic objectives. The request for exercises and instructions in software has now been met as mentioned in Sect. 3.

The other issue was how much time students should devote to the course. Several comments were made about time. Analysis of students' responses to the time taken

Table 4 Categorised excerpts from anonymous comments provided in 2011

CAST
- I found that the course is definitely helpful, especially the use of graphs to demonstrate the points that make it easy to understand
- I found the exercises helpful, but if struggling with something it would have been good to have a "show working" option rather than just the correct answer via "Tell Me"

Quizzes
- Some of the quiz questions need better wording, they were confusing
- The quizzes, particularly towards the end of the course, could have covered the material more comprehensively
- The only critique I would offer about this course would be about having to perform calculations in the quizzes. I think understanding the concepts behind statistical techniques is very important … but memorising formulae or having to calculate them unnecessary

Other aspects
- It would be better if there are more exercises using GenStat or other statistical software so that we can learn how to use the program and also how to interpret the result from the program
- The course objectives did not match any of the text given for learning, nor did it match the questions in the quiz—I did not feel I had satisfactorily met each criteria for each section
- I wish I could have taken this course over the summer break, when I had the time
- My comment would be that potential students don't under-estimate how much time they'll need for it
- I really appreciated the effort made by the tutor. It certainly made me feel more comfortable asking for help if I needed it

question showed that topic 2 was very time consuming. This was the topic in which the first analysis of a case study was made.

To get a complete picture, one should also gather information on non-usage. Early in the life of this course, 30 Ph.D. students who had enrolled, but not completed the course in 6 months, were encouraged to give reasons either by attending a meeting, completing an anonymous feedback or by direct email. Excerpts from non-anonymous comments made by 13 respondents are shown in Table 5. The overwhelming reason given was competing priorities and one student made the general point that students would not do the course if it was not perceived to have a high priority. This illustrates the need to engage supervisors so that the course becomes an important priority as discussed in Sect. 3.

4.3   Qualitative Effect

There is a reasonable amount of anecdotal evidence that this course is having an effect on the statistical skills of research students. Several supervisors have commented on how much their students have benefited. The statistical consultants in the SCU have also reported that students who have taken the course have a much better appreciation of the statistical process in their research.

Table 5 Some reasons for not continuing with the course, collected May 2011

Time issues
- ...as it was my first semester as a Ph.D. student I thought I would have a bit of time when in fact I very quickly became swamped with my own work and didn't find time for the course...
- As a fulltime worker and mother I sometimes get overwhelmed by other things
- I've been quite busy, doing a Ph.D. at the moment... I put this course on the backburner
- I've been wanting to continue on with the course, but I've been pressed for time...
- I put aside 1 weekend to sit and finish it, assuming that I will have the memory from last course I had, but that didn't work as the material was quite intense and needed more time

Course unsuitable
- I prefer the interaction of live lectures though, I think I learn better in that type of environment so I can ask questions in real time and discuss them in person...
- I guess that my initial enrolment was tentative and based on limited understanding of what was offered, and once I found that the course was slightly different to what I had expected, I decided not to continue...

General comment
- Given that this course is not compulsory, if students don't have real determination or motivation to take this course and they have some other important tasks to carry out during the course period.[sic]

5 Conclusions and Lessons Learnt

Several important lessons have been learnt about engaging students in online courses. They include the way that statistical computer packages can be incorporated into such a course, the importance of engaging research students' supervisors and coordinators, some possible dangers of minimum maintenance of such a course and obtaining feedback.

MacGillivray (2003) emphasised that teaching graduate students is very different from teaching undergraduates. She commented that whereas undergraduates benefit from learning experience sessions, graduates prefer to do this in their own time. The graduates in this course have their own research projects and the emphasis needs to be on the concepts and demonstrations of how to put them into practice with succinct examples.

In the first two iterations of ISO, students were asked at the beginning to nominate which statistical package they would use for the practical exercises in five of the topics. A few guidelines were offered but the decision still caused angst among some students, who did not know which one to choose. Others found learning to use the package, for which only minimal instruction was given in the course, distracted them from learning the statistical concepts. While a minority of students enjoyed the computing exercises, others were disengaged. Eventually this choice and these exercises were dropped. Instead there are now optional exercises with detailed instructions for carrying them out in SPSS and in GenStat. Students can just look through the solutions if they wish.

It is not vital to be able to use a computer package in this course because the CAST e-books are so interactive.

Lancaster (2010) comments that developing e-learning courses is time and resource intensive. However, CAST was already available and so this cut the development workload.

Maintenance is a separate issue. Minimum maintenance was one property initially set for ISO. It has been pointed out in Sect. 4 that many more students can be accommodated in an online course than was previously possible with four short courses presented live each year. However, it is a mistake to think that one could devise a course and then sit back and let students go it alone. Perry and Pilati (2011) in a review of online learning literature point out that a sense of community, timely feedback, clear expectations and a reasonable chance f success are just as important in online learning as in the classroom. Some teachers (Da Silva Nascimento and Dos Santos Vaz Martins 2007; Salmon 2002; Tudor 2006) expend considerable effort on moderating discussions and other e-tivities. My experience is that a period of looking through quiz results and activities, and emailing students to offer help or praise, generally results in an increased level of activity by the students.

Another aspect of minimum maintenance is the use of software-assessed quizzes. Manly (2010), in contrast, manually marks assignments in his online courses. An assignment was initially included in this course but only two were submitted and it would be too time intensive if large numbers of students did this. A variety of quiz-type questions is available in Moodle, including, multiple choice, true/false, matching properties in one list with answers in another and calculations with a numeric answer. There are others that have been more difficult to use in the Statistics context. Some students have been put off by the perceived simplistic nature of such questions. At each version of the course, the more simplistic questions are improved.

An area of development for this course would be to model questions on those in ARTIST (Garfield et al. 2002) but apply a local context.

Various ways to engage students' supervisors, and through them their students, were discussed in Sect. 3. Research students generally take the advice of their supervisors and are more inclined to persist with non-compulsory online course if they perceive that their supervisor or research coordinator thinks it is valuable.

Initially anonymous feedback was sought at the end of the course and judging by the numbers, only those students who completed the course, provided feedback. Getting students to send non-anonymous comments by email proved useful, as did incorporating an extra anonymous feedback form halfway through the course. Students' estimates of how long they spent on each topic assisted in gauging the length of the course. We were aiming for 30 h in total for the introductory course.

We have placed a lot of emphasis on flexibility. Perhaps these courses are too flexible. Placing each student in a monthly group with a fixed starting time would make the use of e-tivities, as discussed in Sect. 3, possible. This may improve student engagement but the loss of flexibility may decrease the enrolments. The main aim is that research students should become familiar with some statistical concepts and terminology.

6 Note

Developed from a paper presented at Eighth Australian Conference on Teaching Statistics, July 2012, Adelaide, Australia.

This chapter is refereed.

References

Adobe Connect. Retrieved February 26, 2013, from http://www.adobe.com/au/products/adobeconnect.html

American Statistical Association. (2005). Guidelines for assessment and instruction in Statistics education. Available from http://www.amstat.org/education/gaise/index.cfm

Bloom, B. S. (Ed.). (1956–1964). *Taxonomy of educational objectives*. New York: David McKay Company Inc.

Da Silva Nascimento, M. M., & Dos Santos Vaz Martins, J. A. (2007). *Let us do it in a different way, an alternative assessment proposal*. IASE/ISI Satellite 2007.

Digital Lecture Delivery at ANU. Retrieved December 3, 2013, from http://itservices.anu.edu.au/online-learning/lecture-recording/

Enders, F. B., & Diener-West, M. (2006). Methods of learning in statistical education: A randomised trial of public health graduate students. *Statistics Education Research Journal, 5*(1), 5–19. http://www.stat.auckland.ac.nz/serj © International Association for Statistical Education (IASE/ISI), May, 2006.

Everson, M. G., & Garfield, J. (2008). An innovative approach to teaching online Statistics courses. *Technology Innovations in Statistics Education, 2*(1), 1–18, retrieved November 28, 2012, from http://escholarship.org/uc/item/2v6124xr.

Finch, S., & Gordon, I. (2010). Lessons we have learned from post-graduate students, ICOTS8. Retrieved March 1, 2013, from http://www.stat.auckland.ac.nz/~iase/publications/icots8/ICOTS8_4H3_FINCH.pdf

Garfield, J., & Ben-Zvi, D. (2007). How students learn statistics revisited: A current review of research on teaching and learning statistics. *International Statistical Review, 75*, 372–396.

Garfield, J., delMas, R., & Chance, B. (2002). The Assessment Resource Tools for Improving Statistical Thinking (ARTIST) Project. NSF CCLI Grant ASA-0206571. [Online: https://app.gen.umn.edu/artist/]

Lancaster, G. (2010). Communicating the value of statistical thinking in research, ICOTS8. Retrieved March 14, 2013, from http://www.stat.auckland.ac.nz/~iase/publications/icots8/ICOTS8_6C4_LANCASTER.pdf

MacGillivray, H. L. (2003). Making statistics significant in a short course for graduates with widely-varying non-statistical backgrounds. *Journal of Applied Mathematics and Decision Sciences, 7*(2), 105–113.

MacGillivray, H., & Croft, T. (2011). Understanding evaluation of learning support in mathematics and statistics. *International Journal of Mathematical Education in Science and Technology, 42*(2), 189–212.

Mallee fire ecology. Retrieved March 1, 2013, from http://fennerschool-research.anu.edu.au/malleefire/patterns/

Manly, B. F. J. (2010). Some different models for interacting with researchers and students in other disciplines, ICOTS8. Retrieved March 14, 2013, from http://www.stat.auckland.ac.nz/~iase/publications/icots8/ICOTS8_6C1_MANLY.pdf

Moodle (Modular Object-Oriented Dynamic Learning Environment). Retrieved November 28, 2012, from https://moodle.org/

Perry, E. H., & Pilati, M. L. (2011). Online learning. New Directions for Teaching and Learning, pp. 95–104. Wiley. doi:10.1002/tl.472

Salmon, G. (2002). *E-tivities: The key to active online learning*. London: Kogan Page.

Stirling, D. (2000–2012). Computer-Assisted Statistics Textbooks (CAST). Retrieved February 26, 2013, from http://cast.massey.ac.nz/collection_public.html

Tudor, G. E. (2006). Teaching introductory statistics online—Satisfying the students. *Journal of Statistics Education*, 14(3). www.amstat.org/publications/jse/v14n3/tudor.html

Wattle (Web Access to Teaching and Learning Environments). Retrieved February 26, 2013, from https://wattle.anu.edu.au/about.php

Wimba products. Retrieved February 26, 2013, from http://www.wimba.com/products/

Engaging Entry Level Researchers in Agriculture in Statistical Communication and Collaboration: Why? and How?

O. Kravchuk and D.L. Rutley

Abstract The Biometry Hub, a statistics research and consulting group in an agricultural science school, has commenced a project to enhance the statistical capacity of graduates in agricultural sciences. The project engages with students undertaking research projects as they are completing their undergraduate degrees and considering their careers, possibly in research. This group is motivated learners focused on delivering outcomes in solving real-life problems, and respond well to opportunities for their broad professional development. The project will help them become familiar with the culture of cross-disciplinary collaboration, an essential component of modern agricultural research.

The teaching and learning framework of the project consists of four elements: (1) group workshops in quantitative methods; (2) individual attention from a statistics consultant throughout the research project; (3) targeted guidance with peer-reviewed resources in statistical methods, experimental design and data management specific to the students' research topics and (4) supervisor and peer support encouraged through the dissemination of 'good statistics practice' in the research group hosting a student.

This chapter summarises the problems addressed by the project, presents the project framework and discusses performance measures for the project's elements and potential impact.

Keywords Multidisciplinary collaboration • Enquiry-based learning • Academic workplace culture

O. Kravchuk (✉)
School of Agriculture, Food and Wine, University of Adelaide,
Adelaide, SA 5005, Australia
e-mail: olena.kravchuk@adelaide.edu.au

D.L. Rutley
Biometry Hub, School of Agriculture, Food and Wine University of Adelaide,
Adelaide, SA 5005, Australia

H. MacGillivray et al. (eds.), *Topics from Australian Conferences on Teaching Statistics:*   345
OZCOTS 2008-2012, Springer Proceedings in Mathematics & Statistics 81,
DOI 10.1007/978-1-4939-0603-1_19, © Springer Science+Business Media New York 2014

1   Introduction

1.1   Project Staff and Collaborators

The current project forms part of an Australia-wide project Capacity Building for Statistics (CBS). The CBS is supported by the Grains Research and Development Corporation (GRDC), who have identified the lack of sufficient biometricians available for research and development as a significant threat to their future research programs (NIASRA 2013). The CBS is led by the University of Wollongong (Prof. Brian Cullis) and has links to the University of Adelaide, Charles Sturt University and the University of Western Australia. Each participating university is fulfilling a specific task in enabling the continuation of biometrical support for agricultural research. The Biometry Hub group in the University of Adelaide is responsible for the task of establishing a program for promoting statistical thinking to agricultural scientists while specifically targeting the new generation of agriculturalists. The target audience of the project is students in their honours year (final, optional year in a 3-year + honours undergraduate degree) in agricultural, animal and food sciences.

Individually or collectively, the authors have many years of experience in:

- Providing statistical and data management support to the research projects of students in agriculture, food science and animal science degrees in Australian universities
- Teaching statistics, experimental design and research methodology to undergraduate students in agricultural and bioscience degrees
- Delivering professional development workshops in statistics to postgraduate students and researchers in biosciences
- Statistical consulting and research in the field of design and analysis of experiments in biosciences

Both authors joined the School of Agriculture, Food and Wine, University of Adelaide in 2012. The educational approach being developed and trialled in this project has been prompted, and is informed by our own professional experience, reports from our students, numerous discussions with statistics colleagues in academia and industries and relevant literature.

1.2   The Diversity of Agricultural Science Students and Its Consequences

This 3-year project is being developed to foster statistical thinking in entry level researchers in agriculture to help them advance in the hierarchy from undergraduate students to research scientist (Allan 2011). The honours year is an important formative year for agricultural science graduates, in which they are shaped as young researchers and industry specialists. It is thus highly likely that, upon graduation, they will transfer to their workplace the data collection, management and analysis skills and habits learned in the honours year.

Currently there is a large diversity in the statistical skills of students in an honours year in agricultural science disciplines across Australia, most likely as a function of the students' backgrounds, influenced by their university, undergraduate degree program and level of formal statistical training. In the authors' experience, typical statistical approaches in honours students' research projects vary greatly, from the basic data manipulation in MS Excel®, through ANOVA's of simple designs to elaborate temporal and spatial models and complex models in statistical genetics.

This non-uniformity in statistical awareness among entry level researchers has major short- and long-term consequences. To appreciate this, let us look at the broader picture of statistics and research skills in the Australian agriculture workforce.

1.3 Needs for Improving Statistical Literacy in Agricultural Researchers

Although there is a general need to improve statistical understanding and reasoning across society (ABS 2010b), it is particularly urgent in the agricultural sciences as a result of two specific needs for:

1. A workforce technically skilled in handling data in line with improvements in data capture, computational advances and statistical algorithms development (GRDC 2012)
2. High quality scientists (Pratley 2012; Ransom et al. 2006) competent to make informed everyday decisions towards the overarching goal of sustainably meeting the rapidly growing global demand for agricultural products (Raney et al. 2009)

We consider the increasing demand for technical skills first. As a result of the rapid evolution of data acquisition systems in agriculture over the last decades (Munack and Speckmann 2001), the ability to collect good quality multifaceted data is increasing, allowing researchers to probe more and more challenging questions (CSIRO 2011). The current level of complexity of the problems posed in agriculture is often accompanied by data that is 'messy', both by design and nature (Silva 2006). Making inferences on the basis of such data goes far beyond basic statistical techniques. At the same time, in a highly competitive research 'market', and under the 'Excellence in Research for Australia' (ERA) evaluation framework (ARC 2013), there are expectations placed on Australian researchers, including those in agricultural sciences, to produce high quality research to inform the industries (DIISR 2011) as well as to publish in high impact journals (Hughes and Bennett 2013). For meeting these expectations of excellence in research, the quality of experimental designs and data analyses is fundamental.

Recent advances in statistical algorithms, easily available to users in software packages such as ASReml (Gilmour et al. 2009), SAS (SAS Institute Inc 2010), GenStat (VSN International 2012) and the ever growing R (R Core Team 2013), allow users to analyse sophisticated experimental studies. Alarmingly, this same ready availability of the power of data acquisition and data analysis increases the risk of the misuse of statistical algorithms by non-statisticians. Thus, at a technical level, there is an increasing demand for competent statistical knowledge.

Access to advanced statistical software packages does not lead to good decisions per se; it is critical that scientists understand the statistical principles and methods they apply for analysing their data. These have to be communicated to the scientific communities by statisticians, either through training, consulting or software support documentation, in a way that other scientists can understand and benefit from (ABS 2010a; Sprent 1970). For this, statisticians, in their turn, need to understand the essence of biological research problems. This clearly implies the need for statistics–science collaboration.

In statistics training and consulting, approaches to communicating statistical methods in the biological context will vary depending on the level of statistical awareness of the users. In the authors' experience, and surprisingly close to the classification by Sprent (1970), several typical user groups are worth mentioning: (1) 'occasional users' with technical skills but who lack competence in selecting and interpreting correct procedures; (2) 'conservative researchers' who recognise their statistical knowledge limitations and so restrict themselves to the use of simple, 'safe and classic', experimental designs and analyses, limiting the potential findings from their studies; (3) 'natural data handlers' who are highly skilled in data mining but may neglect to consult a statistician when their knowledge of statistical modelling is not sufficient and (4) 'genuine collaborators' who appreciate the role of statisticians in research teams and support and promote the exchange of knowledge and ideas between their discipline and statistics. In our opinion, a great outcome for agricultural research in Australia will be achieved if the agricultural researchers, both young and experienced, become 'statistically informed experimenters' (Sprent 1970, p. 144), willing and prepared to collaborate with statisticians in their research.

We consider now the demand for high quality agricultural scientists. There is a current shortage of specialists in many agriculture-related disciplines, due to the worldwide decline in the number of trained professional agriculturists since the late 1980s (Pratley 2012; Ransom et al. 2006). As a consequence of small undergraduate enrolment, agricultural science programs in Australia have been restructured, putting pressure on the curriculum in terms of both subject choice and coverage. For example, at the University of Adelaide, the degree of Bachelor of Agricultural Science was a 4 year degree in the 1980s and is currently a 3 year degree, with the fourth, honours, year being optional. Similarly, there is an undersupply of trained statisticians and biometricians (Hall 2004; Stubbs 2012). The implications of these shortages in conjunction with increasing demand for services are that honours graduates in agricultural sciences are readily employed as qualified scientists and are expected to make, without any assistance, informed decisions based on experimental evidence. However, these recently graduated scientists have had limited undergraduate opportunity to develop critical thinking (Quinn et al. 2009) and cross-disciplinary collaboration and communication skills (Ransom et al. 2006). Thus we are effectively placing these fresh graduates in situations where it is likely that they do not have the professional data handling and interpretative skills required. It is further likely that these emerging professionals do not realise that they lack the necessary skills, and so they become at risk of operating outside their area of expertise and breaching basic professional ethics (AIAST 2013), an area that can often be neglected in favour of an exclusive emphasis on technical skill development (Quinn et al. 2009).

1.4 The Scope of the Project

Thus it is clear that there is an urgent need to develop creative and innovative ways to enable agricultural researchers to become both skilled in data handling and statistically informed.

Encouraging and supporting agricultural sciences honours students to approach their research questions from various angles, and to explore different statistical models and methods, will give the students the confidence to 'step outside their comfort zone' (Schwartz 2008). As biological scientists, it is unlikely that these students will choose to become statisticians, but they will gain new technical- and problem-solving skills and experience statistical reasoning in context. Most importantly, they will become accustomed to collaborating with professional statisticians to develop ways for drawing novel interpretations and conclusions from the data they collect (Olkin et al. 1990).

To provide a grounded description of the project and its context, the remainder of this chapter discusses the fundamental role of cross-collaboration with statistics in agricultural research and issues with current structures and approaches, justifies the choice of entry level researchers as the project target, describes the project components and their rationale, and identifies ways in which project impact will be assessed.

2 Statistics Collaboration in Agricultural Research

2.1 Cross-Disciplinary Collaboration is an Important Feature of Modern Research

Cross-disciplinary collaboration has proven indispensable for developing novel approaches and breadth of understanding to solve complex problems (Davies et al. 2010). The appreciation of, and funding for, multidisciplinary research has changed significantly since the mid-1980s when 'pure science' was considered 'better science' (Molenberghs 2005, p. 6). Even before that era, Sprent (1970) had argued that the full benefit of statistical knowledge could only be made available to research communities through cross-disciplinary collaborations. Two decades ago, the Institute of Mathematical Statistics (IMS) identified the science–statistics collaboration to be a requirement for statisticians and academics of other disciplines to solve difficult problems, such as those encountered in complex agricultural systems, and acknowledged that '… advances in substantive knowledge and in statistical theory and methods are virtually inseparable …' (Olkin et al. 1990, p. 122). The needs for, and benefits of, such cross-disciplinary partnerships (Hall 2004; Lindsay et al. 2004) have since increased, as can be demonstrated through recent examples from Australia.

2.2   Statistics–Agricultural Sciences Collaboration Examples in Australia

Recent successful examples of collaboration between agriculture and statistics in Australia include the development of ASReml (Gilmour et al. 2009) by the NSW Department of Primary Industries (DPI), and the National Variety Trials (NVT) of the GRDC (2008). In the case of the development of ASReml, the NSW DPI is not a business that produces and sells statistical software. Rather, it had a need for an efficient and versatile modelling algorithm to cope with complex designs, and so set out to fill this requirement. Similarly, the GRDC was not interested, per se, in developing complex analytical methods; rather they wanted to invest in plant breeding more efficiently and so set about coordinating state-based NVT programs with standard protocols to allow the data to be combined for analysis across field trials in Australia. This has been possible due to the productive efforts of the Statistics for the Australian Grains Industry (SAGI) program (Cullis 2012) in developing an efficient data compilation, reporting and analysis system.

2.3   Needs for, and Examples of, Cross-Disciplinary Training for Entry Level Scientists

Despite such remarkable examples of collaboration between statisticians and research and industry to adapt statistical methods to needs, many agriculturalists are still using statisticians as service providers, often due to a lack of statistical awareness. Establishing a better culture of communication between statistics and other disciplines may be a remedy for this. Leading statistical professional bodies have recommended that collaborative cross-disciplinary training programs between statistics and other disciplines be developed for entry level researchers to ensure that the thinking of statisticians is imparted to the students (Olkin et al. 1990; SSAI 2005).

Such collaborative initiatives have been promoted as educational priorities (Acker 2008): teaching students' skills to draw from across disciplines to develop complex system-wide solutions (Faulkner et al. 2009; Ransom et al. 2006).

The literature suggests that the efforts in this direction have been mostly concentrated on cross-disciplinary training programs for Ph.D. students in sciences. In her invited paper for the eighth International Conference of Teaching Statistics, Lancaster (2010) discusses several common implementations, including course-work programs, short courses in research training programs for Ph.D. students and master classes for further professional development of researchers. Bishop and Talbot (2001) present a case-study of training agricultural science Ph.D. students in the University of Adelaide. At the undergraduate level, on the other hand, many statistical educators (see, for example, Bidgood 2009) have successfully applied the problem-based learning approach to training statistics students in solving real-world problems. However, there is a current gap in the literature with regard to methodology for training biological scientists in cross-disciplinary collaboration with statisticians prior to Ph.D. research programs.

2.4 Structural Contrasts Between Agricultural Science and Statistics

Both disciplines, statistics and agricultural science, are scientific pursuits, though very different in their history and nature. Agriculture has been developing for thousands of years, initially and for a long time by trial and error, to fill our primary food and fibre requirements across many production environments. Hence agricultural science has developed into many specialised scientific branches, ranging from agronomy to animal production, environmental management, microbiology and soil science. Statistical science, although a relatively new discipline, also has a wide range of branches, with biometrics developing from the requirement for science to produce robust solutions based on the interpretation of experimental evidence of ever increasing biological complexity and variation. However, even with the more recent developments in statistics, a well-trained statistician has sound understanding of the complete core of their discipline which can be applied in any field to all types of applications. This is both the nature and the power of the statistical science.

This contrast between the disciplines explains the widespread rationale behind embedding broadly based statisticians into agricultural university departments and assigning them to a technical role in 'Consulting Services' to all agriculturists. But it also means that statisticians have to learn about particular areas of application. This can present a challenge to the statisticians—a challenge that can be enormous if their diverse groups of 'clients' hold an expectation that the statistician is deeply familiar with their fields. Encouraging both statisticians and scientists to be communicators and collaborators has great potential for mutual benefit. Engaging future agricultural scientists to experience and learn the fundamentals of statistical reasoning and problem solving is necessary to enable them to tackle quantitative problems in agriculture as well as to prepare them to collaborate with statisticians.

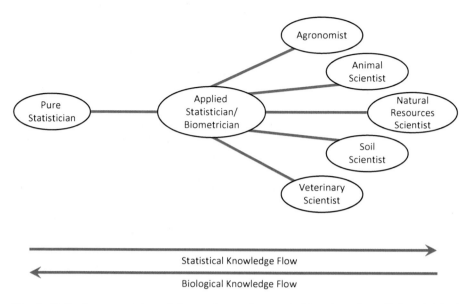

Fig. 1 Stylised representation of the continuum from statistician through applied statistician/ biometrician to agricultural scientist

2.5   Statisticians as Collaborators

In Australia and New Zealand, there are many biometricians who received their undergraduate training in biological sciences followed by higher degree training in statistics, as well as those who hold degrees in statistics and mathematics but have developed an almost expert knowledge in specific agricultural sciences in their work placements. Their different training paths shape them somewhat differently as collaborators with agricultural scientists (Pfannkuch and Wild 2000). There is a vast distance along the continuum from pure statistician to statistically educated agriculturist (Fig. 1), via applied statistician/biometrician, along which distinctions are not always clear. In our experience, those in the applied statistical group, whether they come from the science or mathematical statistics directions, are most efficient collaborators with agricultural scientists.

There is also a separate considerable group of those practising statistics, of whom Silva (2006) speaks as 'self-taught statisticians', who have learnt statistics through their work and practice, rather than via formal course work. Such a term is likely to cause vigorous debate, but it is the variation in this group which presents the challenges. The authors have met throughout their careers representatives of this group who are very successful and others not at all so. We prefer to classify this group as the 'data handlers' mentioned earlier.

The Australian grains industry (B. Cullis, pers. comm., 2012) are interested in recruiting more biometricians and providing more statistical training to Silva's (2006) statistically 'self-taught' practitioners, who will use applied statistics as required for their employment. In addition, a side benefit of projects like ours is that they may encourage bright students in agricultural sciences to develop life-long data analysis learning skills, or even to consider a career path in statistics.

3   Rationale for Targeting Entry Level Researchers

As we mentioned earlier, most efforts in introducing entry level researchers to science–statistics collaboration have been targeted at Ph.D. students. We suggest that such introduction must start earlier

In our project, entry level research students are those undertaking an honours, or equivalent, program containing a significant independent research component. Currently, for entry into an honours program in agricultural sciences, the highest achieving students are selected. This year of 'rounding off' of an undergraduate degree (Kiley et al. 2009) transforms (Allan 2011) them from learners to producers of knowledge (Manathunga et al. 2011). In this year, the students are still in the learning phase of their professional career but developing their preferred and unique ways of solving problems and choosing methods for undertaking tasks. The honours students are adaptable and likely to develop statistical thinking if they find it beneficial and supported by a strong research culture in their placement in an agricultural research group (Lancaster et al. 2013).

As an entry level, an honours research project is a venue for providing learning and developmental opportunities rather than a project that leads to ground-breaking research results. For example, no stress is associated with not obtaining 'conclusive data', which would be a disaster for a Ph.D. project. Typically, students are guided by their academic supervisors in agricultural science in selecting a research topic and then encouraged to explore the topic from different perspectives. In this process, students are gaining skills in independent research. There is also an expectation that they will develop significant quantitative skills in design, analysis and interpretation of experiments. Numerous discussions between the authors and academics supervising honours students indicate that this outcome is, in fact, often an unrealised hope. It is still not an exception when students do not consult a statistician, or the literature, when designing their first experiment and expect that a statistical consultant can salvage the situation with some 'magical' analysis (Bishop and Talbot 2001). Keeping in mind that these students are our future top agricultural researchers and practitioners, it is important to prevent this bad habit from solidifying and carrying over into their careers.

Employers, undoubtedly, value the extra skills, maturity and confidence students develop in undertaking this research year. In the Schools of Agriculture, Food and Wine, and Animal and Veterinary Science, highly motivated and top quality graduates, even those not interested in a research career, usually elect to undertake an honours degree. This extra qualification prepares them for 'real-life' research or helps them compete for 'preferred' employment packages.

Honours graduates often prefer industry employment instead of or before the academic research path due to the continuing shortage of agricultural scientists, discussed earlier. The authors have indeed observed many such cases in the last 10 years across Australia. Many graduates of these honours programs have gone on to become leading decision makers in Australia's agricultural industry. Agreeing with Meng (2009, p. 203), we hold that helping these students develop a strong appreciation of statistics is our chance to contribute significantly to '…preparing whole generations of future scientists and policy makers'.

4 Program and Model for Building Statistics Capacity

4.1 Project Time Frame and Sustainability

The project is being conducted in 2013–2015 in the School of Agriculture, Food and Wine, University of Adelaide, with a typical enrolment of about 10–15 honours students a year. The model of the project will then be made available to other schools in the Faculty of Sciences in the University of Adelaide and to the other universities participating in the CBS project. The model also has potential for inclusion in a new 2-year Masters degree, which is to replace the current honours program in the University from 2017 onwards (UA 2013).

4.2 Venue, Environment and Audience of the Project

The School is research-focused, with consistently high ERA evaluations for its research in agriculture (ARC 2013). The School enjoys a broad spectrum of national and international collaborations and is well supported by research grants from the Government and agricultural industries. There are strong multidisciplinary links within and outside the School, including those with statistics. The choice of topics for research projects for honours students in agricultural and food sciences is fascinating, allowing students to work at the cutting edge of biological sciences as well as to experience working with rural communities in Asia on agronomy extension projects. As a rule, there is funding available for honours projects to support students' direct experimental work, to provide additional training in laboratory skills outside the university and to allow students to attend conferences or professional meetings.

In their honours year, students conduct a year-long research project. In this process, they are required to

- Review the literature to justify their project objectives
- Design and conduct the experimental part, including collecting, analysing and interpreting data
- Report their results in the form of a thesis

In undertaking these tasks, students gain skills in independent research. This set-up presents an ideal opportunity to implement the enquiry-based learning (EBL) approach for teaching statistical principles as well as skills in cross-disciplinary collaboration (Tishkovskaya and Lancaster 2012).

Currently, due to logistics, organisational and other historical reasons, the depth of data analysis in honours projects is limited. An audit of recent honours theses and interviews with supervising academics conducted in the early stages of this project indicated that, alarmingly often, students do not demonstrate a comprehensive understanding of statistical procedures used in their theses. Their data analyses also indicate that little attention is given in the projects to relevant advances in statistics. The majority of agricultural science academics are supportive of changing this situation for the better.

In our project, we benefit from the current supportive and encouraging culture of cross-disciplinary research in the School, and aim to enrich this culture further by disseminating statistical ideas and promoting science–statistics collaboration. The efforts are concentrated on individual honours students, their supervisors and placements.

4.3 Role of the Supervisor

To consolidate the student transformation to a collaborative researcher, the supervisor's behaviour is critical (Lancaster et al. 2013). To encourage transfer and retention of the desired cross-disciplinary collaboration skills, the supervisor will need to be supportive prior to, during and after a student has undertaken the honours training. Supervisors' essential roles, providing physical and emotional support (Lancaster et al. 2013), in the context of our project translate into these activities:

- Allocating time and resources in the honours project to allow the student to study and practise new methods of experimental design and analysis
- Being a good role model in collaborating with statisticians, encouraging the student to share their experience and new knowledge with colleagues in the laboratory, prompting discussions of the student's interpretations of statistical methods and providing the student with constructive feedback

4.4 Role of the Research Group

It has been noted elsewhere that the collegial nature of graduate teaching forms an important non-cognitive aspect (Johnson 1996) and is a prerequisite for research skills training (Cargill 1996; Cargill and Cadman 2005). The research culture in students' placements in their honours year shapes them as young researchers. Often, in addition to supervisors, other members of research groups become mentors of honours students.

Through this mentoring (Kogler Hill et al. 1989b) the students are exposed to the actual culture of science–statistics collaboration and this hugely influences the attitude towards statistics they will develop at the end of the honours year. Thus it is paramount that this culture is healthy. As a way to promote a healthy engagement, one of our goals in the project is to reach out to the mentors (who, in our experience, may themselves be 'occasional users' or 'conservative researchers') and to contribute to their transformation into 'statistically informed researchers'.

4.5   Role of the Statisticians

In a project like ours, the skills of the consulting statistician as both a statistician and a collaborator are crucial for success (Johnson and Warner 2004). It is well recognised that there is a substantial teaching element in statistical consultancy (see, for example, Belli 1998). In our project, the role of the statistician has essential requirements of qualification in statistics; solid knowledge and ongoing experience in biometry, in application to agriculture; good written and verbal communication skills; and small-group university teaching experience.

Formally, one consulting statistician is appointed to work directly with all the honours students (with the exception of those whose projects require statistics co-supervision) during the project. However, the whole Biometry Hub group contributes to presenting talks to research groups and collegially advising the formal appointee on methodologies of analyses and the literature whenever an advanced statistical specialisation is required. The Biometry Hub directly benefit from this opportunity to strengthen the group's support for the research in the School.

5   Methodology of the Project

The methodology comprises elements of corporate (research groups) and individual (honours students) training. Amongst many approaches to corporate training (Buch and Bartley 2002), we only focus on class-room statistics communication to disseminate relevant and efficient statistical methods to research communities in the School. Our main goal in the corporate training is to prepare a more statistically aware and responsive environment for the honours students.

The students' teaching and learning model of the project (Table 1) is transformational (Allan 2011) for enhancing their cross-disciplinary problem-solving abilities (Schuyten et al. 2006). To achieve the desired transformational outcomes, we employ the EBL approach (see, for example, Tishkovskaya and Lancaster 2012). Following Sowey (2006), our approach encompasses the three elements of

Table 1 Model of skills development and student transformation

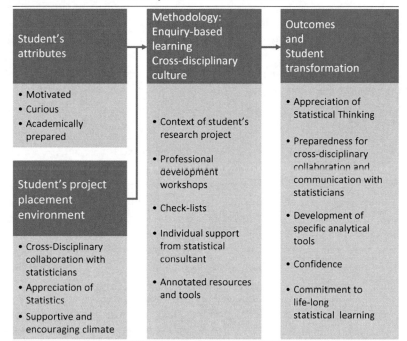

Student's attributes	Methodology: Enquiry-based learning Cross-disciplinary culture	Outcomes and Student transformation
• Motivated • Curious • Academically prepared	• Context of student's research project • Professional development workshops • Check-lists	• Appreciation of Statistical Thinking • Preparedness for cross-disciplinary collaboration and communication with statisticians
Student's project placement environment	• Individual support from statistical consultant • Annotated resources and tools	• Development of specific analytical tools • Confidence
• Cross-Disciplinary collaboration with statisticians • Appreciation of Statistics • Supportive and encouraging climate		• Commitment to life-long statistical learning

developing the sense of worthwhileness of statistics: demonstrating to the students in the context of their research enquiry that statistics is (1) interesting, (2) useful and (3) a discipline of substance.

The project methodology accounts for both the students' attributes and research environment. It is put in practice through the provision of the following teaching and learning interventions: (1) professional development workshops in statistical methods; (2) individual attention from the biometrician; (3) structured check-lists for self-study and preparation for meetings with the biometrician and (4) annotated references from peer-reviewed research and statistical literature and online resources. These elements are discussed in more detail below.

5.1 Professional Development Workshops

Professional development workshops are presented to the cohort of honours students as well as to their laboratories to promote cross-disciplinary links with statistics. The initial workshop is presented to honours students at their induction to ensure they are aware of the intention of the Capacity Building project, its aims and outcomes, and the process and resources available for student support.

Activities to engage with the supervisors and research groups at students' placements are also carried out. Tailored professional development workshops, designed in close collaboration with group convenors, target the research groups with specific biometrical information in the context of their research and useful for their types of projects for honours students. The content of the workshops to which groups they are to be presented and other logistics is determined on the basis of the honours projects. Our hope is that, at the end of the project, such statistical workshops will become an essential part of the research culture in the School.

5.2 Direct Contacts with a Statistician

Good research requires autonomy and independent thinking (Willison and O'Regan 2007). In all direct contacts with the consulting statistician, students are required to exercise their own statistical thinking, to apply, with guidance, the analysis tools they learn, and to consider how these statistical analyses have led to their conclusions. An encouraging atmosphere needs to be maintained in the meetings so that students feel that their efforts in critical thinking, innovation and responsibility for all aspects of their project are genuinely noticed and valued.

Three appointments are scheduled per student to assist with developing their statistical thinking in the areas of (1) experimental design; (2) data quality audit and (3) analysis, interpretation and presentation of results. Each appointment is scheduled at the corresponding stage of the project. These meetings give the student experience in seeing how statisticians 'think statistically' about each stage of the project, as well as in explaining to the statistician their research questions and problems and collaborating on solving them. This helps enhance students' science communication skills and introduce them to the culture of statistics–science cross-disciplinary research.

Students' direct supervisors attend the first and last meetings either in person or via electronic media. The supervisor is to play a mediator role, ensuring that the student's interpretation and presentation of the research problem is correct and collaborating with the statistician on adapting statistical methods to the problem in question. Observing and contributing to the dialogues between the supervisor and the statistician, the student will gain practical experience in cross-disciplinary collaboration. Ensuring the supervisor is part of the meeting presents an excellent opportunity for the knowledge exchange between the scientist and the statistician—an exchange which is likely to be beneficial to the School's research.

Based on the discussions, presentations and findings in these appointments the student is additionally directed towards specific software tools and peer-reviewed published resources to assist their learning and to improve their research practice. Check-lists are designed and distributed to participating students to help them prepare for the meetings.

5.3   Check-Lists

We agree with the comment by Bishop and Talbot (2001) that 'statistical thinking is required ... [to be] intertwined with the contextual thinking throughout the whole project'. Directed sets of questions and check-lists prompt this intertwined thinking and assist students to prepare for consultations with the statistician. The students are prompted to

1. Convert their ideas into quantifiable research questions and consider all the factors and extraneous variables that may influence their study
2. Suggest a robust plan for data collection and auditing
3. Critically consider various approaches to their experiments, from simple to complex analyses and interpretations, and to presentation of their results

Students need to work on the questions prior to the corresponding individual appointment with the statistician, to ensure they are prepared for the discussion and have the necessary information, and to maximise the learning opportunity.

5.4   Annotated Resources

Students are provided with annotated references to (1) peer-reviewed articles and textbooks, (2) online resources and (3) worked examples of computational procedures. These resources will allow the students to learn and develop skills at their own pace between the consultations (Bishop and Talbot 2001).

5.5   Annotated References

Selected relevant references from the biometrical and applied statistical literature are annotated and made available for students.

An efficient and sustainable way for students to develop an appreciation of statistics as a discipline of substance is to become familiar with the statistics research literature. We feel that this focus has been completely missing in the education of honours students. In our experience, confirmed by the low citation indices of statistics journals, it is a common problem that biologists do not read (or cite) methodological statistical papers. We suggest that a possible remedy for this, at least in our project, is to assist the agriculturalists with the language of the statistics discipline as well as with putting the statistical methods in context. This is our approach to annotating references to statistical methods fundamental for students' projects and presented in peer-reviewed statistics and discipline journals and textbooks. With this assistance, the students will be better equipped for communicating and interpreting the methods in their thesis.

5.6   Guide for Online Materials

A plethora of material is available online, targeted at introductory level statistics and generally for educators, with much less targeted towards independent learners, let alone entry level researchers. Thus there is still a need to develop annotated guidelines for online resources for the agricultural sciences honours students, although some resources have already been compiled elsewhere (see, for example, Bishop and Talbot 2001). These resources will assist students in their self-study. Our goal here is to ensure that all the open-source resources we recommend are of good quality.

5.7   Worked Examples of Statistical Procedures

A variety of statistical methods typically used in honours projects are being identified throughout the project. The computational procedures for these methods are explained and supported with worked examples, available directly in the help system of software packages, in the research literature or written by the consulting statistician based on real experimental data from (de-identified) student projects. This resource is strictly educational, and the students are recommended to use it for their self-study only. As the aim and design of an experiment determines the appropriate analysis and its interpretation, the students will be guided in the individual consultations to specific analyses based on their project.

6   Project Impact and Outcomes

The major impact of this project is expected to be cohorts of entry level research graduates who are confident to discuss their statistical questions openly with their colleagues and senior researchers. Confidence will come to students as they are transformed, enabling them to advance in the research hierarchy and in the workplace (Allan 2011). While aiding, with supervisor support, the transformation process, this project is expected to lead to students considering the discipline of statistics as a significant and relevant body of knowledge from which they can draw tools to assist their own research. Thus the project's aim to encourage students to 'think statistically' and to collaborate with statisticians will have been achieved.

The primary outcome of the project will be the training methodology trialled and fine-tuned over 3 years. All the support resources will comprise secondary outcomes: talks to research groups, check-lists, annotated references, worked examples and a repository of evaluated online resources. In developing these essentially self-learning resources we will pay specific attention to one of the most important learning objectives—to enhance students' motivation to develop and exercise statistical thinking (Hussey and Smith 2003).

A comprehensive report on the methodology will be presented to external reviewers from the CBS project for their assessment of its sustainability and efficiency. The metrics for assessing the methodology will include measures of the workload involved, staff and student engagement levels and expert estimates of the potential for resources developed to be used outside the project. The workload will be measured as the number of hours the Biometry Hub staff have spent on the various activities of the project. The staff and student engagement will be measured using the percentage attendance at our statistical seminars in the research groups and the hours of participation by supervising academics in statistical consultations. Other measures will include metrics on the level of resource use (Bishop and Talbot 2001), such as counts of the number of students accessing papers or web pages.

The probability of broader use of the resources in research training and course work will be estimated based on their relevance to common research themes and research student (e.g. Ph.D.) cohorts in the School, as discussed in School staff meetings, and in comments from staff of the other participating universities.

In order to assess the long-term merit of this project, it will be necessary to investigate the change imparted by the project along with the use of project resources more broadly. In the short term, this may require a cross-disciplinary study, in collaboration with a social scientist experienced in higher education research. As cohorts of students are small, we will pay most attention to the analysis of individual cases. Information will be gathered in semi-structured interviews with students, their supervisors and research placement colleagues.

7 Conclusion

In this project, we are utilising the teaching capacity of statistical consulting in an EBL set-up to train researchers in collaboration with statisticians. Our target group of trainees is agricultural students undertaking a research project in the final (honours) year of their undergraduate degree—future leading researchers and decision and policy makers in agriculture in Australia. The major outcome of the project will be the training methodology, developed and fine-tuned, and transferrable to other universities in Australia.

The project engages with these students as well as with their colleagues at their research placements. Our efforts are expected to impact the research culture in the School providing the venue for the project. Part of this culture improvement will be through the development of enhanced communication support via informal mentoring and collegiality, identified as attributes associated with both faculty and faculty member success (Kogler Hill et al. 1989a). As students work through this program, it is expected that they will become motivated to continue applying statistics to their work and so will retain a framework and set of resources to transfer statistical and collaborative skills to their future workplaces. It is also hoped that they will be left with an understanding of the substantial nature of the discipline of statistics (Sowey 2006) and a well-grounded approach to statistical cross-disciplinary collaboration.

8 Note

Developed from a paper presented at Eighth Australian Conference on Teaching Statistics, July 2012, Adelaide, Australia.
This chapter is refereed.

Acknowledgements The authors sincerely acknowledge the support of the Grains Research and Development Corporation in funding the Capacity Building for Statistics project. The assistance of Margaret Cargill in editing and reviewing this chapter has been extremely valuable. We are also grateful to the editors and anonymous reviewers for their numerous constructive comments.

References

ABS. (2010a). *1500.0—A guide for using statistics for evidence based policy, 2010.* Canberra, ACT, Australia: Australian Bureau of Statistics.
ABS. (2010b). *Understanding statistics. Why statistics matter.* Canberra, ACT, Australia: Australian Bureau of Statistics.
Acker, D. (2008). Research and Education Priorities in Agriculture, Forestry, and Energy-Working toward achieving the 25x'25 renewable energy vision, a recent paper by the National 25x'25 Agriculture/Forestry Steering Committee provides an update. *Resource, 15,* 12–13.
AIAST. (2013). *Code of ethics.* Crows Nest, NSW, Australia: Australian Institute of Agricultural Science and Technology.
Allan, C. (2011). Exploring the experience of ten Australian Honours students. *Higher Education Research and Development, 30,* 421–433.
ARC. (2013). *Excellence for research in Australia.* Australian Research Council, Australian Government. Retrieved August 26, 2013, from (http://www.arc.gov.au/era/
Belli, G. M. (1998). The teaching aspect of consultancy. In *Proceedings of the International Conference on Teaching Statistics (ICOTS'5).* Retrieved August 26, 2013, from http://iase-web. org/Conference_Proceedings.php?p=ICOTS_5_1998
Bidgood, P. (2009). Helping students prepare for their future working lives. In *Proceedings of the International Association for Statistics Education and International Statistics Institute Satellite Conference.*
Bishop, G., & Talbot, M. (2001). Statistical thinking for novice researchers in the biological sciences. In C. Batanero (Ed.), *Training researchers in the use of statistics.* International Association for Statistical Education and International Statistical Institute: Granada, Spain.
Buch, K., & Bartley, S. (2002). Learning style and training delivery mode preference. *Journal of Workplace Learning, 14,* 5–10.
Cargill, M. (1996). An integrated bridging program for international postgraduate students. *Higher Education Research and Development, 15,* 177–188.
Cargill, M., & Cadman, K. (2005). Revisiting quality for international research education: Towards an engagement model. In *2005 Australian Universities Quality Forum*: Citeseer.
CSIRO. (2011). *Plant phenomics facilities.* Retrieved August 26, 2013, from http://www.csiro.au/ en/Organisation-Structure/Divisions/Plant-Industry/Phenomics-Facility.aspx
Cullis, B. (2012). Progress Report 2011-2012 for UOW capacity building project. In *Statistics for the Australian Grains Industry Technical Report Series.* Wollongong, NSW, Australia: University of Wollongong.
Davies, M., Devlin, M., & Tight, M. (Eds.). (2010). *Interdisciplinary higher education: Perspectives and practicalities. International perspectives on higher education research* (Vol. 5). Bingley, England: Emerald.

DIISR. (2011). *Focusing Australia's publicly funded research review*. Department of Innovation, Industry, Science and Research, Australian Government. Retrieved August 26, 2013, from http://www.innovation.gov.au/research

Faulkner, P., Gray, B., & Thomas, T. (2009). New skills for a new era: Ideas for preparing professionals for service in twenty first century agriculture. *International Journal of Applied Educational Studies, 4*, 34–46.

Gilmour, A. R., Gogel, B. J., Cullis, B. R., & Thompson, R. (2009). *ASReml user guide release 3.0*. Hemel Hempstead, England: VSN International Ltd.

GRDC. (2008). *Number crunching to yield improved variety information*. Kingston, ACT, Australia: Grains Research and Development Corporation, Australian Government.

GRDC. (2012). *Strategic research and development plan, 2012-2017*. Grains Research and Development Corporation, Australian Government. Retrieved August 26, 2013, from http://strategicplan2012.grdc.com.au

Hall, P. (2004). The sum and the product of our difficulties: Challenges facing the mathematical sciences in Australian universities. *The Australian Mathematical Society Gazette, 31*, 6–11.

Hughes, M., & Bennett, D. (2013). Survival skills: The impact of change and the ERA on Australian researchers. *Higher Education Research and Development, 32*, 340–354.

Hussey, T., & Smith, P. (2003). The uses of learning outcomes. *Teaching in Higher Education, 8*, 357–368.

Johnson, I. H. (1996). Access and retention: Support programs for graduate and professional students. *New Directions for Student Services, 1996*, 53–67.

Johnson, H. D., & Warner, D. A. (2004). Factors relating to the degree to which statistical consulting clients deem their consulting experience to be a success. *American Statistician, 58*, 280–289.

Kiley, M., Moyes, T., & Clayton, P. (2009). 'To develop research skills': Honours programmes for the changing research agenda in Australian universities. *Innovations in Education and Teaching International, 46*, 15–25.

Kogler Hill, S. E., Bahniuk, M. H., & Dobos, J. (1989a). The impact of mentoring and collegial support on faculty success: An analysis of support behavior, information adequacy, and communication apprehension. *Communication Education, 38*, 15–33.

Kogler Hill, S. E., Hilton Bahniuk, M., Dobos, J., & Rouner, D. (1989b). Mentoring and other communication support in the academic setting. *Group and Organization Management, 14*, 355–368.

Lancaster, G. (2010). Communicating the value of statistical thinking in research. In C. Reading (Ed.), *Data and context in statistics education: Towards an evidence-based society. Proceeding of the Eights International Conference on Teaching Statistics (ICOTS8, July, 2010)*, Slovenia.

Lancaster, S., Di Milia, L., & Cameron, R. (2013). Supervisor behaviours that facilitate training transfer. *Journal of Workplace Learning, 25*, 6–22.

Lindsay, B. G., Kettenring, J. R., & Siegmund, D. O. (2004). A report on the future of statistics (with discussion). *Statistical Science, 19*, 387–412.

Manathunga, C., Kiley, M., Boud, D., & Cantwell, R. (2011). From knowledge acquisition to knowledge production: Issues with Australian honours curricula. *Teaching in Higher Education, 17*, 139–151.

Meng, X.-L. (2009). Desired and feared—What do we now and over the next 50 years? *The American Statistician, 63*(3), 202–210.

Molenberghs, G. (2005). Biometry, Biometrics, Biostatistics, Bioinformatics, ..., Bio-X. *Biometrics, 61*, 1–9.

Munack, A., & Speckmann, H. (2001). Communication technology is the backbone of precision agriculture. *Agricultural Engineering International: The CIGR Journal of Scientific Research and Development, 3*, 1–12.

NIASRA. (2013). *Capacity building*. Retrieved October 16, 2013, from www.niasra.uow.edu.au

Olkin, I., Sacks, J., Blumstein, A., Eddy, A., Eddy, W., Jurs, P., Kruskal, W., Kurtz, T., Mcdonald, G. C., Peierls, R., Shaman, P., & Spurgeon, W. (1990). IMS panel on cross-disciplinary research in the statistical sciences. *Statistical Science, 5*, 121–146.

Pfannkuch, M., & Wild, C. J. (2000). Statistical thinking and statistical practice: Themes gleaned from professional statisticians. *Statistical Science, 15*, 132–152.

Pratley, J. (2012). *Professional Agriculture-a case of supply and demand.* Surry Hills, NSW, Australia: Australian Farm Institute. Retrieved February 24, 2012, from www.farminstitute. org.au

Quinn, C., Burbach, M. E., Matkin, G. S., & Flores, K. (2009). Critical thinking for natural resource, agricultural, and environmental ethics education. *JNRLSE, 38*, 221–227.

R Core Team. (2013). *R: A language and environment for statistical computing.* Vienna, Austria: R Foundation for Statistical Computing.

Raney, T., Gerosa, S., Khwaja, Y., Skoet, J., Steinfeld, H., Mcleod, A., et al. (2009). Livestock in the balance. In *The State of Food and Agriculture.* Rome, Italy: Food and Agriculture Organisation of the United Nations. Retrieved August 26, 2013, from http://www.fao.org/docrep/012/i0680e/i0680e.pdf

Ransom, C., Patricka, C., Ando, K., & Olmstead, J. (2006). Report of breakout group 1. What kind of training do plant breeders need, and how can we most effectively provide that training? *HortScience, 41*(1), 53–54.

SAS Institute Inc. (2010). *SAS OnlineDoc® 9.2.* Cary, NC: SAS Institute Inc.

Schuyten, G., Batanero, C., & Cordani, L. (2006). Working cooperatively in statistics education. In *7th International Conference on Teaching Statistics.* Salvador, Brazil: ISI.

Schwartz, M. A. (2008). The importance of stupidity in scientific research. *Journal of Cell Science, 121*, 1771–1771.

Silva, P. L. D. N. (2006). Statistical education for doing statistics professionally: Some challenges and the road ahead. In *Seventh International Conference on Teaching Statistics*, Salvador, Brazil.

Sowey, E. R. (2006). Letting students understand why statistics is worth studying. In *7th International Conference on Teaching Statistics*, Salvador, Brazil.

Sprent, P. (1970). Some problems of statistical consultancy. *JRSS. Series A, 133*(2), 139–165.

SSAI. (2005). *Statistics at Australian Universities, an SSAI-sponsored review.* Brando, ACT, Australia: Author.

Stubbs, E. (2012). The changing skills graduates need to be successful in applying analytics. In *Australian Statistical Conference*, Adelaide, SA, Australia.

Tishkovskaya, S., & Lancaster, G. A. (2012). Statistical education in the 21st century: A review of challenges, teaching innovations and strategies for reform. *Journal of Statistics Education, 20*. Retrieved August 26, 2013, from www.amstat.org/publications/jse/v20n2/tishkovskaya.pdf

UA. (2013). *Honours at Adelaide.* University of Adelaide. Retrieved August 26, 2013, from http://www.adelaide.edu.au/study/honours/

VSN International. (2012). *Genstat.* Hemel Hempstead, England: VSN International.

Willison, J., & O'regan, K. (2007). Commonly known, commonly not known, totally unknown: A framework for students becoming researchers. *Higher Education Research and Development, 26*, 393–409.

Researchers' Use of Statistics in Creative and Qualitative Disciplines

Sue Gordon, Anna Reid, and Peter Petocz

Abstract The aim of this chapter is to open up and explore a little-researched area of statistics education. We investigate how academics and postgraduate students use quantitative approaches to carry out research in creative and qualitative disciplines, such as music, design and art. We describe our method of interviewing 19 participants by email, indicate respondents' research contexts and the role that statistical techniques played in their research. We discuss how interviewees familiarised themselves with quantitative methods and what assistance they received and would have liked to receive. We investigate the researchers' epistemological views underpinning their methodological approaches. Finally, we develop an interpretive tool for situating research approaches where the home discipline is not usually associated with quantitative methods. The findings raise important issues about the training and institutional support of researchers in these fields.

Keywords Statistical support for research • Music • Creative arts • Epistemology • E-mail interviews • Thematic analysis

S. Gordon (✉)
Mathematics Learning Centre, The University of Sydney, NSW 2006, Australia
e-mail: sue.gordon@sydney.edu.au

A. Reid
Sydney Conservatorium of Music, The University of Sydney, Sydney, Australia
e-mail: anna.reid@sydney.edu.au

P. Petocz
Department of Statistics, Macquarie University, Sydney, Australia
e-mail: peter.petocz@mq.edu.au

H. MacGillivray et al. (eds.), *Topics from Australian Conferences on Teaching Statistics:*   365
OZCOTS 2008-2012, Springer Proceedings in Mathematics & Statistics 81,
DOI 10.1007/978-1-4939-0603-1_20, © Springer Science+Business Media New York 2014

1   Introduction

Karlo is a jazz pianist. His income is derived from evening performances in restaurants and clubs and is supplemented through private teaching. Playing jazz piano is his main focus and he has developed such skills that he is in constant demand as a performer. He can imagine innovative and unusual chord progressions, he has a finely developed ear for subtle shifts in nuance and timbre between instruments, he has an intuitive grasp of melodic form and development, he has a Masters degree in Jazz and he also has an insatiable curiosity. Recently, his interests have led him to doctoral-level research where he is investigating the impact of different instruments on the creation of music. He has noticed in his performances that different pianos seem to demand different sorts of responses from his music. Why is this so? He decided he needed to carry out a comparison between two major brands of instruments and study the perceptions of listeners to the music made on each of them. The first stage of his project was easy—play similar pieces on both instruments and see how the audience reacts. He prepared and publicised a series of concerts of his own jazz compositions, an audience filled the hall to listen, and after the show they completed a questionnaire asking about their opinions of the works and the piano sounds created. Having completed the creative component of the research, Karlo was left with the dilemma of analysis.

Jasmine is an artist. She is a social activist who believes that co-created public art works can be means of emancipation for urban youth. For many years she has curated an arts festival where the aim is to enable young people to express their experiences of youth, trauma, illness, race and other critical issues through the construction of artistic installations. Each curated event results in a public artwork that speaks to a specific community. Funding for this form of co-created work is hard to acquire and requires application to various funding schemes. Jasmine is also a university lecturer and her research has been to explore the lived experiences of young people through art. Her Masters degree used 'arts informed practice' as her main theory and method of analysis—this comprises conducting interviews, creating artworks and finally analysing the resulting artefacts. She has been less successful recently in gaining funding for her curated work and has now enrolled in a PhD program to investigate the impact of funding schemes on community art works. To start her exploration she decided to build a questionnaire and send it out using social media sites to all her previous participants and their local communities. But, what goes into constructing such an investigation? And how will Jasmine deal with the responses that she gets?

People like Karlo and Jasmine, researching in creative and qualitative disciplines (CQD) such as art, music, language, literature or design, have usually had a very different research experience to those in a typical science or business degree. Research students in such disciplines will have a strong background in their own area of arts interest and a keen critical knowledge of their craft, but they are unlikely to have encountered anything mathematical or statistical in their undergraduate degrees, not even the standard introductory statistics course of many science or business degrees (and some arts degrees). Yet when students and academics such as Karlo and Jasmine start their research careers, some of them will meet and wish to utilise statistical approaches. In many cases, their supervisors—academics in these disciplines—will have little experience of quantitative research approaches and may not be well placed to help them get started.

Our chapter is about the way that researchers in these CQDs encounter and relate to statistical ideas as part of their research environment in a context where quantitative thinking is often seen as quite peculiar. Why do CQD researchers decide to utilise a quantitative approach and how do they go about using statistics? How do they learn about quantitative methods and what sources of help do they have in this endeavour? What do they think that a quantitative approach will add to their projects, and how does this change the way that they think about their research questions? And what part do we, as professional statisticians and statistics educators, play in the process?

Research in CQDs often takes an epistemological direction that is quite different to the approach that seems natural to statisticians, yet is essential to the artefacts that are created by that discipline. Continuing our music example, researchers might investigate the artistry involved in playing an instrument. This artistry involves an appreciation of the sort of music that can be played on the instrument, the sorts of sounds that can be produced by the instrument, the situations in which the music is performed and the way in which the musicians interact. Each of these orientations leads to different sorts of research questions, for instance: When and why did the clarinet come into existence, and what were its antecedents? What particular sounds is it capable of making, and are there different sound qualities associated with different contexts (such as jazz or chamber music)? Is the instrument associated with a specific style of music? What impact has the development of the instrument had on compositional practice? How does the instrument feel to the performer and what connotations does it have for the listener?

All these questions are typical of the search for knowledge that may be associated with clarinet playing. Some of them may be addressed by standard qualitative methodologies, such as examination of texts via critical theory, content analysis of interview transcripts, or the investigation of everyday life via ethnomethodology (Garfinkel 1967). In music, the notion of 'historically informed performance'—using knowledge of the period when the music was written as a basis for its contemporary presentation—captures many of the questions raised above (Butt 2002). Art-based research (McNiff 1998, 2003) illustrates how practitioners can include creative arts in their profession's approach to research and has been extended into cognate areas such as early-childhood education. Researchers in CQD will be familiar with some or all of these approaches, but may be in a quite different position when it comes to quantitative enquiry; the most common position in the area of music research, for instance, seems to be a complete lack of knowledge of all but the simplest aspects of statistical method.

de la Harpe and Peterson (2008) investigated the key competencies, content and practices in the undergraduate education of several creative arts fields (architecture, art and design). While research qualities such as criticality, reflection, innovation, problem solving and knowledge building (among many more) were mentioned as core aspects of the disciplines, there was no mention of research activity that could, even remotely, be called statistical. It is possible to conjecture from their study that in CQD quantitative research preparation is not considered necessary for undergraduates and is rated as being of limited or minimal importance for post-graduate level research. Statisticians might disagree, pointing to various benefits of addressing questions in creative contexts using quantitative approaches. Yet this process of 'cross fertilisation' can work in both directions; a recent study (Liu et al. 2013) investigates patients' experiences following a medical procedure using the 'interpretive lens' of music, concluding that the solo concerto is a useful and appropriate metaphor.

The scenarios we have depicted above and the questions raised set the context for our project. In the next section we outline our aims for the study and then describe our methods of data collection and analysis.

2 Methodology

Our purpose in this project is to investigate the ways in which researchers (academics and research students) use quantitative approaches to carry out research in disciplines not commonly associated with statistical methods. Since this is an initial investigation in an area in which there is limited information and experience, we formulated the study using a qualitative approach, aiming to explore and open up the area by utilising rich descriptions of experience of a small group of participants. We planned to recruit participants with diverse backgrounds and in different contexts, to cover a broad selection of situations where researchers in CQD were including (or planning to include) quantitative approaches in their research. A qualitative design of this type has the potential to reveal aspects of experience that can form the basis for later research (including quantitative enquiry). The results can indicate areas for future investigation in more targeted areas of creative arts practice.

There are many examples of research projects that have successfully utilised qualitative approaches that focus on participants' own experience collected through interview studies. For example, far-reaching and ground breaking early studies of how students went about reading and making sense of a written text (Marton and Säljö 1976) were carried out with a small number of participants and led to the identification of surface and deep approaches to learning (Marton and Booth 1997). Such an initial study is able to identify a wide range of themes that could form the basis for a quantitative tool such as a questionnaire, in this case the many questionnaires on 'approaches to learning', such as the *Study Process Questionnaire* (see Biggs et al. 2001). Fielding and Schreier (2001) and the other papers in the same volume of the journal give an incisive discussion of the issues concerning qualitative and quantitative approaches, and their combinations, to research problems. The findings of our qualitative study can provide the information needed to construct a questionnaire with which institutions can survey their incoming CQD research students regarding the forms of research understanding and assistance that they may require.

Although there are generally accepted ideas of what constitutes 'creative disciplines' (Carey and Matlay 2010) and even some agreement that other disciplines are generally 'qualitative', there is no definite line that can be drawn between CQD and quantitative disciplines, and indeed we were interested in the overlap—researchers in traditionally CQD who were considering or actually carrying out research that had at least a quantitative component. Accordingly, we left it to potential participants to decide for themselves whether they were 'researchers in creative and qualitative disciplines'.

We invited participation by post-graduate and academic researchers. Recruitment notices were posted on bulletin boards in arts and education faculties and conservatoriums of music in Australia and circulated through international e-discussion groups and lists (such as music-and-science@jiscmail.ac.uk). In these notices, we stated that we were investigating the use of quantitative approaches for research in CQDs. We asked the following questions of potential participants: *Have you thought about whether or how you could use statistics in your research? Would you be interested in participating* via *e-mail in a research project investigating strategies and preparation for using statistics in research?* Our university ethics committee approved the project, and all our participants gave their informed consent.

Nineteen researchers volunteered to participate in the e-interviews. Most (11) were working or studying in Australia (one jointly in New Zealand), with seven in the United Kingdom and one from Greece. They were working on diverse research projects that ranged from those that included very little statistics to those that were essentially quantitative studies with a qualitative component. PhD candidates made up the majority (14 out of 19) of the respondents, including some who were employed as lecturers by tertiary institutions; the others were academics not enrolled in higher degrees, including one head of department and two retired 'honorary associates'.

Our protocol of interviewing by email was developed over more than 5 years (as described in Petocz et al. 2012) and 'fine-tuned' in the course of carrying out diverse projects, including pedagogical approaches in statistics service courses (Gordon et al. 2007), ways of introducing professional disciplines to students (Reid et al. 2010) and teachers' awareness of diversity in their classes (Gordon et al. 2010). This protocol consisted of an initial series of six open-ended questions with a further two rounds of questions probing initial responses and seeking clarification and further depth. The first question asked about participants' work and/or study and background. Other first-round queries included these: *What benefits could (or does) a quantitative approach have in your research? What formal support is available to you to develop statistical skills for research? Please tell us about a current (or proposed) research project that uses (or might utilise) a quantitative approach.* Second- and third-round questions were more discursive and specific to each participant, such as: *Can any research question be answered using either a quantitative or a qualitative approach? What sorts of problems arise when a research student with a predominantly qualitative background asks a statistician about a proposed research project?*

Participants were strongly engaged in the project, with 18 out of the 19 respondents completing all three e-interviews (one withdrew after one round). Answers to first-round interview questions were sometimes quite succinct. However, in response to further and more individually targeted questions, participants tended to describe their research and associated challenges and issues in considerable depth, so that the total length of the final interviews was around 64,000 words (questions and responses), ranging from 1,000 words to one (exceptional) interview of over 19,000 words.

Our analytic method was to identify themes in response to each of our main questions (Boyatzis 1998). Thematic analysis is considered a fundamental form of qualitative analysis and is located in the 'constructionist paradigm within psychology' (Braun and Clarke 2006, p. 78); it aims to identify common ideas established from shared emphases and common phrases put forward by participants. These common ideas are then grouped and labelled by a descriptor. The method requires iterative reading of the texts with the aim of comparing them to the developing descriptions of themes. The iterative process enables the emergence of nuanced outcomes for the final thematic set. In our analysis we also took an inductive approach in which the texts are viewed as research data that generate and test the themes. Using this approach, the data are collected specifically for the research and the researchers anticipate that they will learn from that data the experiences and concerns that are of importance to the participants. Once the themes are identified from the data texts, the context of each participant's text is subjected to analysis using the themes.

In this chapter we illustrate how and why participants used quantitative methods in their research and describe how they familiarised themselves with these methods and what support they were able to access. We investigate participants' views about the nature and status of quantitative approaches to research, and the effects that these views had on the research questions that they addressed. The transcripts were considered in full to understand, amplify and contextualise individual responses. The data—records of the interviews—are summarised by selection of quotes that illuminate various aspects of the analyses. We illustrate the primary ideas that arose in these areas with short excerpts, reported under pseudonyms chosen by participants themselves—in line with our ethics approval and with the aim of enabling respondents to track their own responses in publications. Rather than representativeness, which would be the aim in a quantitative study, we planned to cover a broad range of experience, sometimes summarised using the term 'saturation' (e.g. Mason 2010), in order to uncover a significant avenue of exploration for future quantitative research. Given the heterogeneity of the target group, it is likely that we did not achieve saturation, although our sample size was within the generally accepted range for qualitative studies of this type.

3   Participants' Use of Statistics in CQD

Participants had backgrounds in music (4), language/linguistics (4), creative arts (2), education (2) or other fields such as architecture and design, cultural studies and physiotherapy. A few had a quantitative background (one in engineering, one in psychology), and one even had a background in statistics, though she was now focusing predominantly on educational research. The research projects on which they were working included studies on injuries sustained by musicians, the performance of sign language interpreters, cultural policy in Australia and New Zealand and the design history of cars. Short extracts from several interviews will illustrate how our participants used (or planned to use) statistical methods in their projects. These extracts are presented in the form that our participants sent them, and we have used ellipses (…) to indicate when we have left out parts of a quotation and square brackets [] if we have inserted any text.

The first quote comes from Emma, who is undertaking doctoral research at an Australian university. Although she originally planned a mixed-methods study, she has been frustrated in her efforts to include a quantitative component in her essentially qualitative study.

> Emma: My PhD research is about language teachers' experiences with digital technologies in different contexts (everyday life, work, study) and how they inform teachers' perspectives on the role of technology in language education. … It's a qualitative study, which employs a case study methodology. The research methods are participant-generated photography, individual interviews and on-line shadowing. … Initially, I planned a mixed method study. To investigate teachers' perspectives on the role of technology in language education I wanted to include a survey. … However, I spent the first year of my candidature on theory and literature review. It appeared to be quite challenging to find the topic I am interested in and develop an appropriate theoretical framework. As a result I did not have enough time to design a survey.

A retired teacher of music, dance and drama in the UK, Mermaid is undertaking post-doctoral research on artistic communication during rehearsals. She has taken a qualitative phenomenon and subjected it to a quantitative and 'objective' analysis.

> Mermaid: I created a methodology to break down types of communication between director and company in the rehearsal period that leads to an artistic production. … Additionally, in order to determine if the system and its use were valid, I also conducted an inter-judge concordance test to check the reliability of the schedule. … As far as myself and my professors are aware (and my publishers), there is no other type of objective analysis in existence to study theatrical directors' styles of communication.

Grace works as an academic in an Australian university department of International Studies. She describes her doctoral research project that builds on her skills as a teacher of German.

> Grace: For my PhD I am investigating whether high school students can develop commitment towards a form of a learning plan. As part of the study, I am developing an instrument to measure three types of commitment. This involves quantitative as well as qualitative methodologies. The quantitative approach makes use of an already developed instrument from another research field. The qualitative methodology includes interviews to identify further potential commitment factors and other factors that have an influence on commitment.

With a first degree in music, Ironchicken is now working on a doctoral study at a UK university. He first sketches the background to his qualitative study, and then explains how his new context is encouraging him to include a quantitative dimension.

Ironchicken: Since around 2000, significant developments have been made in software for analysing sound. Such software could be applied to the sound recordings of musical performance and thus give musicologists an analytical handle on musical performance. However, it turns out to be the case that very few musicologists are actually doing this. My work, then, seeks to address why this is case. I'm currently putting together some case studies of practicing musicologists to discover their understandings of computer techniques for working with music.

... I'm now working on my PhD in a computing department and have been exposed to some very different modes of research. My colleagues now include data scientists, computer scientists and psychologists and I have begun to learn about empirical and experimental approaches to research. As a result, the PhD which I began as yet another critical project is now acquiring an empirical slant: I am performing a data-based analysis of the musicologist community and its research practices as well as carrying out case studies of individual musicologists (using semi-structured interviews).

A final extract from Rieba, a UK doctoral student with a background in experimental psychology, shows the use of qualitative aspects in an essentially quantitative study to find out how singing affects well-being.

Rieba: Since I am not a musician or singer, and had no experience in group singing, I used qualitative methods to understand what is going on. You can only set up an experimental study or develop a quantitative method when you have some idea of what is going on. So I asked singers to describe anonymously what they experienced while they were singing (open-ended questions) and joined a choir to experience it for myself. From the collection of answers I got and from previous research I developed the items for a questionnaire on the experiences of choir singing which was aimed to be quantitative.

Applied statisticians will be familiar with the use of statistics in a diverse range of projects, most commonly in the 'hard' or human sciences. The five examples above illustrate how statistical approaches are utilised in some less-traditional contexts of CQDs in language, drama, education and music. A range of other studies will be outlined in the context of the following sections.

4   Reasons for Including Statistical Approaches in CQD

Our respondents put forward a variety of reasons to explain why they had decided to make use of quantitative approaches, and in this section we explore these reasons. We do this by first giving several examples from our interviews, and then drawing out some common threads.

Working with sign language interpreters, KWSN, a research student of linguistics, recognised that she would learn things from statistical tests that she could not learn from a qualitative approach. For instance,

KWSN: T-tests will tell me whether there is significant difference between the two groups [native and non-native signers of Auslan] in terms of their English working memory span ... I will learn things from the statistical test that I can't learn from a qualitative approach.

Another respondent, CD, carrying out a project on car design history, concurred. He explained that a quantitative approach

CD: provides a realistic time line and identifies dates and events, showing when significant innovations took place, what they were, how long they took to become accepted, and provides some real, hard data that can be investigated to give trends and overviews over this long period.

Some respondents, such as Emma, believed that using quantitative and qualitative approaches together would help 'to confirm and cross-check the accuracy of data to achieve greater validity and reliability'. Archie, too, found that numbers as objective data were reassuring. In his project exploring how music preferences evolve, he said that he would be 'adopting a mixed approach to understand why choices are made in the real social world' and goes on to state in a self-contradictory way 'I know all data (quantitative and qualitative) is subject to bias and open to interpretation, but the data provides an objective, measurable scientific front to my arguments'.

Sally Farrier had been increasingly involved in teaching Alexander Technique for performance to musicians and actors during the past 8 years. It was in the context of this work that she discovered 'the appalling level of playing-related injuries in musicians and set out to do something about it'. Her project included both quantitative and qualitative aspects.

Sally Farrier: My study started as a cross-sectional survey and continued on as a longitudinal prospective study using online questionnaires. I also completed a separate qualitative study on posture last year using structured interviews. These research approaches are those most commonly found in the performing arts literature.

Sally Farrier had not seen quantitative methods as relevant to becoming a physiotherapist until she needed to apply them to her research. She said:

... Much as I hate to admit it, I suspect that most people aside from maths nuts simply would not absorb formal stats training unless they were using that area of stats at the time. My desire to produce high quality work has driven my pursuit of stats knowledge over the past five months, but without that motivation and the immediate practical application of the information I found, I would not have been able to learn it in the way that I have.

Becky considered that 'statistical results are often seen as a more conclusive output than, say, narrative analyses' while Rieba pointed out that 'for the people who provide the funds and society as a whole, it's quantitative research that is more credible'. By contrast, after submitting an article on design history to a journal, CD observed that the editor's response 'indicates that numbers don't seem to be understood by some disciplines'.

The above responses illustrate respondents' perceptions that quantitative approaches can provide information that is not available using qualitative methods or can be used to support qualitative results in a mixed-methods approach. Quantitative results may be seen as more credible or conclusive than qualitative ones, and indeed in some situations may be perceived as necessary for the project to be funded and the outcomes to be seen as 'high quality' or 'scientific'.

5   Support for Learning and Using Statistics in CQD

Given that many of the researchers in CQD were unfamiliar with quantitative methodology, we now look at the ways in which participants tried to acquire enough knowledge and experience to utilise quantitative approaches. Respondents had varied backgrounds in statistical methods, ranging from virtually no experience to one with previous training as a statistician. For most of them, a current task was to increase their knowledge of quantitative methods, an undertaking that they carried out in different ways and with varying degrees of success. One common approach was attending a course in research methodology and/or using statistical computer packages. However, such courses were often described as too broad or too advanced. Francis expressed a common concern: 'it needs to be related to a data set that you are actually using or it is not relevant and becomes a waste of time and no one has time to waste'.

Emma's experience, below, is an Australian example that shows how the researcher often has to fall back on her own resources.

> Emma: I attended [a] couple of seminars on quantitative research but they appeared to be too complicated for me as they assumed certain level of knowledge. My 'beginner' level was not enough. When I was doing my Masters degree, I did a unit on methodology, the first part of which was general for qualitative and quantitative. I did not find it very helpful for me because of the lecturer. The references suggested for reading were not helpful too. Most knowledge I developed later when I started my research and was reading the books I found myself or suggested by supervisor and fellow research students as well as from references in journal articles.

The same training approach was experienced in a UK university, as Mermaid relates.

> Mermaid: All PhD students were bound to undertake a 1-term course in SPSS (one session per week) … The university itself gives out SPSS software to all its PhD students and encourages them to use it. One of my own professors used to lecture in SPSS and she helped me with the concordance data. … During the obligatory, first-year courses in stats, it was announced by the course tutor that anyone could book a personal meeting if they were experiencing difficulty, but that time was rather limited. This is the nearest I can say that 'met' the idea, but I wouldn't say that it really qualifies. Anyone who might have the expertise to help would also invariably have their own workload to contend with.

Beyond the formal courses, statistical help may be hard to find. Jack, professor and head of a department of Creative Arts at an Australian university, was asked whether statistical advice was available and adequate for his researchers' needs. He seems to put his finger on the main problem for researchers in CQD explaining that 'there doesn't seem to be a middle ground between the very heavily science-oriented quantitative research methods and the qualitative approaches. What's needed is more support for mixed methods research'.

Bourbaki, working in the area of writing support for doctoral candidates (at another Australian university) as well as undertaking her own doctorate, points out the consequences of such a lack of support.

Bourbaki: There is, however, no structure for statistics teaching and learning and for statistics support. This has had sad consequences. Students avoiding statistics from lack of knowledge and support – not from discounting a quantitative approach because it is not appropriate.

Some participants reported that their colleagues were able to provide useful support for their statistical learning. Ironchicken, quoted in an earlier section, indicated that his computer science colleagues had been broadening his quantitative horizons. Also in the UK, working in a department of Engineering and Design, CD talked about the help that he obtained for his project on car design history.

CD: There is significant support from those in other disciplines who carry out statistical research. Although my skills are not perhaps what they ought to be in this area, people from our Arts and Human Sciences and Health Faculties have been extremely helpful in developing not simply statistical skills but statistical thinking processes and how research questions might be answered. We also have some support in our faculty and in the university generally, but these have (strangely) been less helpful. … The faculty research doesn't cover very much statistical work … and the university general statistics help tends to be for the undergraduates and how to use (say) SPSS to obtain particular results.

Asked whether challenges associated with quantitative methods were different from those associated with qualitative methods, Kate answered 'Yes and No':

Kate: Yes, in terms of the statistical analysis. Quantitative methods need a professional support and advice in some cases. You need to have certain amount of knowledge about the statistics for the quantitative methods. Also, it seems that there are sometimes difficulties to collect enough number of data to enable statistical analysis. No, in terms of research design and question items. To obtain the outcome that you aim to investigate, questions need to be carefully designed/selected to lead the participants to reveal the issues related to your research questions.

A further point raised by respondents who are PhD candidates is that their supervisors are not necessarily experienced with statistical methods. KWSN lamented this situation.

KWSN: I urgently need to learn how to use SPSS. But neither my supervisors nor my colleagues can teach me how to use it. … Besides, I'd like to have a statistician to turn to. I wish my supervisors would know how to use quantitative research methods. As I said before, they usually do qualitative research.

Some participants, particularly among the Australian interviewees, were very relieved to have found a helpful statistician who was able to give them advice on their specific project. Grace found such a person from her university website, while Sally Farrier identified an expert through family connections.

Grace: I used the university homepage to identify statistics experts and found an expert who was kind enough to help me with my work. This has turned out to be of tremendous support for my work. However, not all students were that 'lucky'. Many exclusively rely on 'learning by doing' techniques. I have used about four applied statistics books myself but still required help from an expert.

Sally Farrier: The stats advisor assigned to Health Sciences is a very nice man, but he appeared not to know what he was doing when it came to my study. My daughter's friend gave me better advice in ten minutes than I had had over weeks from the other person.

Other participants had the benefit of institutional access to statisticians. Archie explains how this works at his UK university.

> Archie: We are allowed five hours with the stats department per year. There are courses available but no basic courses are available. As such an understanding of statistics is required for these courses, which is difficult for a novice such as myself. … I had two sessions with a postgraduate statistics student who helped me understand my data and assisted with technical SPSS issues. As far as I know he was funded by the stats department who in turn claimed the money back from the university.

We have included a wide range of quotes in this section, as the problem of learning about quantitative approaches and getting support with quantitative research problems was the greatest concern for many of our respondents in CQD. This is not surprising as most participants had limited backgrounds in statistics, and so, without support, were unable to progress with quantitative investigations. Although our study was designed to uncover issues rather than draw numerical conclusions, our CQD researchers in Australia tended to find their institutions lacking in terms of support for quantitative research, while respondents in the UK seemed to have greater access to more formal institutional support. The intriguing apparent difference between these two respondent groups suggests a direction for further research.

6 Epistemological Views of Researchers in CQD

Earlier, we investigated our participants' practical reasons for including quantitative methods in their research. We left that section with the notion that quantitative approaches might be necessary in order to achieve 'credible' and 'scientific' results. What do such statements say about the underlying epistemological views of those who make them? In this section, we broaden our exploration by looking at our participants' ideas about the philosophical and theoretical dimensions underpinning their use of statistics in CQD.

As an academic in the field of music, Panda grappled with an aesthetic problem: 'how can music 'mean', when it doesn't (as many authors argue) mean anything in a referential way'? He considered that 'experimental methods are both narrow and unreliable—it is only possible to test one narrow faculty at a time, and when you do, it is usually uncertain whether the result is due to one underlying mechanism or several'.

> Panda: However, I think I am drawn to the epistemology of science; and I tend to see analytic philosophy, computational modeling, and experimentation, as a continuum. … I enjoy the fact that nothing is necessarily excluded from a theory; that there is quite a lot of instinct in accounting for a set of phenomena (such as those associated with the experience of music listening); and, importantly, that this involves challenging and reworking one's own assumptions.

This tension between some researchers' perceptions of quantitative methods as 'scientific' and their inherent distrust of these methods is echoed by Stark.

Stark: I am intensely suspicious of quantitative surveys and investigations and data – because when I'm at the giving end, they never give me the opportunity to pass on all the information, which I would like to convey, in my words and way. ... I don't think modelling is more authoritative than well-argued analytical philosophy; and I think quantitative research can be prey to any number of fallacies, so that conclusions from e.g. experimental psychology or sociology are often less authoritative in my view than philosophy, anthropology, or ethnography. ... Honestly, I think what mostly drives method is people's predilections, and their background and training.

At the other end of the spectrum, Rieba asserted 'that any research question, by nature requires quantitative methods. The qualitative methods are of a great importance to complement and inform the numbers but the numbers provide the answer'. This claim corresponds more closely to the standard view of statistics and its role in the scientific approach. She expanded her views:

Rieba: I work under the scientific assumption of cause-effect and only an experimental study (randomly assigned participants to control and experimental conditions) will help us determine the effects of singing on wellbeing. ... [However] numbers are just that and sometimes not sensitive enough to capture the nuances or the details in individual experience.

Many participants supported a view that the research question itself determines the approach and methodology. Ironchicken explains the elements of this idea, and Becky continues along the same path. She is only momentarily perturbed by the interviewer's deliberate reversal of expectations about methods for evaluating drug effects and voting preferences and finds refuge in the combination of qualitative and quantitative approaches.

Ironchicken: I believe the question determines the approach. It may be the case that quantitative methods could be applied in almost all cases, but qualitative methods not. For example, drug trials must be conducted quantitatively in order to obtain a significant result to demonstrate the suitability of the drug for its intended purpose. I can't imagine how a qualitative approach would suit a drug trial; even if subjects were asked to self-report on the effectiveness of the drug, their responses would still be used to contribute to a quantitative data set. As another example, literary criticism is a well established and valid approach to analysing literature, but quantitative techniques can also be applied in analysing literature, even for similar questions.

Becky: You could probably try to investigate any research question using any method but there would definitely be some cases where a particular approach is more suitable than another. If you want to understand the reasons people vote why they do, you are going to need a different method than if you want to determine the side effects of a particular drug.
[Interviewer: So I'm thinking: the side effects of a drug would have to be investigated qualitatively – maybe by interviewing people about their experience – whereas understanding the reasons why people vote as they do could be investigated quantitatively – maybe by adding a short questionnaire to their voting paper?]
That's funny – I would think if you were going to use those examples in that way that the qual/quant distinction would be around the other way! Probably this is another example where a mix of qualitative and quantitative methods would be appropriate in each case, but exactly which methods and how they are combined would depend on the research question and objectives.

Having a background in both qualitative and quantitative research, CD was able to pinpoint some of the difficulties of combining methods, and he suggests utilising a design approach. Mermaid expands on the benefits of such a combination in her project on communications in rehearsals.

CD: Sometimes they are to do with the ways in which the two relate technically to each other with things like sequencing, sorting out what happens with null returns and lack of information. At other times it is to do with the fact that describing what the components mean has significant difficulties in that people feel that I should be measuring something tangible rather than something that is the result of a statistical process using categories, and these have no units, for instance (which is something engineers would expect). ... It could be worth framing the question in several different ways to see what would be the most natural way of answering it and starting off with a very vague question, or even with several ones that start to define an area. It starts to become more of a design-type method than a research one and ends up with a sort of diamond process with a broadening of the question followed by a narrowing down (which may imply a specific method).

Mermaid: The quantitative data, if used alone, would be significantly impoverished without the back-up of observational input, and especially interviews, along with the written notes I made during each observation. On the other hand, the patterns that emerged (from Excel analyses) were sometimes quite extraordinary and would probably not have been identified without this (quantitative) approach to expose them.

Bourbaki also discussed the benefits of combining quantitative and qualitative approaches. She considered that viewing 'concepts and phenomena through only one lens, there seems to be – for me – something missing'.

Bourbaki: Quantitative research can challenge and change qualitative insights in exciting and rapid ways. ... The juxtaposition of the specific with the broadly conceptual can produce rich results. And when quantitative observation and analysis concur with conceptual understanding – then it seems that there are frequently additional sometimes difficult-to-explain learnings that occur at a level just below consciousness that may influence still further observations. ... The integration of quantitative results may therefore involve an integration of non-numerical information – such as the settings in which observations occurred, a method of observation or the manner in which a particular question arose.

Finally, Bourbaki articulated how beliefs could be transcended by research.

... This is the exhilarating part when you start to purposefully dump some of your beliefs (formerly construed as knowledge), and start to realise that your research could be thought of, thought about and understood in a way that does not match what you believed to be so just a few days or months before... and perhaps you had understood things like that for years.

We have outlined, in this section, a range of epistemological understandings of quantitative methods by our CQD researchers, indicating how their research approaches both shape and are shaped by their developing projects. In their quotations we see participants grappling with the traditional view of the nature and benefits of science, interpreted almost entirely using the quantitative paradigm. We see many of them putting forward the notion that a specific question determines the methodological approach to its investigation. And we see respondents accepting, sometimes uncritically, the epistemological 'solution' of combining quantitative and qualitative research approaches.

7 Emerging Themes

Statistical advice for CQD researchers is not essentially different from statistical help for research in any field. However, researchers in the more traditionally quantitative areas such as the sciences have an a priori expectation that they will need statistical support, and in many cases they will have received some prior training in statistical thinking and methods. This is not generally the case in CQDs. As Jack summed up: '... most artists struggle to build a methodology around their practice let alone understand the different approaches to research. Hence, to get them to apply quantitative research methodology is usually a big achievement'!

Our findings raise some important themes. Firstly, respondents indicated a variety of reasons for including or planning to include quantitative approaches in their research in CQD, despite the generally qualitative context in which they worked. These reasons included their perception that quantitative methods would enhance the objectivity of their findings, could be used to identify cause and effect, or were needed to augment qualitative or non-empirical methods alone. However, our participants' responses suggest that researchers in CQD may be hampered in these endeavours by a lack of knowledge of and support for quantitative skills and techniques, and associated analytic tools, such as statistical software.

Secondly, training in statistics may be inadequate for researchers in CQD; in our group of participants, this was especially the case in Australia. The common short course in research methods, or introduction to statistical software such as SPSS, was generally rated as inadequate or inappropriate for researchers' needs by our participants. The types of courses described by Harraway (2010) do not seem to be accessible to many CQD researchers. Although such courses are often described as not needing any mathematical background, nevertheless, they are aimed at scientifically trained researchers and would be unlikely to meet the needs of most of the participants in our study.

Thirdly, our findings show that institutional support for researchers in CQD in the form of competent advice or assistance in statistics is limited at best. Where respondents did find help and expertise in their university the assistance tended to depend on the goodwill of a statistician or colleague. The general lack of such support was a matter of major concern to our respondents; many of them had trouble even finding out what they needed to know.

We leave the summary of these issues with Isabel, a trained statistician working in a School of Technology at a university in the UK, and carrying out predominantly educational research. Isabel's views of students' problems of learning statistics and getting statistical research help are particularly important, coming as they do from someone straddling the divide between the quantitative and the qualitative.

Isabel: Statistics is a massive subject with so many pitfalls people commonly fall into – such as confounding. Really if you are doing quantitative research, you need the input of a statistician. ... I am always happy to offer advice to research students, regardless of their subject; however, I am one person and there are not many other statisticians about either. I am sure that if their supervisor doesn't know a 'proper' statistician, and also if they don't understand the subject of statistics sufficiently, the student won't even know they have to look for a statistician, let alone where to find one.

8   Transgression, Confession and Border Crossing

In the course of undertaking this research we have noted diverse aspects related to the manner in which our participants speak about research in their home discipline and the role of statistics in such research. In our research design, we encouraged participants to think of their home discipline first and then focus on the use of quantitative approaches in relation to that home discipline. The relational aspect enables us to see more acutely how participants experience the link between them.

Some CQD researchers experience an epistemological dilemma which we might call 'transgression'; their epistemological perspective is confronted by an alternate paradigm that then places them in some sort of cognitive dilemma. Other researchers make a form of apology for their home discipline—we call this 'confession'; the researcher apologises for their primary discipline as lacking something (often, quantitative methods) and decides that something needs to be done to rectify the situation. Yet other CQD researchers move freely from one epistemological position to another—we refer to this as 'border crossing'—in order to accommodate the procedures and values of different disciplines. Each respondent may illustrate one or several of these experiences depending on his/her own perception of the situation.

Using the quotations given in this chapter, Emma provides us with an example of confession. She describes her basic approach to research as using case-studies that comprise three elements: photographs, interviews and online shadowing. In the field of arts-based research, these data collection methods are suitable for a robust analysis of the situation. But Emma suggests that there is something incomplete in the validity of these methods with her comment 'I wanted to include a survey'. She gives no information at all about the sort of survey that she could use, or how she might carry out the analysis of it, simply that she wanted to include one. However, time passes and she can't quite work out how to include a survey in her research project.

An example of transgression comes from Grace, who adopts the use of a previously devised survey to form the basis of her qualitative interview analyses. The survey is not directly tied to her research question, as it was 'an already developed instrument from another research field', but because it has been used before it is imbued with rigour and validity. The interchange with Becky about her epistemological views provides a further example of transgression; in this case the cognitive dilemma is resolved, or at least put off, by an appeal to 'mixed methods'.

Ironchicken furnishes us with an example illustrating the notion of border-crossing. The world of sound analysis has progressed due to technological changes, and forms of musical analysis that are common to musicians are deemed inadequate for contemporary studies. He finds that computer scientists are more in touch with the sound wave analysis that he seeks to carry out. There is an element of surprise as 'what began as yet another critical project [based on semi-structured interviews] is now acquiring an empirical slant'. While few other musicologists are using this approach he is forging ahead into a new world.

Some participants show more than one of these experiences, maybe at different stages of their research. Mermaid, working in the world of theatre, has often wondered about the social interplay between director and actors. Like Ironchicken, she suggests that her statistical analysis of communication activity is new to her home discipline and hence worthy of exploration. There is a form of confession in the implied inadequacy of theatre discourse, a transgression when she suggests that the communication analysis will lend validity and reliability to an otherwise subjective world, and then a border crossing when she proceeds with her quantitative analysis of communication between theatre professionals. In her case there is an intellectual disjunction between the world of subjective practice and the potential world of statistical objectivity.

9   Pedagogical Implications

This study initiates exploration into an area of statistics education where there is a dearth of data and builds on our previous and long-term research about students' experiences of learning statistics in more usual contexts such as mathematics, science or psychology (including Reid and Petocz 2002; Gordon 2004; Gordon et al. 2009). The findings reveal implications for further pedagogical research and practice.

In Petocz and Reid (2010) we identified three levels of students' conceptions of statistics, summarised from narrowest to broadest by their focus on the techniques of statistics, on the data used in statistical analyses, and on the meaning that can be developed through statistical approaches. Clearly, researchers in CQD disciplines have a range of different ways in which they experience and understand the utility of statistics for their research and a range of attitudes to learning statistics. Some of our CQD participants talked about the possibility of utilising statistics in their study, but were unable to articulate in any way how this might illuminate their research question. Such a view of statistics may represent a conception that is even narrower and more restricted than the ones we previously identified with statistics students, based as it is on these researchers' very limited experience with quantitative data.

For educators working with CQD researchers this may be a key to pedagogical development. Our findings show that the CQD researchers in our pilot study were generally rather nervous about the use of statistics, from the identification of a sensible statistical design for their research through to analysis and presentation of any numerical data that they might have collected. Moreover, many of them had little idea about where to find statistical help for their research. This is a challenge for research institutions as many groups of students in creative disciplines have not had any previous exposure to quantitative research methodology and come to the problem of the use of numerical data completely naively. A just-in-time approach to developing statistical skills is one possible solution, maybe in the form of statistical consultancy for research students; an alternative could be an introductory statistics (or quantitative research methods) course appropriate for CQD researchers. A problem with the former might be that students only acquire patchy knowledge about statistical approaches, or even none at all if they just rely on a consultant to give them the appropriate answers. However, any introductory course is likely to be too generic and miss the specific questions that the students have: recall Francis's comment that 'it needs to be related to a data set that you are actually using or it is not relevant and becomes a waste of time and no one has time to waste'!

Perhaps a solution to identifying CQD researchers' statistical needs could be the development of an instrument that canvasses the scope of their research question, the various approaches to their research that may need support (in qualitative and quantitative fields) and their level of comfort with a range of research methodologies. In this way a fluid, yet targeted, approach to support may be identified early on in CQD research activity. Further, our identification of the concepts of transgression, confession and border crossing in our participants' interview records provides a theoretical tool for investigating how CQD researchers come to think of working with statistical methods, and how appropriate pedagogical approaches could be tailored to their research contexts.

10   Conclusion

Our findings concur with the observation of MacGillivray (2010) that, as for communication skills, no discipline is immune to the need for statistical thinking, although forms, contexts and applications may vary. Researchers in music and other creative or qualitative disciplines may be familiar with a diverse range of research approaches for evaluating or appreciating artistry, yet our findings highlight a neglected area of their preparation and education—their training in quantitative methods. The problems of learning statistics and getting statistical help for projects in CQD are exacerbated for researchers in areas where few colleagues or even supervisors have expertise in statistical methods.

The responses from our interviews demonstrate that participants' needs for knowledge in quantitative methods were, in most cases, accompanied by a sense of frustration at their own lack of skill and their limited access to assistance in this area. Where respondents utilised statistics as part of a mixed-methods approach, they often reported searching for self-help in the form of a textbook or online material. Some researchers were fortunate to find a statistician to provide expert advice, but this generally voluntary assistance has implications for the workload of such a 'lone statistician'. The outcome of the lack of knowledge and/or support in quantitative methods is that researchers in CQD may be constrained from utilising the best approach to answer their research questions—surely a fundamental criterion of good research.

At the beginning of this chapter we introduced Karlo and Jasmine, who provoked our thinking about this area of research. Karlo's current work is in the transgression stage. He knows that his habitual methods of musical analysis are sound, but he has started to utilise quantitative approaches in his current project and he is now in need of statistical advice. His faculty is not statistically literate, as for many of our research participants, and so he is looking for advice from someone with statistical expertise to carefully explain his results from his intuitively developed questionnaire. He hopes that this may help him make best use of the material he collected during his performances. Jasmine is now a border crosser. She is still an artist, working and researching using a range of arts-informed methodologies, but she is now also enrolled in a statistical research methods course to build up her skills in quantitative approaches. Auspiciously, she has also become her faculty's resident quantitative analyst of various forms of student data.

11 Note

Developed from a paper presented at Eighth Australian Conference on Teaching Statistics, July 2012, Adelaide, Australia.

This chapter is refereed.

Acknowledgment We gratefully acknowledge the time and insights of the participants in this project.

References

Biggs, J., Kember, D., & Leung, D. (2001). The revised two factor study process questionnaire: R-SPQ-2F. *British Journal of Educational Psychology, 71,* 133–149.

Boyatzis, R. (1998). *Transforming qualitative information: Thematic analysis and code development.* London, UK: Sage.

Braun, V., & Clarke, V. (2006). Using thematic analysis in psychology. *Qualitative Research in Psychology, 3*(2), 77–101.

Butt, J. (2002). *Playing with history: The historical approach to music performance.* Cambridge, UK: Cambridge University Press.

Carey, C., & Matlay, H. (2010). Creative disciplines education: A model for assessing ideas in entrepreneurship education? *Education + Training, 52*(8/9), 694–709.

de la Harpe, B., & Peterson, F. (2008). A model for holistic studio assessment in the creative disciplines. *ATN Assessment Conference 2008 Proceedings.* Retrieved from http://ojs.ml.unisa. edu.au/index.php/atna/article/view/339/238

Fielding, N., & Schreier, M. (2001). Introduction: on the compatibility between qualitative and quantitative research methods. *Forum Qualitative Sozialforschung/Forum: Qualitative Social Research, 2*(1). Retrieved from http://www.qualitative-research.net/index.php/fqs/article/view/965

Garfinkel, H. (1967). *Studies in ethnomethodology.* Cambridge, UK: Prentice Hall.

Gordon, S. (2004). Understanding students' experiences of statistics in a service course. *Statistics Education Research Journal, 3*(1), 40–59. Retrieved from http://www.stat.auckland.ac.nz/serj

Gordon, S., Petocz, P., & Reid, A. (2007). Teachers' conceptions of teaching service statistics courses. *International Journal for the Scholarship of Teaching and Learning, 1*(1). Retrieved from http://www.georgiasouthern.edu/ijsotl/v1n1/gordon_et_al/IJ_Gordon_et_all.pdf

Gordon, S., Petocz, P., & Reid, A. (2009). What makes a 'good' statistics students and a 'good' statistics teacher in service courses. *The Montana Mathematics Enthusiast, 6*(1/2), 25–39. Retrieved from http://www.math.umt.edu/TMME/vol6no1and2/

Gordon, S., Reid, A., & Petocz, P. (2010). Educators' conceptions of student diversity in their classes. *Studies in Higher Education, 35*(8), 961–974.

Harraway, J. (2010). Statistics for postgraduates and researchers in other disciplines: case studies and lessons learned. In C. Reading (ed.), *Proceedings of the Eighth International Conference on Teaching Statistics (ICOTS8)*. Retrieved from http://www.stat.auckland.ac.nz/~iase/publications/icots8/ICOTS8_6C3_HARRAWAY.pdf

Liu, L., Lapum, J., Fredericks, S., Yau, T., & Micevski, V. (2013). Music as an interpretive lens: Patients' experiences of discharge following open-heart surgery. *Forum Qualitative Sozialforschung/Forum: Qualitative Social Research, 14*(1). Retrieved from http://www.qualitative-research.net/index.php/fqs/article/view/1812/3485

MacGillivray, H. (2010). Abstract for session 6C: Statistics training for researchers in other disciplines. In *Scientific Programme for the Eighth International Conference on Teaching Statistics (ICOTS 8)*. Retrieved from http://icots8.org/session.php?s=6C

Marton, F., & Booth, S. (1997). *Learning and awareness*. Hillsdale, NJ: Lawrence Erlbaum.

Marton, F., & Säljö, R. (1976). On qualitative differences in learning: I. Outcome and process. *British Journal of Educational Psychology, 46*, 4–11.

Mason, M. (2010). Sample size and saturation in PhD studies using qualitative interviews. *Forum Qualitative Sozialforschung/Forum: Qualitative Social Research, 11*(3). Retrieved from http://www.qualitative-research.net/index.php/fqs/article/view/1428/3027

McNiff, S. (1998). *Art-based research*. London: Jessica Kingsley Publisher.

McNiff, S. (2003). *Creating with others: The practice of imagination in life, art, and the workplace*. Boston: Shambhala Publications.

Petocz, P., Gordon, S., & Reid, A. (2012). Towards a method for research interviews using e-mail. In C. S. Silva (Ed.), *Online research methods in urban and planning studies* (pp. 70–85). Hershey, PA: IGI Global.

Petocz, P., & Reid, A. (2010). On becoming a statistician: A qualitative view. *International Statistical Review, 78*(2), 271–286.

Reid, A., & Petocz, P. (2002). Students' conceptions of statistics: A phenomenographic study. *Journal of Statistics Education, 10*(2). Retrieved from www.amstat.org/publications/jse/v10n2/reid.html

Reid, A., Petocz, P., & Gordon, S. (2010). University teachers' intentions for introductory professional classes. *Journal of Workplace Learning, 22*(1/2), 67–78.

Part F
Papers from OZCOTS 2012

Evaluation of the Learning Objects in a Largely Online Postgraduate Teaching Program: Effects of Learning Style

Imma Guarnieri and Denny Meyer

Abstract One of the challenges of providing an online course is the development, maintenance and evaluation of teaching resources provided for students. The purpose of this study is to evaluate, on the basis of student preferences, the diverse learning resources provided in a postgraduate course of Applied Statistics, taught predominately in an online mode, and to explore how students' learning styles impact the evaluation of these resources. In a survey of 57 students 74 % selected more than one preferred learning style, with the "Doing" (75 %) learning style being most favoured in comparison to Reading (63 %), Looking/Watching (42 %) and Hearing (37 %). While online tutorials, face to face class, assignments and quizzes were helpful to students across all learning styles, those students who preferred to learn by "Doing" or "Looking/Watching" attached greater importance to the prescribed text/notes while those who learn by "Looking/Watching" also found lecture recordings and practice exams particularly helpful. This paper discusses how information about students' learning styles and preferences for resources can assist in the prioritisation of resource development.

Keywords Online course evaluation • Learning styles • Statistics education • e-Learning • Learning resource preferences • Binary logistic regression models • Conventional learning theories

1 Introduction

Over the last decade the educational paradigm has changed its focus from being teacher-centred to being more student-centred, and along with this shift has also come a change in the delivery of many educational programs from being solely face

I. Guarnieri • D. Meyer (✉)
Faculty of Health Arts and Design, Swinburne University of Technology,
Hawthorn, VIC 3122, Australia
e-mail: iguarnieri@swin.edu.au; dmeyer@swin.edu.au

H. MacGillivray et al. (eds.), *Topics from Australian Conferences on Teaching Statistics:*   387
OZCOTS 2008-2012, Springer Proceedings in Mathematics & Statistics 81,
DOI 10.1007/978-1-4939-0603-1_21, © Springer Science+Business Media New York 2014

to face to largely online. Larson (2002) argues that whereas traditional course delivery is centred on the facilitator, online education is more student-centred, translating into the development of diverse teaching resources such as online quizzes, narrated powerpoints, online classes and discussion board posts.

As noted by Mestre (2010) the core characteristics of resources which support learning are efficiency (i.e. cost and time savings), reusability, interoperability, durability and accessibility. However, a learning object's effectiveness may vary in response to different learning styles. This study considers this question within the context of a postgraduate program for Applied Statistics taught predominantly in an online mode. The program serves students from a wide range of disciplinary and ethnic backgrounds, suggesting considerable variation in learning style.

2 Literature Review

There are many models for learning styles that have been developed to allow learners to be categorised into a specific learner type, and some of these models have looked at patterns for diverse populations in the online environment (Mestre 2010). Previous studies have used several of these models to examine the association between students' learning style and the selection of course delivery format. However, the research outcomes have provided mixed results (Zacharis 2011; Kolb 1984; Witkin et al. 1977).

There is some evidence that learning styles have more impact in web-based courses than in traditional face to face learning. Manochehr (2006) compared online learning versus traditional instructor-based learning in regard to preferred student learning styles, showing that learning style was irrelevant for knowledge performance in traditional learning but important in online learning. However, the literature regarding student *satisfaction* with online courses is less clear about the impact of a student's learning style on their preferences for types of learning objects (Glass and Sue 2008), and it is hoped that this research will make a contribution in this area.

Learning styles are not fixed and learners can adopt a different learning style depending on the subject matter and current learning environment. Mestre (2010) refers to a 2007 study which suggested changes in learning preferences as people aged. In particular it was found that older people tended to have a higher preference for a single learning approach (43 %) than younger people (36 %). For most learners, one or two styles are preferred above the others (Pritchard 2009). However, students do typically have one learning style that is preferred over others and can be motivated by learning material compatible with this preference (Larkin and Budny 2005). Zacharis (2011) argues that "being aware of … students' learning styles, instructors can design online modules and activities, or redesign sequence of events and interventions, to accommodate effectively their different academic skills and interests".

The goals of this study are twofold; to identify the preferred learning style of the postgraduate students of an applied statistics course and to determine if there is an association between learning style and preference for types of learning resources for this group of students. It is hoped that answers to these questions will help to inform future development of learning resources, especially for online courses. However, it is known that students change their learning style preferences across different disciplines (Jones et al. 2003) and, presumably, also across different deliveries, so this study should be regarded as a useful case study rather than as a more general model.

3   Research Method

In this study we consider a questionnaire completed by postgraduate students of applied statistics. The students surveyed were enrolled in a nested postgraduate Applied Statistics program; a Graduate Certificate, consisting of four units; a Graduate Diploma, including an additional four units; and a Masters, including a further four units. This is a largely online program and the students come from all over Australia with a few overseas students as well. The program is designed assuming that the students do not have a strong background in mathematics, with emphasis placed on the learning of statistical software skills (e.g. SPSS, SAS, R, SAS Enterprise Miner and AMOS), statistical modelling and reporting of research results. Most students study part-time completing only 2 units per semester. This means that most students who complete the masters need to study with us for 3 years, allowing them to apply their learning directly to their work situation.

The students were invited to complete the anonymous Opinio survey by email at the end of semester 2 2011, with a second reminder email sent 2 weeks later. In total 57 responses were obtained providing a 52 % response rate. Almost half (28) of these students were enrolled in at least one Graduate Certificate unit. Each of eight learning resources was rated on a scale with 1 = not at all helpful, 2 = a little/some help, 3 = very helpful, 4 = extremely helpful, with not applicable responses, which occurred when a resource had not been accessed, treated as missing values. The eight resources named were: text/notes, lecture recordings, online tutorials, face-to-face classes or workshops, assignments, weekly quizzes, practice exams and the discussion board. In addition students were asked their level of study (Graduate Certificate, Graduate Diploma or Masters) and they were asked to indicate on a Yes/No scale about their preferences for each of the four learning styles; *Reading, Hearing, Doing* and *Looking/Watching*. This classification of learning styles relates to the Index of Learning Style Inventory (Felder and Spurlin 2005) developed by Richard M. Felder and Linda K. Silverman from North Carolina State University (NCSU).

The *Doing* preference relates to the NCSU Active category. As explained by Mestre (2010), Active learners, often including Latinos, African Americans, and Native Americans, tend to retain and understand information best by doing something active with it—discussing or applying it or explaining it to others. "Let's try it out and see how it works" is an active learner's phrase. Students who prefer this learning style tend to prefer group work over individual work and this clearly poses a challenge for online study. Non-active or reflective learners, often including Asian

Americans according to Mestre (2010), prefer to think about the material before working with it and they tend to prefer working alone. Mestre describes the NCSU Verbal learners as students who get more out of words, both written and spoken explanations, suggesting the *Hearing* and *Reading* learning preferences used in this study. The NCSU Visual learners remember best when they see pictures, diagrams, flow charts, time lines, films and demonstrations, relating best to our *Looking/Watching* preference.

Descriptive statistics were derived for the learning style preferences and for the satisfaction ratings for the eight learning resources. On account of the small frequencies for some cells, these ratings were binary coded to identify "very or extremely helpful" responses and Fisher Exact tests were used to test for relationships between the four learning styles. Then binary logistic regression was used to predict the helpfulness for each resource on the basis of the preferred learning styles of each student. This was done while controlling for the level of study, using a binary indicator variable for students enrolled in at least one of the Graduate Certificate units. In all cases Hosmer Lemeshow tests suggested that these models described the data well. However, the small sample size meant that the odds ratio confidence intervals were too wide to be useful.

4   Results

The *Doing* learning preference was the most popular method of learning for these students with 75 % of students selecting this option. The *Reading* preference was also popular among the students with 63 % of students selecting this option. However, *Looking/Watching* was selected by only 42 % of the students and *Hearing* was selected by only 37 % of the students. No significant relationships were found between *Reading, Hearing* and *Looking/Watching* preferences. However, the results did show an interesting relationship between *Looking/Watching* and the *Doing* preference ($p = .012$) with only 33 % of students who selected the *Doing* option also selecting the *Looking/Watching* option while 71 % of the students who did not select the *Doing* option did select the *Looking/Watching* option. As Mestre (2010) comments in relation to Active learners, sitting through lectures without getting to do anything physical is particularly difficult for students with a preference for *Doing* as opposed to *Looking/Watching*.

In our relatively young sample, mostly aged between 25 and 35, 26 % of the students selected only one learning approach with 39 % selecting two approaches, 26 % selecting three approaches and 9 % selecting all four approaches. These results suggest that a combination of learning approaches may work best for our students. Watson (2004) has provided some examples of the benefits of combining audio and visual components into tutorials, but our priority must be the combination of *Doing* with *Reading* and *Hearing* learning styles because of the popularity of the *Doing* approach among our students. No significant relationships were detected

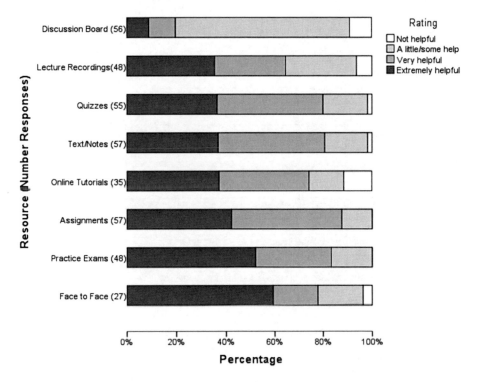

Fig. 1 Resource helpfulness

between level of study and learning preference, suggesting that, for this program, level of study does not need to be taken into consideration when designing resources to accommodate different learning styles.

Our goal in this study is to link resource satisfaction with preferred learning style in the case of the eight specific resources given in Fig. 1. The summary results shown in Fig. 1 suggest that the students tended to find all the resources very or extremely helpful, except for the discussion board. However, the relatively small number of students who accessed online tutorials (35) and face-to-face classes (27) should be noted.

Figure 2 suggests that there are interesting differences in the perceptions of resource usefulness that relate to preferred learning style. For example, 83 % of students who prefer to learn by *Looking/Watching* found recorded lectures very or extremely helpful, but this percentage dropped to 59 % and 58 % for students who preferred to learn by *Reading* or *Doing*, respectively.

Finally, binary logistic regression was used to model the Fig. 2 relationships using a very/extremely helpful response as a positive response while controlling for the level of study. The results again show no significant level of study effect for any of the resources, suggesting that any conclusions from this study can be generalised

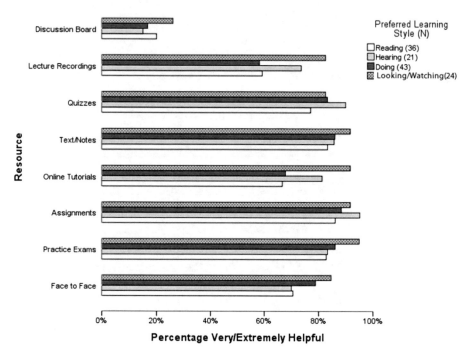

Fig. 2 Relationship between resource helpfulness and preferred learning style

across all levels of study in this program. There are also no significant effects for the *Reading* and *Hearing* learning styles suggesting that people with these preferred learning styles do not have any special preferences in regard to learning resources, but there are some significant effects for *Doing* and *Looking/Watching*.

The odds ratios for the binary logistic regression analyses are illustrated in Fig. 3. Students with a preference for learning by *Doing* appear to find text/notes 18.8 times more helpful than other students on average (Wald = 5.43, df = 1, p = .020). Students who prefer to learn by *Looking/Watching* also find text/notes particularly helpful. These students find text/notes 29.12 times more helpful than other students on average (Wald = 5.90, df = 1, p = .015). In addition students who prefer to learn by *Looking/Watching* find practise exams 24.63 times more helpful than other students on average (Wald = 3.805, df = 1, p = .051) and lecture recordings 6.83 times more helpful than other students on average (Wald = 6.83, df = 1, p = .025). However, as indicated in Figs. 2 and 3, for all the other learning resources there is consensus. Regardless of learning style preference, students find face-to-face classes/ workshops, assignments, online tutorials and quizzes very or extremely helpful, while discussion boards are found to be of limited help. This is in agreement with the findings of who also established a clear preference for practice exercises and low ratings for online discussions among online students.

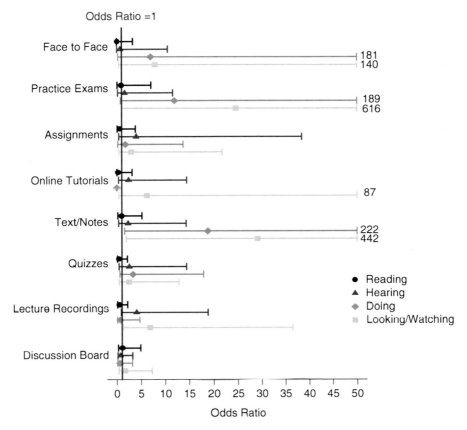

Fig. 3 Odds ratios for learning preferences with 95 % confidence intervals and upper limits above 50 given as the number at the end of the confidence interval

5   Conclusions and Future Work

This research has shown that for a small sample of applied statistics postgraduate students learning style does affect the resources that students find most helpful. This means that, ideally, the content of online courses should be stored in learning objects with a variety of formats in order to accommodate the different learning styles of students. As explained by Shaoling (2011), the sequence in which such learning objects are presented to students needs to be carefully designed, but he does not suggest how this should be done. Future work therefore needs to consider the design and sequencing of appropriate learning objects in the context of the conventional learning theories (e.g. Behaviourism, Cognitivism or Constructivism). Ertmer and Newby (1993) explain that Behaviourism is all about examples of the correct answer or way to do something with repetition and reinforcement leading to the correct response, while Cognitivism focuses more on equipping learners with effective learning strategies to process the information that they are given, and Constructivism constructs new ideas or concepts based upon current and past knowledge or experience. It is expected that the performance of resources in addressing each of the above learning approaches will differ depending on preferred learning style. For example, whereas assignments may be an appropriate learning object for producing Constructivism learning in students who prefer to learn by *Doing* this may not be true for students with other preferred learning styles. For these other students online tutorials, on-campus laboratory classes, practice exams or text exercises may be a more appropriate way to teach Constructivism learning. The challenge for educators may be in modifying learning objects in such a way that they address more than one type of learning approach. For example, it would probably improve the helpfulness of discussion boards if they could be designed in a way that facilitates Constructivism and Cognitivism learning as well as Behaviourism learning. Discussion boards are ideally suited to emphasise collaboration between peers and teachers within a supportive context; however, to support the constructivist perspective the learning needs to be personally meaningful. Gilly Salmon's (2002) e-tivities, defined as "frameworks for enhancing active and participative online learning by individuals or groups", provide the appropriate structure to achieve this as well as describing a scaffolding approach to support constructivism learning. At the same time it would be helpful if learning objects could be designed to have appeal for students with a variety of preferred learning styles. For example, the development on the discussion board of e-tivities which cater to the different learning styles by incorporating sound clips, additional reading material, videos and special exercises may be one way to achieve this.

Apart from its small size this study has several limitations. In particular it uses a simplistic "yes/no" response to determine preferred learning style. More accurate scales such as those developed by Kolb (1984) should be used in future studies. Secondly, it is possible that non-response bias has affected the results in regard to both learning preferences and resource helpfulness. Future research should investigate the likelihood of such bias.

6 Note

Refereed contributed paper presented at the Eighth Australian Conference on Teaching Statistics, July 2012, Adelaide, Australia.

This chapter is refereed.

Acknowledgments This research was funded by the Faculty of Life and Social Sciences at Swinburne University of Technology. The authors also wish to thank Cathy Pocknee and the referees for their excellent help.

References

Ertmer, P. A., & Newby, T. J. (1993). Behaviorism, cognitivism, constructivism: comparing critical features from an instructional design perspective. *Performance Improvement Quarterly, 6*(4), 50–72.

Felder, R. M., & Spurlin, J. (2005). Application, reliability and validity of the index of learning styles. *International Journal of Engineering Education, 21*(1), 103–112.

Glass, J., & Sue, V. (2008). Student preferences, satisfaction, and perceived learning in an online mathematics class. *MERLOT Journal of Online Learning and Teaching, 4*(3), 325–338. Retrieved from (http://jolt.merlot.org/vol4no3/glass_0908.htm)

Jones, C., Reichard, C., & Mokhtari, K. (2003). Are students' learning styles discipline specific? *Community College Journal of Research and Practice, 27*, 363–375.

Kolb, D. A. (1984). *Experiential learning*. Englewood Cliffs, NJ: Prentice Hall.

Larkin, T., & Budny, D. (2005). Learning styles in the classroom: approaches to enhance student motivation and learning (pp. F4D/1–F4D/8). *Proceedings of the Sixth International Conference on Information Technology Based Higher Education and Training*, Juan Dolio, Dominican Republic.

Larson, P. D. (2002). Interactivity in an electronically delivered marketing course. *Journal of Education for Business, 77*(5), 265–269.

Manochehr, N. N. (2006). The influence of learning styles on learners in E-learning environments: An empirical study. *Computers in Higher Education Economics Review, 18*, 10–14. Retrieved from http://www.economicsnetwork.ac.uk/cheer/ch18/manochehr.pdf

Mestre, L. S. (2010). Matching up learning styles with learning objects: What's effective? *Journal of Library Administration, 50*(7/8), 808–829.

Pritchard, A. (2009). *Ways of learning: learning theories and learning styles in the classroom* (2nd ed.). London: David Fulton.

Salmon, G. (2002). *E-Tivities: The key to active online learning*. London: RoutledgeFalmer.

Shaoling, D. (2011). Using learning with styles to implementing personalized e-learning system (pp. 1–4). In *Proceedings of the 2011 International Conference on Management and Service Science (MASS)*, Wuhan, China.

Watson, J. (2004). Going beyond screen captures: Integrating video screen recording into your library instruction program. *Feliciter, 50*(2), 66–67.

Witkin, H. A., Moore, C. A., Goodenough, D. R., & Cox, P. W. (1977). Field-dependent and field-independent cognitive styles and their educational implications. *Reviews of Educational Research, 47*, 1–64.

Zacharis, N. Z. (2011). The effect of learning style on preference for web-based courses and learning outcomes. *British Journal of Educational Technologies, 24*(5), 790–800.

Problem-Based Learning of Statistical Sampling Concepts Using Fantasy Sports Team Data

Robert Brooks, Ross Booth, Jill Wright, and Nishta Suntah

Abstract This paper explores a case of using data from a fantasy sports competition (the AFL Dream Team competition) to teach the core concepts in statistical sampling and the central limit theorem as they apply to problems of inference regarding the population mean.

Keywords Problem-based learning • Statistical sampling • Fantasy sports

1 Introduction

The core curriculum of the undergraduate business degree typically includes at least one unit in introductory business statistics or quantitative analysis. A summary of the unit offerings in the US setting is provided by Haskin and Krehbiel (2012) who document that all of their top US schools require a course in statistics, although less than half require a second course in statistics. Despite the important role of base statistics knowledge in the undergraduate business curriculum, the challenge remains in teaching an introductory statistics course that is engaging and interesting to students who do not wish to major/specialise in statistics while also developing

R. Brooks (✉)
Department of Econometrics and Business Statistics, Faculty of Business and Economics, Monash University, PO Box 197, Caulfield East, VIC 3145, Australia
e-mail: robert.brooks@monash.edu

R. Booth
Department of Economics, Monash University, Clayton, VIC 3800, Australia
e-mail: ross.booth@monash.edu

J. Wright • N. Suntah
Department of Econometrics and Business Statistics, Monash University,
Berwick 3806, Australia

H. MacGillivray et al. (eds.), *Topics from Australian Conferences on Teaching Statistics:*   397
OZCOTS 2008-2012, Springer Proceedings in Mathematics & Statistics 81,
DOI 10.1007/978-1-4939-0603-1_22, © Springer Science+Business Media New York 2014

their understanding of key statistical concepts and improving their statistical literacy. Yilmaz (1996) discusses these issues at a broad level as a way of informing course design.

Sampling distributions are a pivotal concept in any elementary statistics course, in the following sense. An understanding of sampling distributions is essential in order for a student to comprehend how inferential statistical techniques work. However, a number of more elementary concepts must be mastered and combined in order to understand sampling distributions themselves. Thus, as is discussed comprehensively by Chance et al. (2005), the teaching of sampling distributions must be considered in the context of: required prerequisite knowledge; the sampling distribution ideas that the student must absorb and what students can then do with that knowledge. They provide tables listing items under each of these headings and also include a table of common student misconceptions.

Various approaches have been adopted to facilitate a more engaging and interesting approach to statistics education. A range of interactive simulation activity-based and problem-based learning (PBL) approaches have been used in teaching in the sampling distributions area. Interactive simulation approaches are detailed in del-Mas et al. (1999) and Aberson et al. (2000). Activity-based approaches are detailed in Dyck and Gee (1998), Schoenfelder et al. (2007) and Matz and Hause (2008). PBL approaches are detailed in Budé et al. (2011). In essence these approaches involve combinations of computer simulation work, students working in small groups and student working on an inquiry or PBL approach. The present paper outlines an approach adopted in the introductory business statistics unit taught in a PBL-based approach in the Bachelor of Business at the Peninsula campus of Monash University. The paper focuses on an approach to the understanding of sampling distributions and the central limit theorem in this course and makes use of data from the Australian Football League (AFL) in which to provide a real world context for the problem. The use of sports statistics is a common way of improving the relevance of teaching introductory statistics courses and Albert (2002) details an entire statistics course taught using baseball statistics.

The AFL data is an example of a fantasy sports game. Fantasy sports games involve the use of actual sports statistics to create games for fans and other players. In the US setting this is managed by the Fantasy Sports Trade Association (FSTA). According to the FSTA website there are over 32 million fantasy sports players in the US and Canada. In the US context the existence of fantasy sports leagues has been found to impact match attendance for American football (see Nesbit and King 2010a) and American baseball (see Nesbit and King-Adzima 2012), and television viewing for both American football and baseball (see Nesbit and King 2010b), thus the problem has a real business context for sporting leagues.

In the context of Australian Rules Football the AFL runs a competition called Dream Team in which players select a fantasy team and players score points based on actual match statistics. The plan of this paper is as follows. Section 2 details the PBL approach adopted for the teaching of this problem in this unit. Section 3 contains some concluding remarks.

2 Problem-Based Learning Using AFL Dream Team Data

We now focus on the first implementation of PBL in the core statistics unit of the new PBL degree program at Monash University, Peninsula campus. Students were enrolled in a core business statistics unit in the first semester of their course. One full day per week for one 12-week semester was devoted to this core unit. Each week, the day began with a presentation by the lecturer lasting up to 1 h. This consisted of an introduction to the statistical topic for that week and a description of the problem that the students were to tackle. The students then worked on the problem in groups of four for several hours, during which time they had access to a tutor when questions arose. The day concluded with each group giving a joint presentation of its findings.

Where possible, one set of data was used over several weeks, so that students did not use too much time becoming familiar with the context. In particular, the AFL Dream Team data was used for the discussion of the normal distribution in week 5, sampling distributions and the central limit theorem in week 6, confidence intervals and hypothesis testing in weeks 7 and 8, respectively.

The 'real' business problem in this setting consists in understanding the statistical properties of the data underlying a fantasy sports game, namely the AFL Dream Team competition. In this game, players select teams consisting of 33 players, drawn from the whole pool of 742 players in the 16 clubs of the league. The players score points based on their actual performance in the games that occur throughout the season. The fantasy teams are then ranked according to the total points scored by the 33 players.

The discussion of sampling and the central limit theorem was undertaken in week 6 of the semester. In earlier weeks, the prerequisite concepts of populations and random samples, parameters and statistics, random variables and their distributions and normal distributions had been covered. In the instructor's introduction to the day's activities, the concept of the distribution, for a given sample size, of all possible sample means had been introduced. The planned learning activity was for students to repeatedly select random teams and build up the sampling distribution of the average player scores for these random teams. To enable an understanding of central limit theorem the students simulated random teams of different sizes, even though the AFL Dream Team competition is a team size of 33. Students were encouraged to observe and comment on characteristics of these distributions. By observing the actual drawing of the samples, they could obtain an intuitive understanding of the mechanism whereby variability of the sample means became less for larger sample sizes.

The population for analysis is the player scores for the 2010 AFL Dream Team competition. A histogram of this data is shown in Fig. 1, and it exhibits certain unusual characteristics. The histogram shows a distribution that is non-normal with a large number of player scores in the lowest class. In fact most of these scores are zero, as 161 of the 742 players in the league did not play an AFL game in 2010 and thus did not score any points. The histogram is right skewed with a small number of

Fig. 1 Population of AFL dream team player scores for 2010 (mean = 751.09, standard deviation = 711.29)

higher scoring players in the right tail. From this population students then draw random samples of various sizes using the random number generation functions in Excel. Hardy (2002) details an approach for generating repeated random samples in Excel and using these repeated samples to teach the central limit theorem and confidence intervals. While Hardy (2002) uses particular statistical functions in Excel to define the underlying population distribution, we make use of real Dream Team data to represent the underlying population distribution. Real data are seen as desirable in statistics classes because they introduce students to the real statistical problems of skewness and multi-modality as identified in Gould et al. (2006). This is a significant advantage over simulated data which produce smoother underlying distributions.

Each student in the class can then use a random sampling approach to produce a unique collection of random samples. Further, by then pooling the random samples across students this approach enables students to cumulatively build up the sampling distribution of the sample mean across the class. In order to illustrate the behaviour of sampling distributions for varying sample sizes, students were asked to find samples of sizes $n = 16, 25$, and 36. For our small class size of eight students aggregating across the class, they obtained 100 samples of each size. The histograms for different sample sizes are presented in Figs. 2 and 3, respectively. These histograms show a greater peakedness about the mean as the sample size increases consistent with the theoretical predictions around variability falling as sample size increases.

The figures also report the mean and standard deviation of the population as whole, as well as the mean and standard deviation of the 100 samples for each of the three sample sizes. These values can then be compared to the theoretical values expected in a repeated sampling case. We find that the mean value for the 100 samples differs from the overall population mean. This is due to both the sample size and the number of samples taken. In the context of our real AFL Dream Team data the skewness is quite extreme because of the concentration of zero observations and the exploration of different sample sizes illustrates how larger samples are needed in the case of more extreme population data. The PBL setting also enables students to grapple with complications in the problem. These are more tractable because the

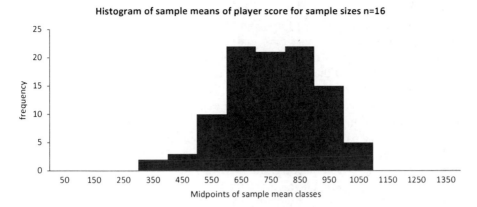

Fig. 2 Histogram for average player score for 100 randomly selected player scores (sample size: $n = 16$, mean = 756.94, standard deviation = 154.88)

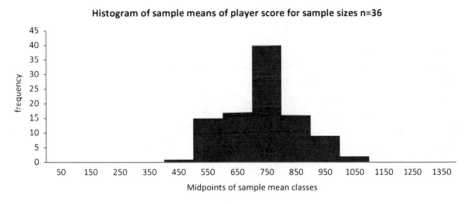

Fig. 3 Histogram for average player score for 100 randomly selected player scores (sample size: $n = 36$, mean = 744.49, standard deviation = 124.58)

real data setting makes them more concrete. For instance, in the present problem students may wish to move to larger sample sizes where the convergence to normality would be clearer. However, from a population of 742 player scores in the AFL Dream Team moving to larger sample sizes potentially raise the need to make finite population correction factors in the calculation of the standard deviation.

Overall our results are comparable to those reported in Vaughan (2003), and a larger number of samples than 100 is needed to approach the true population mean. We also find the standard deviations differ from their theoretically expected values, again due to both the sample size and the number of samples taken. Thus, we again find that a larger number of samples than 100 are needed to approach the theoretically expected values. With a larger class size a greater number of samples could be pooled across students to also illustrate the importance of the number of samples.

In an assessment setting, students were required to deal with a range of group and individual learning tasks. Their formal end of semester examination then presented the histograms from Figs. 1 through 3 and the summary statistics and asked:

Based on the histograms and summary statistics explain what is happening to the sampling distribution of the sample mean as the sample size increases.

In addition to using the 2010 AFL Dream Team data to study sampling distributions the students and staff involved in the unit and program were participants in the actual 2011 AFL Dream Team competition. Some participants played the game fully aiming to make player trades and optimise their teams each week, while others picked favourite players or teams, and other players used randomly generated teams from the random selection function built in to the game website. The top performing participants are those who played the game with the aim of optimising teams thus showing the returns associated with the 'skill' in the game. The randomly selected teams are generally in the middle performing group of participant teams.

3 Conclusion

This paper explores how a PBL activity can be used in teaching one of the technical components of an introductory business statistics unit. The paper illustrates how combining group work and sample simulation-based approaches can produce an interactive learning approach to the teaching of a 'real' problem. With a larger group of students it would be possible to present a formal comparison of student statistical literacy and understanding of key concepts comparing across students taught in a conventional lecture/tutorial approach versus a PBL approach. This task is planned for future semesters.

4 Note

Contributed paper presented at the Eighth Australian Conference on Teaching Statistics, July 2012, Adelaide, Australia.

Acknowledgment The authors wish to thank Ian Weeding and the Australian Football League for the provision of the Dream Team data for use in the teaching program.

References

Aberson, C., Berger, D., Healy, M., Kyle, D., & Romero, V. (2000). Evaluation of an interactive tutorial for teaching the central limit theorem. *Teaching of Psychology, 27*, 289–291.

Albert, J. (2002). A baseball statistics course. *Journal of Statistics Education, 10*. Retrieved from http://www.amstat.org/publications/jse/v10n2/albert.html

Budé, L., van de Wiel, M., Imbos, T., & Berger, M. (2011). The effect of directive tutor guidance on students' conceptual understanding of statistics in problem-based learning. *British Journal of Educational Psychology, 81*, 309–324.

Chance, B., delMas, R., & Garfield, J. (2005). Reasoning about sampling distributions, chapter 13. In D. Ben-Zvi & J. Garfield (Eds.), *The challenge of developing statistical literacy, reasoning and thinking.* Dordrecht, the Netherlands: Kluwer Academic.

delMas, R., Garfield, J., & Chance, B. (1999). A model of classroom research in action: Developing simulation activities to improve students' statistical reasoning. *Journal of Statistics Education, 7.* Retrieved from http://www.amstat.org/publications/jse/secure/v7n3/delmas.cfm

Dyck, J., & Gee, N. (1998). A sweet way to teach students about the sampling distribution of the mean. *Teaching of Psychology, 25*, 192–195.

Fantasy Sports Trade Association. http://www.fsta.org/blog/fsta-press-release/fantasy-sports-participation-sets-all-time-record-grows-past-32-million-players

Gould, R., Kroutei, F., & Palmer, C. (2006). Towards statistical thinking; making real data real. *Refereed paper to the international conference on teaching statistics ICOTS-7 2006.*

Hardy, M. (2002). Repeated simulated sampling in Excel as a tool for teaching statistics. *Journal of Computing Sciences in College, 17*, 167–174.

Haskin, H., & Krehbiel, T. (2012). Business statistics at the top 50 US business programmes. *Teaching Statistics, 34*, 92–98.

Matz, D., & Hause, E. (2008). "Dealing" with the central limit theorem. *Teaching of Psychology, 35*, 198–200.

Nesbit, T., & King, K. (2010a). The impact of fantasy football participation on NFL attendance. *Atlantic Economic Journal, 38*, 95–108.

Nesbit, T., & King, K. (2010b). The impact of fantasy sports on television viewership. *Journal of Media Economics, 23*, 24–41.

Nesbit, T., & King-Adzima, K. (2012). Major league baseball attendance and the role of fantasy baseball. *Journal of Sports Economics, 13*, 494–514.

Schoenfelder, E., Olson, R., Bell, M., & Tom, K. (2007). Stop and smell the roses: An activity for teaching the central limit theorem. *Psychology Learning and Teaching, 6*, 80–84.

Vaughan, T. (2003). Teaching statistical concepts with student-specific datasets. *Journal of Statistics Education, 11.* Retrieved from www.amstat.org/publications/jse/v11n1/vaughan.html

Yilmaz, M. (1996). The challenge of teaching statistics to non-specialists. *Journal of Statistics Education, 4.* Retrieved from http://www.amstat.org/publications/jse/v4n1/yilmaz.html

BizStats: A Data and Story Library for Business Statistics

Howard Edwards, Sarah Edwards, and Gang Xie

Abstract Undergraduate degrees in Business and Commerce usually have a compulsory paper in Business Statistics or similar, and many students who enrol in these papers are apprehensive or sceptical about studying statistics and may be unaware of the need for and relevance of statistical thinking and statistical methodologies in business. This paper describes the results to date of an ongoing project to identify and gather data sets which are relevant to business planning and decision making within a New Zealand context and which can be used to: (a) illustrate the types of statistical methods that are commonly introduced in an introductory statistics paper; and (b) provoke statistical thinking within a business setting. The project is envisaged to be an ongoing one and the data and story library will be made freely accessible on a website.

Keywords Data • Business • Statistics education • New Zealand • Data-and-story library • DASL

H. Edwards (✉) • S. Edwards
Massey University at Albany, North Shore Mail Centre,
PO Box 102904, Auckland, New Zealand
e-mail: h.edwards@massey.ac.nz; sarahjoyedwards@gmail.com

G. Xie
CRC for Infrastructure and Engineering Asset Management (CIEAM),
Queensland University of Technology, O413 O Block, Gardens Point campus,
2 George Street, Brisbane, QLD, Australia
e-mail: masseyxie@gmail.com

H. MacGillivray et al. (eds.), *Topics from Australian Conferences on Teaching Statistics:*   405
OZCOTS 2008-2012, Springer Proceedings in Mathematics & Statistics 81,
DOI 10.1007/978-1-4939-0603-1_23, © Springer Science+Business Media New York 2014

1   Introduction

Most undergraduate degree programmes in Business, Management and Commerce have a compulsory requirement in Quantitative Methods such as Business Mathematics and/or Business Statistics. For example, the Association to Advance Collegiate Schools of Business (AACSB) accreditation process requires that undergraduate business majors should possess a basic understanding of mathematics and statistics, and all eight New Zealand universities have a compulsory paper (or its equivalent) in Quantitative Methods or Business Statistics in their core Business/Management/ Commerce degree. In most cases this topic is covered in the first year of degree study.

Many students entering these programmes may have not studied much Mathematics or Statistics at high school level. For example, a survey of all students enrolled in the Statistics for Business paper offered at Massey University's Auckland campus in Summer 2011/2012 reported that 36 % of students rated their background knowledge of Statistics on entry as poor—18 % had not studied mathematics beyond Year 11 in New Zealand and 19 % had studied no Statistics at high school level overseas. These students are typically over-represented in self-help tutorial or clinic classes.

Even among students who have recent background study in these areas, however, there may be a lack of understanding or appreciation of the importance of statistical thinking and statistical literacy in business and decision making. McAlevey and Stent (1999) reported that students who had completed an introductory course believed that the first year was too early in their Business course to appreciate the relevance of Statistics. Coleman and Conrad (2007) found that students in their study were significantly less likely to rate a Statistics or research methods course as helpful to their career compared to non-Statistics-based courses.

One way to assist students in perceiving relevance is to provide examples, exercises, questions and case studies based on real business data. Cryer (2002) reviewed the principal conclusions of 17 annual U.S. Conferences called Making Statistics More Effective in Schools and Business, which included using real data (preferably hands-on), and that students are more effectively motivated by seeing statistics at work in solving real problems. John and Johnson (2002) state that "…students are most effectively motivated by seeing statistics at work in real applications, problems, cases, and projects". Although there are many textbooks on Business Statistics available which use real data, the authors of this paper chose to focus on collecting business datasets in a New Zealand context.

2   Background to the Project

Massey University is New Zealand's only multicampus and multimodal university, offering degree and diploma programmes from campuses in Auckland, Palmerston North and Wellington together with distance mode teaching. The core undergraduate Business programme is the Bachelor of Business Studies which includes a compulsory course in Business Statistics which is offered at all three campuses and in distance mode. Up until 2010 various texts in Business Statistics had been used in this course, but in 2010 it was decided to abandon the traditional published textbook from 2011 onwards and replace it with the ebook CAST (Stirling 2001). See Edwards et al. (2011) for a discussion of CAST and its usage in this course.

CAST is able to be customised for a particular course (the CAST website lists ten publicly available customised ebooks), but the versions for introductory courses all share a common set of examples and exercises spanning a range of disciplines. Therefore, it was decided to instigate a project to gather a range of business-related datasets and cases that could be used in conjunction with CAST. Copyright and confidentiality considerations meant that locating data that was publically available was seen as a priority.

The first places to be searched were the online Data and Story Libraries (DASL for short). A DASL houses datasets (usually in text format) together with information about the dataset: a description of the data including the variables in the data set, the context in which the data was collected and a listing of statistical methods that might be used to summarise or analyse the data (the story). Webpages and links allow the user to list datasets by subject or by statistical application. The original DASL was developed at Cornell University and is currently hosted at Carnegie-Mellon University as part of the StatLib project (see e.g. Witmer 1996), and the Australasian version OzDASL is hosted as part of the StatSci project courtesy of the Walter and Eliza Hall Institute of Medical Research (Smyth 2011). While both sites have some business or business-related datasets, there are only one or two datasets that were sufficiently relevant and up to date to be considered for inclusion.

This led to the idea of creating our own DASL (perhaps a library branch is a better description!) of New Zealand business datasets that could be used by anyone but targeted at Zealand Business Statistics courses.

3 The BizStats Data and Story Library

The BizStats DASL currently contains 40 datasets, collected from various sources including data websites, published reports and journal articles and individuals. Subject areas include Communications, e-Commerce, Economics, Finance, Management, Marketing, Real Estate and Tourism and application areas include descriptive statistics, categorical data analysis (tests of independence and goodness of fit), simple and multiple linear regression and time series analysis.

Figure 1 below shows a typical story file. Each "data-and-story" group has a title (note there may be more than one dataset associated with a given story), a description which includes descriptions of each of the variables in the dataset(s), the name(s) of the dataset(s) to be downloaded, a source for the data and a summary of possible analyses that could be carried out using the data.

The first stage of data collection focussed on webpages of government departments and private organisations such as banks and insurance companies. The second stage (still in progress) is focussed on scholarly articles and publications in Business, Economics and Management journals. Much of the data is longitudinal (i.e. time series data) which will require updating on an ongoing basis.

Title: The 2001 Household Savings Survey (HSS).

Description: The 2001 Household Savings Survey (HSS) was a cross-sectional nationwide survey that collected information on the net worth (assets minus liabilities) of New Zealanders. It was a one-off survey commissioned by the Retirement Commission and conducted by Statistics New Zealand. The survey results provide factual information to feed into the public debate on superannuation, and raise public awareness of retirement issues. This activity was first released in 2008.

The survey population for the HSS was the usually resident population of New Zealand, aged 18 years and over, living in permanent private dwellings on the North, South, and Waiheke Islands. The total achieved sample size for the HSS was 5374 households. This sample was made up of a core sample (from 6600 households that were approached) and an additional 6600 households that were screened for Maori residents, to provide a Maori booster sample. In both cases all households were randomly selected. The overall response rate was 74 percent. This data set contains 300 records and 10 variables based on data from the 2001 Household Savings Survey (HSS). These records are not about real people, but were generated using statistical techniques to have similar characteristics as respondents to the survey. This modification helps Statistics New Zealand prevent any unintentional disclosure of data about an individual's personal information. This data set is prepared in the form of a data frame of 301 rows by 11 columns.

ID:	Personal ID index.
Gender:	Male or female.
Employ:	Status of employment, E = individual is currently employed; NotE = individual is not currently employed.
Qualification:	The highest qualification achieved, None = No qualification; School = School qualification; Vocation = Vocational or trade qualification; Degree = Bachelor's or higher degree.
Ethnicity:	European, Maori, or Other.
Age:	Age of individual.
AgeP:	Age of individual's partner (where applicable). AgeP = 0 implies that no partner.
Income:	Total annual income from all sources (includes income of partner where applicable). Rounded to nearest $1000.
WageSalary:	Income from wages and/or salary for the individual. Rounded to nearest $1000.
Debt:	total debt from all sources (includes debt of partner where applicable). Rounded to nearest $1000.
Networth:	Total net worth (includes debt of partner where applicable). Rounded to nearest $1000.

Download: savingsurvey.txt

Source: http://www.stats.govt.nz/tools_and_services/services/schools_corner/surf for schools.aspx, 2010.
Analysis: Descriptive data analysis, regression models, ANOVA, hypothesis tests, contingency table data analysis. This is a valid random sample data set containing both categorical and numeric variables. This data set is suitable for a range of statistical analyses and the results may be used to infer the population in New Zealand.

Fig. 1 The "Story" file for the dataset savingsurvey.txt

4 Massey University Experience: To Date

Some of the datasets have been introduced to the Statistics in Business paper at the Auckland campus, firstly over the Summer School offering in 2011/2012 and secondly in the Semester 1 2012 offering. Material production deadlines meant that it was not possible to incorporate datasets into the custom version of CAST; however, datasets were used in teaching materials as examples and exercises, and also in workshops where students work through a directed exercise in groups of two or three. Figure 2 below shows part of the workshop relating to linear regression and time series modelling.

The scatterplot below shows the monthly New Zealand median house price from January 1992 to October 2010. Using the scatterplot, describe the relationship.

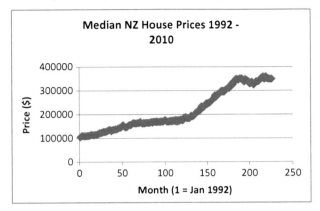

The last page gives the relevant simple linear regression model output.

Write down the least squares regression equation for predicting NZ median house price based on the month and draw it on the scatterplot. (You may round each coefficient to the nearest whole number).

Interpret the values of the y-intercept (b_0) and the slope (b_1).

Examine the residual plots – have the assumptions of regression been met?

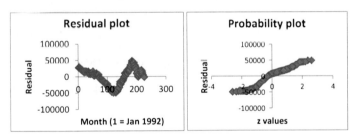

Fig. 2 Workshop exercises using the dataset houseprice.txt

Anecdotal evidence suggests that some students have found these datasets to be helpful in aiding their motivation for and understanding of statistical thinking and statistical methodologies in Business Studies. An online questionnaire will be made available on the class Moodle site at the completion of Semester 1 2012.

5 BizStats Website

The current working version of the BizStats DASL is available at http://bizstats. massey.ac.nz and it is intended that the contents of the DASL will eventually be available on OzDASL.

6 Note

Contributed paper presented at Eighth Australian Conference on Teaching Statistics, July 2012, Adelaide, Australia.

References

Coleman, C., & Conrad, C. (2007). Understanding the negative graduate student perceptions of required statistics and research methods courses: implications for programs and faculty. *Journal of College Teaching and Learning, 4*, 11–20. Littleton, CO: The Clute Institute.

Cryer, J. D. (2002). A review of the lessons learned at the conferences on making statistics more effective in schools and business. In *ICOTS6: Proceedings of the sixth international conference on teaching statistics*, Cape Town, South Africa. IASE.

Edwards, H. P., Fitch, A. M., Heath, J. M., Jones, B. M., Jones, G., Leader, D. (2011). The use of the interactive e-book CAST in a first year Business Statistics paper (pp. 86–92). In *The eighth southern hemisphere conference on the teaching and learning of undergraduate mathematics and statistics: Conference proceedings*. Christchurch, New Zealand: University of Canterbury.

John, J. A., & Johnson, D. G. (2002). Statistical thinking for effective management. In *ICOTS6: Proceedings of the sixth international conference on teaching statistics*, Cape Town, South Africa. IASE.

McAlevey, L., & Stent, C. (1999). Undergraduate perceptions of teaching of a first course in business statistics. *International Journal of Mathematical Education in Science and Technology, 30*, 215–225. London, England: Taylor & Francis.

Smyth, G. K. (2011). *Australasian data and story library (OzDASL)*. Retrieved from http://www.statsci.org/data

Stirling, W. D. (2001). *CAST: Textbooks for learning statistics*. Retrieved from http://cast.massey.ac.nz

Witmer, J. A. (1996). DASL—The data and story library. *The American Statistician, 51*, 97–98. Retrieved from http://www.jstor.org/stable/2684709

Statistics Training for Multiple Audiences

Emma Mawby and Richard Penny

Abstract Statistics New Zealand is the official statistics agency in New Zealand. Within Statistics NZ there are a diverse range of statistical knowledge needs ranging from basic statistical knowledge to advanced statistical skills. Also Statistics NZ as the leader of the Official Statistics System (OSS) has been given the task of coordinating the development of better statistical skills within all of government. Finally it is in the interests of Statistics NZ to increase the statistical skills of the users of its data outputs. While diverse, there are many commonalities in terms of the needs within parts of each audience. The methodology section of Statistics NZ is working on how we can develop training resources for methodological skills that can be easily modified and reused for different audiences. Also how can we effectively find, assess and use the increasing amount of online training resources available and collaborate on their development?

Keywords Statistics teaching • Statistics learning • Training resources • Official statistics

E. Mawby
Statistical Methods, Statistics New Zealand, P.O. Box 2922,
Wellington 6140, New Zealand
e-mail: emma.mawby@stats.govt.nz

R. Penny (✉)
Statistical Methods, Statistics New Zealand,
Private Bag 4741, Christchurch 8140, New Zealand
e-mail: richard.penny@stats.govt.nz

H. MacGillivray et al. (eds.), *Topics from Australian Conferences on Teaching Statistics:*   411
OZCOTS 2008-2012, Springer Proceedings in Mathematics & Statistics 81,
DOI 10.1007/978-1-4939-0603-1_24, © Springer Science+Business Media New York 2014

1   Introduction

Statistics New Zealand is involved in improving the statistical skills of its own staff, other government departments and its clients. This means that Statistics NZ has to manage a diverse set of statistical training needs. These arise from the variety of skills and knowledge to be acquired and the range of learning styles, prior knowledge and perceived outcomes for learners. Statistics NZ staff are based in three cities, Wellington, Auckland and Christchurch, across the two main islands of New Zealand.

Statistical Methods (SM) is the group within Statistics NZ that provides the methodological statistical support. Most SM staff have mathematics or statistics degrees, though very few have much knowledge of data collection and analysis before they join SM. SM needs to provide ongoing training for all SM staff. SM is not responsible for more general statistics training which is co-ordinated by a separate Statistical Education team (SE).

Few SM staff have much experience in providing training, though all have consumed large amounts. Also there is not much time set aside to assess available training resources and develop courses. We tend to provide training that is similar to what was done in the past or which reflects how the trainer was trained, as discussed in Martin (2006).

In the next section of the paper we discuss training specifically for SM staff. We provide the results of a small survey of SM staff which elicits their approaches to learning different topics to see what they did or preferred. The following section covers the issues associated with providing statistics training to a range of clients within Statistics NZ and externally. Next we discuss our view that there are more similarities in teaching statistical ideas and concepts to any person, irrespective of their prior experience, than differences, and the potential this gives to building and acquiring teaching resources. We end the paper with a summary and some tentative conclusions.

2   Training the Statistical Methods Division

Since June 2011, the number of roles in SM has increased from 45FTE to 56FTE. This increase is due mainly to the need to support an organisation-wide programme of development called Statistics 2020. Also the training needs of new recruits have changed as discussed in Bagherzadeh and Giovannini (2006). Previously most new recruits entered SM shortly after completing undergraduate study in statistics or a statistics-related field. Recently analysts have been recruited from other parts of Statistics NZ, other New Zealand organisations and overseas.

Analysts who are new to SM undertake IN2SM. This two-part 7-day training course is intended to equip them to complete standard methodology tasks independently. IN2SM is delivered mainly in lecture style, with trainer and learners in the same location. Practical exercises make up a small part of the course. The cost of travel means that the choice of trainer is sometimes restricted and turnover of trainers leads to some revision of the course material every time it is delivered. The material for these courses is decided upon and maintained by the methodological "networks" within SM; groups of analysts who have knowledge and interest in areas such as survey design, time series and editing and imputation.

Some networks offer more advanced courses, such as the Introduction to Design and Estimation in Practice (IN2DEEP). This is a 5-day course that has been run by the Business Survey Methodology Network. These courses tend to be run irregularly and infrequently as they are needed less often and can be of a more practical nature (and therefore costly compared to other methods of learning). We estimate that an SM analyst spends around 5 % of their first 2 years in SM completing training activities.

SM also offers weekly methodological seminars, conducted by videoconference. These are usually presented by SM analysts who spend up to an hour describing an interesting aspect of their recent work. They are open to learners from outside SM. A 3-day Professional Development Offsite course for all SM staff is run every 18 months. This involves some presentations from people external to SM. The format is predominantly lecture style with some workshop-style practical activities and all learners are gathered in the same location.

In April 2012 we conducted a survey of analysts working in SM (see "Appendix" for details). We received 25 responses from the 61 members of SM, a response rate of 41 %. About two thirds of the respondents had worked in SM for less than 5 years. We asked how they used the sources of learning that were available to them, not what sources they would like to use if they were available.

The three most used methods of learning in SM were:

- "Another written source (e.g. book, internet)"
- "Analyst who previously did the task" and
- "Another analyst"

Most analysts preferred to learn about existing statistical methodologies from another methodologist;

> I...like learning from other people, in a coach-like arrangement. People with experience can provide advice and support that can't be found in other sources of information, particularly if the task or problem is unique to Statistics NZ business. I think it would be good if this kind of thing was a bit more formalised than it is now. (Analyst with <5 years experience)

Notes from Statistics NZ courses were used as a source of reference by about half of all respondents, mainly for finding out about new (to the analyst) subjects. However, this method was not favoured for learning about existing statistical methodologies. Analysts preferred to do this by working with colleagues or by referring to another written source, such as survey documentation.

> Group discussions (e.g. in network meetings discussing a paper everyone has read ahead of time) or group problem-solving workshops to learn something new (e.g. how to do a particular task in R or SAS) I find more interesting and engaging than the lecture format that we tend to use for seminars and training courses. (Analyst with ≥5 years experience)

SM staff members seemed to favour multiple different ways of learning about software and respondents were more likely to use online learning (such as YouTube clips) for this type of task than any other task.

> I've used online courses…to learn how to do different things in SAS (analyst with <5 years experience)
>   We need more online stuff, both the best from external as well as much we create ourselves in Statistics NZ. (Analyst with ≥5 year's experience)

Respondents commonly reported their preference for "learning at their own pace" through repeatable training material, e.g. examples or exercises from in-house courses or through coaching or less formal discussion.

Few analysts used people outside of Statistics NZ as a source of learning, although one experienced analyst commented;

> If I have a specific question which no-one in SM (or that I know in wider Statistics) can help with, then I would contact someone from ABS or ONS if relevant. (Analyst with ≥5 years experience).

3 Training Outside Statistical Methods

The main responsibility for general in-house and external technical training belongs to SE. SE organises courses to raise statistical capability. Some of these, such as *Public Sector-Decision Making with Official Data,* which involves a workplace-based statistical research project, lead to qualifications through standards-based assessment. In 2011, an Official Statistics postgraduate paper was offered at multiple New Zealand universities. Practising experts from around the country (including an SM analyst) gave lectures via videoconference. Students were assessed on five written assignments that were not just doing official statistics, but also writing about them. These and other external training initiatives are discussed in Forbes et al. (2014).

SM is "contracted" by SE to provide in-house courses on various official statistics methods (e.g. confidentiality, editing & imputation) as well as a course on Statistical Thinking (using the Wild and Pfannkuch definition of Statistical Thinking). There is overlap between the material in these courses and IN2SM. However, the audience for these courses generally consists of learners from survey output areas, IT personnel and field interviewers, most with little or no university statistical training. As a result we have had many difficulties in getting this training to work as effectively as it should. SM's current lecture style presentations with some learner interaction means we tend to have a "convoy approach" to training. SM trainers present a single standard course designed for a particular level of prior knowledge and a particular style of learning. Some learners struggle to understand while others find the presentations too simple. The timing of these courses means that learners nominated for a course cannot always attend due to work demands (e.g. imminent data release). While group instruction is efficient for SM and its trainers, it does not always work well for learners.

Learners are often highly motivated to learn if they can see the immediate use for the material. However, we have observed that learners taking courses run by SM do not always absorb what we would regard as the key learning. Instead they absorb a set of rules and processes to apply to their work. It is often difficult for learners and trainers to assess the value of learning until some months after the course.

SM has limited knowledge in developing, maintaining and enhancing training courses. The people within SM are not specifically employed for their previous teaching experience, so presenting training outside SM can be the first time the trainer has provided statistical training. We also cannot guarantee that any person will present a course more than three or four times.

The methods of eliciting feedback about training from learners do not always evaluate the effectiveness of the mode of training. The World Wide Web holds some very useful resources to improve and enhance SM courses, but finding and integrating them into our training requires knowledge and time. It is necessary to have a range of resources available so that there is something that matches a learner's knowledge, skill level and learning style with the level technical knowledge the learner requires. It may benefit both learner and trainer if we could have some formal assessment and accreditation for this learning, as training is often perceived as a more valuable activity if it results in accreditation.

The key is building a suite of resources that learners can use when they wish, on topics they want to learn and with the approach to learning that best suits them. With this set of resources SM can use trainers for those aspects of training where one-on-one is best, perhaps more acting as a guide, coach and mentor. As can be seen from our small survey consulting someone seen as knowledgeable is still the preferred way to learn. This suite of statistical training resources could also be used as a basis to improve statistical knowledge externally, both in government and for our data users. It would assist SM, as well as statistical decision makers in government and elsewhere, to have a deeper and broader set of learning opportunities.

4 A New Approach

We should find or develop resources that enable the less experienced trainers to deliver effective training using different modes of delivery. SM trainers can then focus on providing assistance to learning rather than feeling that they need to be "the expert" as discussed in Jorner (1990) and Martin (2006).

Another benefit of having a range of resources is that it will enable more "just in time" training. So when a learner needs to learn something and has the time to learn, the resources are available as described in Talbot et al. (1998). We need to ensure that we have mentors and coaches available to support learners. Many people could help, as much of this learning is provided "on the job". How many of us work out most of our firm's administrative procedures from the instructions? We generally ask someone who knows a lot when we get stuck!

We can see an opportunity to develop resources that meet the needs of many groups of learners. Much statistical training in formal training institutions is "bridging training", targeted at learners in the workforce. SM should use the knowledge being built up in this area as well as supporting it. SM is committed to raising levels of statistical knowledge. We could be a useful supporter and willing party to testing and evaluating resources on people other than full-time tertiary students.

5   Summary and Conclusions

We believe that the mode of delivery is as important as course content. We need to find a better way of delivering improved courses for learners who are at the start of their careers by completely re-engineering our approach to work-based training. IN2SM in its present form seems to fulfil the need to "get people going" but the value of having notes to refer to in the future seems limited. Also "just in time" learning can be excellent for fulfilling an immediate need to complete an operational task, but is less suitable if knowledge needs to be retained and cross-linked. As we move towards new methods of collecting data, e.g. using administrative data and online collection of survey data, making cross-links will become even more important.

We are looking to use resources developed for universities and schools to improve our training using the co-operative approach suggested by Wild (2006). This benefits both groups as we can provide input into developing practical training resources that can be seen to be relevant to students.

6   Note

Contributed paper presented at Eighth Australian Conference on Teaching Statistics, July 2012, Adelaide, Australia.

Appendix: Questionnaire About Training Habits Sent to SM Analysts in April 2012

Information gathering for OZCOTS paper—"Statistics Training for Multiple Audiences"

We are interested in how SM analysts fill the gaps which exist in our present training programme: i.e. what does an analyst do when they begin a new task and find that they need some new knowledge to complete it?

Please complete the [6x11] matrix below, ticking all situations that apply

Method of learning	Talk to					Refer to					
Type of task	Analyst who previously did the task	Another analyst	Senior/principal methodologist	Someone outside SM but inside Statistics NZ	Someone outside Statistics NZ	Network homepages (and associated links)	Notes from IN2SM	Notes from another Statistics NZ course	Refer to notes from an external course (e.g. uni)	Refer to another written source (e.g. book/internet)	Refer to another non-written source (e.g. YouTube clip)
A Statistics NZ process—e.g. how the R&D survey works											
A new subject—e.g. confidentiality, E&I											
How to do something in a piece of software											
Software not previously used											
A existing methodology as used in Statistics NZ—e.g. ARIMA modelling											
A new skill—e.g. doing cool graphs											

Please list any methods of learning that would like to use in your work but do not currently use in Statistics NZ:

Please list any other sources we have missed that you use (and the type of task you use them for):

References

Bagherzadeh, M., & Giovannini, E. (2006). Towards a more integrated international statistical system: The role of training. In A. Rossman, & B. Chance (Eds.), *Proceedings of the seventh international conference on the teaching of statistics*. Bahia, Brazil: IASE, ISI. Retrieved from http://www.stat.auckland.ac.nz/~iase/publications.php?show=17

Forbes, S. D., Harraway, J. A., Chipperfield, J. O., & Siu-Ming, T. (2014). Raising the capability of producers and users of official statistics. In H. MacGillivray, M. Martin, & B. Phillips (Eds.), *Topics from Australian conferences on teaching statistics: OZCOTS 2008–2012*. New York: Springer Science + Business Media.

Jorner, U. (1990). Vocational training and university education—is there a difference? In *Proceedings of the third international conference on the teaching of statistics*. Dunedin: IASE, ISI. Retrieved from http://www.stat.auckland.ac.nz/~iase/publications.php?show=18

Martin, P. (2006). Achieving success in industrial training. In *Proceedings of the seventh international conference on the teaching of statistics*. Salvador, Bahia, Brazil: IASE, ISI. Retrieved from http://www.stat.auckland.ac.nz/~iase/publications.php?show=17

Talbot, M., Horgan, G., Mann, A., Badia, J., & Dieter Quednau, H. (1998). SMART: Introducing specialist statistical techniques via the Web. In *Proceedings of the fifth international conference on the teaching of statistics*. Singapore: IASE, ISI. Retrieved from http://www.stat.auckland.ac.nz/~iase/publications.php?show=2

Wild, C. (2006). On co-operation and competition. In *Proceedings of the seventh international conference on the teaching of statistics*. Salvador, Bahia, Brazil: IASE, ISI. Retrieved from http://www.stat.auckland.ac.nz/~iase/publications.php?show=17